Genetic Recombination

Genetic Recombination

Understanding the Mechanisms

HAROLD L. K. WHITEHOUSE

*Fellow of Darwin College, Cambridge,
and Reader in Genetic Recombination,
University of Cambridge, UK*

A Wiley–Interscience Publication

1807 1982

175 YEARS OF PUBLISHING

JOHN WILEY & SONS

Chichester · New York · Brisbane · Toronto · Singapore

Library of Congress Cataloging in Publication Data:

Whitehouse, Harold L. K.
　　Genetic recombination.
　　'A Wiley–Interscience publication.'
　　Bibliography: p.
　　Includes index.
　　1. Genetic recombination.　I. Title.
QH443.W48　　574.87'322　　81-21981

ISBN 0 471 10205 9　　　　AACR2

British Library Cataloguing in Publication Data:

Whitehouse, Harold L. K.
　　Genetic recombination.
　　1. Recombinant DNA　2. Deoxyribonucleic acid synthesis
　　I. Title
　　574.87'3282　　QP624

ISBN 0 471 10205 9

Typeset by Pintail Studios Ltd., Ringwood, Hampshire.
Printed in the United States of America.

Contents

Preface

Genetic recombination seems to be of universal occurrence in living organisms. Knowledge of its mechanism is of basic importance because it is the process by which segments of corresponding chromosomes of paternal and maternal origin are exchanged. Thus, knowledge of the molecular basis of heredity, and of how that molecular structure allows exchanges to occur, underlie any understanding of the process.

I realized the need for a book on recombination when I received over 800 requests, originating in more than 40 countries, for copies of a review published in 1970. Knowledge of recombination mechanisms has lagged behind understanding of the other two basic features of the hereditary material, namely, how it functions in the cell and how it replicates. I suspect, however, that many biologists may not be fully aware of how far research in this subject has progressed in recent years, nor appreciate the considerable consensus of opinion that has been reached on formerly controversial issues. Certainly, I have learned much in the course of preparing the book, and I hope others will benefit from my attempt to describe the main discoveries that have been made and the evidence for them.

Within the last few years it has become apparent that living organisms possess a second method of bringing about genetic recombination. It differs fundamentally from the 'normal' mechanism in not requiring correspondence of nucleotide sequence between the two DNA molecules. In the final chapter of the book I have written a brief introduction to this rapidly advancing field of non-homologous recombination.

The book is addressed not only to those with a particular interest in recombination, but to all geneticists, and indeed to biologists in general and anyone with an interest in genetics. Because of their diversity of structure and life-cycle, I discuss individually the results obtained with the various bacterial viruses that have contributed so much to understanding of recombination. This approach was not necessary, however, with fungi and other higher organisms (eukaryotes), where the evidence points to considerable uniformity in the recombination process. A handicap for anyone reading publications on recombination is the lack of standardization in the use of the terms 'crossover' and 'conversion', and the need to be familiar with the terminology both of the geneticist and the molecular biologist. I have therefore included a glossary of technical terms.

The biennial recombination workshops, organized by Dr Robin Holliday and Professor Neville Symonds and supported by the European Molecular Biology

Organization, have been of great value in helping me to obtain a broad view of current recombination research. I have tried in this book to convey some of the excitement, so evident at these meetings, at what that research has revealed.

I thank Dr Philip Oliver and Dr Paul Markham for their critical reading of Chapters VI and VII, respectively, and Mr G. J. Clark for help with many of the diagrams and with the references and index. I am grateful to Dr Patricia J. Pukkila for permission to refer to unpublished work by her and associates. I thank Mrs Ruth Hockaday and Mrs Anita Bennett for typing the manuscript.

Cambridge, October 1981   HAROLD L. K. WHITEHOUSE

I.   Introduction

The term *re-combination* seems to have been introduced into genetical literature by Bateson (1909), and was used initially for pairs of character differences showing independent inheritance. He and associates had discovered partial linkage (see Chapter VII, Section 1) and it is the mechanism underlying this phenomenon that forms the subject of this book. This recombination of linked characters was called *crossing over* by Morgan and Cattell (1912). Much later, it was found that recombination of closely linked mutants could occur by a non-reciprocal process that was given the name *conversion* (Chapter VII, Section 4).

Understanding of the mechanism of recombination had to await knowledge of the molecular basis of heredity, and even then progress has been slow. The reason why the recombination mechanism has proved so difficult to resolve is because the most favourable organisms for genetic analysis do not coincide with those most suitable for molecular studies. Fungi such as *Ascobolus* and *Sordaria*, with the products of meiosis retained in a single cell, the ascus, and with mutants available affecting the colour of individual spores within the ascus, have given detailed genetical information about recombination. The bacterium *Escherichia coli* and its phages have been the prime sources of molecular information. With organisms as widely separated in evolution as eukaryotes and prokaryotes, it could not be assumed that the mechanism of recombination was the same in both. Furthermore the phages of *E. coli* that have been most extensively studied from the point of view of recombination, namely φX174, f1, T4 and λ, have proved to be so diverse, and to have such bizarre life-cycles, that it was not possible to generalize from the conclusions drawn from one of them. A further complication was the discovery of two or more pathways for recombination in some of these prokaryotic organisms. On the other hand, the evidence points to uniformity of recombination mechanism in eukaryotes, with differences between species of a minor character only.

In view of the diverse nature of the *E. coli* phages, a chapter is devoted to each of the main categories of phage, and another to *E. coli* itself. On the other hand, there is considerable similarity between the mechanism of transformation in several different bacteria, so these are treated in one chapter. Eukaryotic recombination, for the reasons indicated, is likewise treated as a whole. As becomes apparent from the reviews of the current state of knowledge of recombination in these organisms (Chapters II–VII), there has been considerable convergence of opinion in recent years about the mechanism, despite the great diversity in structure and life-cycle. It has therefore been possible to survey some of the

1

molecular events in recombination (Chapter VIII) in the knowledge that they are of wide applicability, even though in none of the recombination pathways that have been studied is the sequence of events fully understood.

Recombination was discovered in the chloroplasts of *Chlamydomonas reinhardii* by Sager and Ramanis (1963) and in the mitochondria of baker's yeast (*Saccharomyces cerevisiae*) by Thomas and Wilkie (1968). Both of these organelles contain DNA and their genetic analysis has been reviewed by Sager (1977) and Birky (1978). There are indications of heteroduplex formation, conversion, and reciprocal exchange, both in *Chlamydomonas* chloroplasts and yeast mitochondria: see Birky (1978). Thus it seems likely that the recombination mechanism resembles that of nuclear DNA. But the difficulties encountered in studying organelle recombination have meant that understanding of the mechanism has lagged behind that of the prokaryotes and eukaryotes described in the following chapters.

A recombination mechanism of a different kind from the normal one has recently been discovered. It occurs in both prokaryotes and eukaryotes and is responsible for the transfer of DNA segments to new positions in the genome without dependence on homology of nucleotide sequence. In consequence, it can cause high frequency mutation. This non-homologous recombination by transposable elements is reviewed briefly in Chapter IX.

II. Genetic transformation

1. INTRODUCTION

The phenomenon of transformation holds a special place in genetics as the process in which the molecular basis of heredity was first shown to be DNA. Transformation had been discovered by Griffith (1928) in the course of experimenting with *Diplococcus* (or *Streptococcus*) *pneumoniae* (the pneumococcus). He had found that the smooth capsule of polysaccharide surrounding the cell wall of a virulent strain could be inherited by an avirulent non-encapsulated strain with which it was mixed, even though the cells of the virulent strain had first been killed by heat. Subsequently it was shown that cell-free extracts of the virulent strain could bring about transformation of the other strain, and ultimately Avery, MacLeod and McCarty (1944) purified the substance responsible and showed that it was DNA.

2. THE INCORPORATION OF DONOR DNA INTO THE GENOME OF THE RECIPIENT CELL

Investigation of the process of transformation in the pneumococcus was hampered by the fact that cells are receptive to external DNA only during a brief period in the cell cycle. Fox and Hotchkiss (1957) devised a technique for preserving cells in the receptive state. The method depended on freezing them in 10% glycerol after growth to the appropriate physiological state. With receptive cells available when required, Fox and Hotchkiss were able to analyse the kinetics of the reaction between these cells and the external DNA. They found there was an initial adsorption that was reversible, followed by an irreversible fixation of the donor DNA in the recipient cells.

The wild-type pneumococcus is able to utilize maltose as a source of carbohydrate by virtue of the presence of the enzyme amylomaltase. By ultraviolet irradiation Lacks and Hotchkiss (1960*a*) obtained eight mutants deficient in the enzyme and hence unable to utilize maltose. They investigated the transformation of the mutants by wild-type DNA. To their surprise they found (Lacks and Hotchkiss, 1960*b*) that the mutants differed in the level of enzyme production that they attained, relative to total protein increase, but this rate was proportional to the extent of permanent genetic change, that is, the fraction of cells that were transformed. They concluded that, in order to elicit enzyme synthesis, the donor DNA must not only be incorporated in the recipient cell but also be integrated into its genome. This idea was supported by the observation that the enzyme is

3

rapidly formed even when the donor DNA carries a second amylomaltase mutant at a closely linked but different site from that in the recipient cells. In this experiment enzyme formation required the donor DNA to interact with that of the recipient, presumably by some form of recombination.

Fox and Hotchkiss (1960) incubated pneumococci sensitive to streptomycin with DNA labelled with ^{32}P isolated from streptomycin-resistant cells. The incubation was for 15–40 min at 30 °C before DNA remaining outside the cells was destroyed by the addition of pancreatic deoxyribonuclease. The cells were sampled at intervals up to 2 h in order to measure the quantity of ^{32}P incorporated and the number of cells transformed. DNA was extracted from the cells and used for transformation of a new culture of streptomycin-sensitive bacteria, the same physical and biological measurements being made. It was found that transforming DNA re-isolated a short time after fixation in the recipient cell had suffered neither physical nor biological dilution. It was evident that the transforming DNA increased at the same rate as the DNA of the recipient bacteria during their growth.

Using the same technique but paying particular attention to the first few minutes after exposure to the external DNA, Fox (1960) discovered that the ability to transform further cells was lost if the DNA was extracted immediately after it had entered the recipient cells. If a few minutes had elapsed, however, before the DNA was extracted, it was capable of bringing about further transformation. These results were comparable to those of Lacks and Hotchkiss (1960b) with enzyme production, which likewise went through a period of eclipse until the donor DNA had been integrated into that of the recipient cell.

The speed with which the donor DNA seemed to be integrated suggested that it had not replicated. Fox (1960) found that transformation was normal in cells treated with fluorodeoxyuridine, which largely inhibits DNA synthesis. He concluded that little or no synthesis was required for transformation.

Lacks (1962) used column chromatography to investigate the fate of the donor DNA on entering the recipient cells. DNA labelled with ^{32}P was prepared from streptomycin-resistant pneumococci and used to transform a streptomycin-insensitive strain. DNA was extracted from the recipient cells at various times and fractionated on methylated albumin-coated kieselguhr columns. From the position in the column of the radioactivity, he found that immediately after entry about half the donor DNA was converted to single-stranded form, the other half being degraded to oligonucleotides. These results would explain the eclipse just after entry, both of information transfer and of transforming ability, if transcription and cell entry in transformation require duplex DNA. If the DNA was extracted at slightly later times after the donor DNA had entered the cells, Lacks found that the radioactivity was rapidly incorporated into duplex structures. It seemed therefore that the single-stranded DNA was quickly integrated into the recipient genome.

Fox and Allen (1964) confirmed these conclusions by studying pneumococcal transformation with the aid of density-gradient centrifugation. Both strains were marked genetically, the donor being resistant to streptomycin and the recipient to *p*-nitrobenzoic acid. The donor bacteria were allowed many generations of growth in a medium containing ^{32}P as a radioactive label, and deuterium (^{2}H) and ^{15}N as density labels, before the DNA was isolated. After not more than 15 min incubation following cell entry, and hence before DNA replication had occurred, the DNA was extracted from the recipient cells and subjected to density-gradient centrifugation in caesium chloride. It was found that both the streptomycin resistance and the radioactivity of the donor DNA were now associated with DNA only slightly heavier than that of the recipient cells. It was evident that the donor DNA had become associated with the genome of the recipient cells.

When the size of the DNA molecules extracted from the recipient cells was reduced, by sonication, from an estimated molecular weight of 2×10^7 daltons to about 10^6 daltons, the streptomycin resistance and the radioactivity became associated with heavier molecules, approaching a density halfway between that of the two parents. Denaturation, that is, separation of the strands of the duplex, using heat or alkali, revealed that most of the ^{32}P was still associated with a large amount of DNA of light density. Denaturation of sonicated extracts, on the other hand, revealed that the bulk of the radioactive label was associated with heavy single strands. Fox and Allen (1964) concluded that the donor material formed part of one strand of the duplex DNA of the recipient genome and was covalently linked to it. They favoured the hypothesis that either strand of the donor DNA could form a hybrid with the complementary strand of the DNA of the recipient. If it were one unique strand, after one bacterial doubling none of the cells receiving the other strand would carry donor information. This is contrary to Fox and Hotchkiss's observation (see above) that the transforming activity multiplies in synchrony with that of the recipient DNA.

The steps in the process of recombination in pneumococcal transformation, to which Fox and Allen's results pointed, were illustrated by Fox (1966) and are shown in Fig. 1. One strand, chosen randomly, of the duplex donor molecule is inserted into the corresponding region of the recipient genome, the other strand of the donor and the displaced strand of the recipient being broken down. Finally, the inserted donor strand is joined covalently to the neighbouring parts of the corresponding recipient strand.

Miao and Guild (1970) showed that the donor DNA does not have to be duplex for pneumococcal transformation to take place, but the efficiency of single strands is only about 0.5% of duplex, probably because of inefficient cell entry. Their demonstration depended on the use of the light strands only as donor. The light and heavy strands of the pneumococcal DNA were separated by denaturation, and fractionated by reaction with the ribopolymer poly(G,U).

Lacks, Greenberg and Neuberger (1974) obtained mutants of the pneumococcus lacking a deoxyribonuclease necessary for transformation. They used a strain that was deficient in two major deoxyribonuclease activities known not to affect transformation, and selected mutants defective in transformation, following treatment with the mutagen 1-methyl-3-nitro-1-nitrosoguanidine. One group of mutants showed no detectable transformation and failed to bind or take up DNA. Their deoxyribonuclease content was unchanged. A second group of mutants

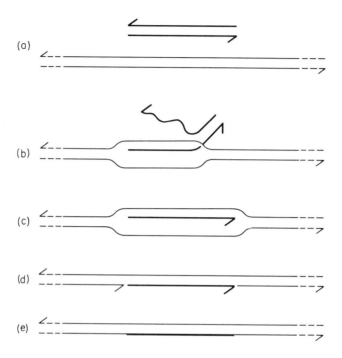

Fig. 1. Possible intermediates in the integration of donor DNA into the recipient genome in the pneumococcus, according to Fox (1966). (a) Donor segment (thick lines) and recipient genome (thin lines). (b) One strand of the donor pairing with the complementary strand of the recipient in the homologous region of the genome. (c) Pairing completed, the other donor strand having separated. (d) The segment of recipient strand that has been displaced is lost through nuclease action. (e) Covalent links seal the donor strand segment into the recipient molecule.

showed some residual transformability, were able to bind DNA to the outside of the cell, and lacked a deoxyribonuclease present in wild type. Lacks *et al.* (1974) suggested that this enzyme may break down one strand of the donor duplex, the energy from this hydrolysis being used to pull the other strand into the cell.

Another bacterial species in which transformation has been investigated is *Bacillus subtilis,* which belongs to a different family from the pneumococcus. Szybalski (1961) showed there was an association between the transforming DNA and that of the recipient cells, and Bodmer and Ganesan (1964) made a detailed investigation. They used donor DNA that carried several genetic markers and also ^{15}N and ^{2}H as density labels and ^{3}H-labelled thymidine as a radioactive label. The recipient strain was also genetically marked and carried ^{32}P as a radioactive label. The recipient cells were exposed to the DNA extracted from the donor strain for 10–30 min, before pancreatic deoxyribonuclease was added to destroy the external DNA. Transformation was studied by density-gradient centrifugation of the DNA extracted from the recipient cells, in conjunction with measurement of the differential radioactivity of the tritium from the donor and the

^{32}P from the recipient. The effects of fragmenting the DNA by shearing and of separating the strands by denaturation with heat or alkali were investigated in relation to buoyant density and radioactivity. Clear evidence was obtained that parts of the donor DNA were built into the recipient genome and held by bonds resistant to shearing and denaturation. It was not possible to distinguish simplex from duplex integration with certainty. The mean molecular weight of the inserted DNA was estimated to be 2.8×10^6 daltons, equivalent to 4200 nucleotide pairs if duplex or 8400 nucleotides in one strand. From the frequency of transformation of linked markers, single-strand integration was favoured.

Evidence supporting single-strand integration was obtained by Bodmer (1966) from study of the kinetics of the inactivation of transforming DNA by pancreatic deoxyribonuclease. This enzyme is believed to make single-strand breaks at random positions in duplex DNA, with the result that the viscosity of the molecule is not affected until two breaks in opposite strands occur close enough together to break the whole molecule. Nevertheless, Bodmer found that transforming activity declined rapidly with the nuclease treatment, for example, to 20% of its initial value when the viscosity had fallen only to 85%. A direct relationship was found between the transforming activity of the DNA and the single-strand molecular weight, but not with the duplex molecular weight. Takagi, Ando and Ikeda (1968) discovered that the inactivation of transforming activity brought about by pancreatic deoxyribonuclease could be reversed by treatment with polynucleotide ligase of phage T4 of *E. coli*. This showed that the cause of the decreased transforming activity following the nuclease treatment was single-strand breaks, that is, nicks, in duplex DNA. Similar results were obtained by Laipis, Olivera and Ganesan (1969) using the ligase of *E. coli* and also of *B. subtilis* itself. The single-strand molecular weight increased after the ligase treatment.

Chilton and Hall (1968) discovered that *B. subtilis* DNA that had been denatured with alkali and neutralized with ethylenediaminetetraacetate (EDTA) showed a high level of transforming activity. By investigating its buoyant density in caesium chloride and its sedimentation in a sucrose gradient, they showed that the DNA that brings about transformation in the presence of EDTA was single-stranded. They pointed out that this finding supported the view that single-stranded DNA was an intermediate in transformation by duplex DNA. Chilton (1967) found that the two single strands, which could be separated by differential complexing with polyadenylic acid, had identical transforming activity.

Dubnau and Davidoff-Abelson (1971) followed transformation in *B. subtilis* both molecularly, using density (^{2}H) and radioactive (^{3}H) labels of the donor DNA, and genetically, using a tryptophan-requiring recipient and a histidine-requiring donor strain. They found that a donor–recipient complex formed faster when measured by the transfer of radioactivity into the low density fraction than it did when the appearance of recombinant transforming activity was monitored. This activity was the ability to transform doubly mutant cells, that is, requiring both tryptophan and histidine, to a requirement for neither amino acid. The tryptophan independence derived from the original donor strain and the histidine independence from the original recipient one. That transformation lagged behind the appearance of radioactivity in the donor–recipient complex implied that this complex existed transiently in a form with low biological activity. Dubnau and Davidoff-Abelson (1971) suggested that this recombination intermediate had the donor and recipient parts held together only by non-covalent bonds.

Harris and Barr (1969) discovered that at the particular stage of the cell cycle when transformation can occur – the state termed competence – the DNA of *B. subtilis* became denser in a caesium chloride gradient as the pH was raised from 8 to 11. Such a density shift with increasing alkalinity is characteristic of partially single-stranded DNA. It was absent from non-competent cells. From the magnitude of the density shift, they estimated that not more than 5% of the DNA could be single-stranded. The sedimentation pattern of this dense DNA on alkaline sucrose gradients indicated that each molecule could possess only a very limited number of simplex regions. If there was only one, as seems likely, it would be about 2000 nucleotides in length. Subsequently (Harris and Barr, 1971), caesium chloride density gradient experiments at pH 11 were carried out using ^{32}P-labelled donor DNA, and recipient cells in which the DNA was labelled with ^{3}H. The DNA of competent recipient cells was extracted 3, 10 and 30 min after a 3 min exposure to the donor DNA. It was found that at these successive times the ^{32}P-labelled DNA showed a progressive loss from the position in the gradient corresponding to single-stranded DNA and a progressive gain at the position of the ^{3}H-labelled DNA. They concluded that the donor DNA was associating with the recipient molecules by annealing with the pre-existing single-stranded regions where these happened to be homologous, that is, a complementary gap.

Haemophilus influenzae is a third species of bacterium in which transformation has been studied. It is a member of a different family from either the pneumococcus or *B. subtilis*. Voll and Goodgal (1961) used DNA extracted from a cathomycin (novobiocin)-resistant strain of *H. influenzae* to transform a streptomycin-resistant recipient strain. In other experiments the genotypes were reversed. At various times after uptake of the external DNA, the DNA of the recipient cells was assayed for transforming ability using a strain sensitive to both cathomycin and streptomycin. DNA capable of transforming cells of this strain to resistance to both antibiotics could be extracted from the recipient cells within a few minutes of uptake of the original donor DNA – half the ultimate level was reached within 15 min at 36 °C, following a 5 min period of uptake. Experiments of the same kind were performed in saline, where growth is much reduced. Under these conditions there was little if any increase in transforming activity of the recipient character but nevertheless within 30 min a considerable amount of linkage of donor and recipient transforming factors had occurred. It was estimated that no more than 15% of the DNA could have undergone synthesis. Purification of the DNA carrying donor and recipient factors failed to separate them. It was evident that recombination of donor and recipient DNA occurred rapidly after uptake of the donor DNA, and that this could take place in the absence of cell growth or of appreciable DNA synthesis.

Transformation in *H. influenzae* was investigated by Stuy (1965) using a variety of techniques. The donor strain was resistant to streptomycin and the recipient to cathomycin. DNA isolated from recipient cells that had been incubated at 37 °C for 10 min following a 1 min exposure to the external DNA, showed only a quarter of the frequency of transforming ability for streptomycin resistance compared with that immediately after cell entry. Judging by the ability of the extracted DNA to transform sensitive cells to both streptomycin and cathomycin resistance, incorporation of donor DNA into the genome of the recipient cell began after 10 min incubation.

Similar results were obtained when the donor DNA had about 10% of its thymine residues replaced by bromouracil. Density-gradient centrifugation in caesium chloride showed the DNA carrying the streptomycin resistance factor shifting from heavy to near to light buoyant density between 10 and 30 min after the start of incubation. This is the result expected if, during this time, small pieces of the donor DNA (carrying the heavy bromine atoms) are incorporated into the recipient genome. Moreover, all the DNA fractions carrying the factor for streptomycin resistance showed the sensitivity to ultraviolet light expected of DNA containing bromouracil. There was no evidence that streptomycin resistance had been transferred to a bromine-free copy, such as recombination by copy-choice would imply. When the fate of donor DNA was traced by labelling it with ^{32}P, about half the radioactivity appeared soon after uptake as small fragments. Chromatography over methylated albumin kieselguhr columns failed to reveal single-stranded DNA with either the radioactive or the biological markers. Stuy tentatively concluded that in transformation in *H. influenzae* duplex donor DNA is incorporated into the genome of the recipient cell. In other respects his results were in agreement with those for *D. pneumoniae* and *B. subtilis*.

Duplex integration in *H. influenzae* was not confirmed by Notani and Goodgal (1966). Admittedly, they found no evidence that the donor DNA became single-stranded after uptake by the recipient cells, in the way that it had been shown to do in *D. pneumoniae* and *B. subtilis*. Thus, ^{32}P-labelled donor DNA extracted from transforming cells of *H. influenzae* at 0, 15 and 40 min after a 3 min uptake period was found by density-gradient centrifugation in caesium chloride to remain in the duplex condition. It was concluded that if any of the donor DNA became single stranded during these 40 min after uptake, it was immediately either broken down or integrated into the recipient genome.

Notani and Goodgal (1966) demonstrated single-strand integration by labelling the donor DNA, which carried a genetic marker for resistance to cathomycin, with ^{2}H and ^{15}N to give a heavy density, and with ^{32}P to give radioactivity. The recipient genome carried the genetic marker for resistance to streptomycin and it was labelled with tritium (^{3}H) as a radioactive marker. DNA extracted from the recipient cells at various times after they had been exposed to the donor DNA was subjected to density-gradient centrifugation in caesium chloride. It was found that there was a progressive transfer of the ^{32}P to a density intermediate between that of donor and recipient, and this DNA of hybrid density included recombinants, that is, with the ability to bring about transformation to resistance to both cathomycin and streptomycin. It was evident that donor DNA was being integrated directly into the recipient genome, and the fact that the recombinant material did not contain more than 50% of the heavy donor DNA pointed to single-strand integration. This was confirmed by two additional experiments.

In the first experiment the fate of recombinant DNA, with resistance to both antibiotics, was followed through a replication cycle, which takes about 30 min at 36 °C. If the genome contained double-stranded donor, that is, heavy, material, it will become lighter on replicating, whereas a genome containing only single-stranded donor material will stay at the same density when it replicates. A density shift was not found.

In the second experiment the DNA of hybrid density extracted from the recipient cells was denatured and centrifuged again in a caesium chloride gradient. If the donor material were duplex, the separated strands would have a predictable buoyant density, but if the donor material were simplex the single strands that carried the ^{32}P would be denser than if it had been duplex. This was the result obtained. Furthermore, these strands had a buoyant density less than that expected if all their atoms were heavy, implying that the integrated donor segment was covalently joined to recipient material of lighter density. The integrated piece of donor DNA had a mean molecular weight estimated to be 6×10^6 daltons.

Notani and Goodgal (1966) concluded that the mechanism of integration of donor DNA into the recipient genome was similar to that of *D. pneumoniae* and *B. subtilis*. The difference from those species seemed to be merely in the timing of the breakdown of one strand, which appeared to be delayed in *H. influenzae* until immediately before integration.

Support for the conclusion that integration of DNA in *H. influenzae* transformation is single stranded was obtained by Goodgal and Postel (1967). They had shown (Postel and Goodgal, 1966) that single-stranded DNA was taken up by recipient cells and caused transformation as efficiently as duplex DNA, provided that the pH was initially low and was then raised to neutral. The fate of denatured, that is, single-stranded, DNA extracted from cathomycin-resistant cells and used to transform a streptomycin-resistant strain was investigated, both biologically and physically, by Goodgal and Postel (1967). Lysates of transforming cells were tested for cathomycin resistance. As a function of time this character decreased in denatured (that is, single-stranded) DNA but increased in native (that is, duplex) DNA. Furthermore, linked cathomycin and streptomycin resistance increased with time and the characters failed to separate on denaturation, showing that recombination had occurred and covalency had been established between donor and recipient DNA. Labelling the donor DNA with density (^{15}N, ^{2}H) and radioactive (^{32}P) markers and the recipient DNA with a different radioactive marker (^{3}H), enabled the physical fate of the donor DNA to be followed by density-gradient centrifugation in caesium chloride. A progressive transfer of the donor DNA to a hybrid density took place, concomitant with the appearance of duplex DNA carrying both cathomycin and streptomycin resistance. The integrated piece of donor DNA was estimated to have a molecular weight of about 6×10^6 daltons. Goodgal and Postel (1967) concluded that the integration process was quite similar whether the donor DNA was initially simplex or duplex, as expected if only one strand is integrated when the donor is duplex.

The conclusions reached about the process of transformation are similar for *D. pneumoniae*, *B. subtilis* and *H. influenzae*. It appears that in all three species DNA can enter cells at a particular stage of the cell cycle and that this DNA is normally duplex. Under special conditions, however, single-stranded DNA can penetrate. There are differences in the timing of events subsequent to cell entry, but in all three species one strand of the donor DNA, if it is duplex, is broken down. A single strand of the donor becomes incorporated in the recipient genome in place of its homologue. The association is believed initially to be by hydrogen bonds between the bases, covalent joining of phosphodiester links occurring subsequently.

3. MISMATCH REPAIR

(a) Marker effects in the pneumococcus

Lacks and Hotchkiss (1960*a*) had obtained eight amylomaltase-deficient mutants of the pneumococcus by means of ultraviolet treatment (see Section 2). They compared the transformation frequencies using wild-type DNA as donor, with each mutant in turn as recipient. The wild-type DNA was obtained from a streptomycin-resistant strain. The streptomycin-resistance marker provided a standard of reference. They found that there was a 15-fold range in the frequency of transformation to maltose utilization. This integration efficiency was measured by the ratio of maltose factor transformation to streptomycin factor transformation. Recombination between the maltose mutants was studied by using one as donor and another as recipient. Three of the mutants were multisite, showing no recombination with others that themselves recombined when crossed. One of the multisite mutants overlapped all the others. It was concluded that all the mutant sites were in the same region of the genome – the *maltose* (*mal*) locus. The differences in integration efficiencies seemed to be characteristics of the individual mutants. The multisite mutants had lower transformation frequencies than the others, and Lacks and Hotchkiss (1960*a*) suggested that the variation in frequency depended in part on the size of the genetic region involved.

A 30-fold range in integration efficiency was found by Iyer and Ravin (1962) with five mutants of spontaneous origin conferring resistance to erythromycin and mapping at the *ery* locus. They concluded that this variation depended on the nature of the mutated region being integrated.

Mutants of the pneumococcus that were resistant to aminopterin were found by Sicard (1964) not to grow in a culture medium rich in isoleucine. It appeared that the aminopterin resistance was associated with sensitivity to an imbalance in the relative concentrations of branched amino acids (isoleucine, leucine and valine). This made it possible to score transformations of wild-type cells with mutant DNA by selecting for resistance, and transformations of mutant cells with wild-type DNA by plating in the appropriate medium. The two properties of the mutants were inseparable either by reverse mutation or recombination, and the sites appeared to be linked. It was concluded that the mutants were all in one functional region of the genome, the *amiA* locus. Ephrussi-Taylor, Sicard and

Kamen (1965) investigated a total of 73 *amiA* mutants, some of spontaneous origin and others induced with ethyl methane sulphonate (EMS) or with nitrous acid. For individual mutants the efficiencies of reciprocal transformations, that is, to wild type and to mutant, were the same. The mutants fell into two classes, however, as regards this efficiency, differing by a factor of about 10. It was found that all the mutants induced by EMS and by nitrous acid were of low efficiency, whereas the spontaneous mutants were distributed about equally between the low and high categories.

A detailed investigation of integration efficiency was made by Lacks (1966) with mutants at the *maltose* (*mal*) locus. Altogether, 76 mutants were analysed, including the eight already referred to. Various mutagens were used, either on wild-type cells or wild-type DNA, to obtain the mutants, some of which proved to be multisite, probably deletions. As already mentioned, the evidence for this was that they failed to show recombination with two or more *mal* mutants that showed recombination with one another. The tests for recombination were made by using DNA from one mutant to attempt to bring about transformation to wild type when the recipient cells carried a second *mal* mutant. Further evidence that a mutant was multisite was provided by a failure to obtain spontaneous reversion to wild type, unlike the single-site mutants. Some of the deletions were extensive and overlapped others. From the recombination frequencies, and with the help of the deletions, the mutants were mapped.

Each mutant in turn was transformed by wild-type DNA and the integration efficiency determined with the aid of a particular sulphonamide-resistance marker in the donor DNA as a standard of reference. Five important results were obtained.

(1) The single-site mutants, that is, those whose sites did not overlap two or more recombining sites, fell into four classes as regards integration efficiency, and for some mutagens there was a well marked relationship with efficiency: nitrous acid and hydroxylamine gave low efficiency mutants only (Table 1). Members of this lowest class showed higher reversion rates than members of the other classes.

(2) The multisite mutants showed a gradation in integration efficiency, with no discrete classes, but with a clear relationship to the length of the mutant site. The longer the deletion the lower, in general, was the integration efficiency.

(3) There was no relationship between the integration efficiency of a mutant and its position on the map of the *mal* locus.

(4) Single-site mutants at the same site sometimes differed in integration efficiency, and a short deletion with a high integration efficiency could span the sites of single-site mutants that were integrated only with low efficiency.

(5) With two closely linked mutants in the recipient cell and wild-type DNA as donor, or with the converse situation (double mutant DNA as donor), a low efficiency marker reduced the integration frequency of a high efficiency marker, but never the converse, that is, no enhancement of low by high. The extent to which a high efficiency marker was excluded by a low one depended inversely on the distance between the markers.

Table 1. Data of Lacks (1966) for the integration efficiencies of 57 single-site mutants at the *amylomaltase* (*mal*) locus in the pneumococcus. The integration efficiency is the ratio of *mal*$^+$ transformants to sulphonamide-resistant transformants when the *mal* mutant is the recipient, and wild-type DNA containing a particular sulphonamide resistance marker is the donor. The table shows the number of mutants obtained with various mutagens in each efficiency class.

Integration efficiency	Mutagen						
	Nitrous acid	Hydroxyl- amine	Ethyl methane sulphonate	Triethylene melamine	Hydrogen peroxide	Ultraviolet light	Proflavin
Low, 0.026–0.085	13	3	4	4	3	6	3
Moderately low, 0.17–0.26						5	
Intermediate, 0.49–0.64			1	1	1	3	2
High, 0.86–1.06			1	1	1	5	

It was evident that the molecular nature of the mutation was important in determining integration efficiency, and that the exclusion occurring with low efficiency mutants extended along the DNA probably for hundreds of nucleotides. Lacks (1966) attributed the low integration efficiency of long deletions to an end effect, that is, the end of the donor segment of wild-type DNA occurring in or near the region deleted in the mutant. He found much higher integration efficiencies when these lengthy multisite mutations were in the donor DNA and the recipient cell was wild type, in agreement with this hypothesis.

Ephrussi-Taylor and Gray (1966) reported further results with the *aminopterin* (*amiA*) mutants. They found that the 85 mutants studied fell into two non-overlapping efficiency classes, as previously reported with a smaller sample. Ephrussi-Taylor (1966) discovered that the low efficiency (LE) *amiA* mutants were transmitted to daughter cells at the second cell division after uptake, and the high efficiency (HE) markers usually at the third division. She had previously found a similar difference between an LE optochin-resistant mutant and an HE streptomycin-resistant mutant (Ephrussi-Taylor, 1960, 1961), and she now found this was a constant feature distinguishing LE from HE *amiA* mutants. The technique was as follows. Wild-type cells were exposed to the mutant DNA for 10 min. The cells were then transferred to fresh medium and the divisions synchronized by thawing after storage at −70 °C. Samples were taken after each division and the cells separated from one another by sonication. The division at which donor and recipient characters segregated was determined by testing colonies derived from single cells for the parental characters. The testing was done by further transformation experiments.

The delayed transmission of HE donor mutants to progeny, compared with LE, was accounted for by Ephrussi-Taylor and Gray (1966) by the following hypothesis, which also explained the diversity of integration efficiency, and the over-riding effect of an LE mutant on the integration efficiency of a closely linked HE mutant.

They made three postulates:

(1) After the donor strand has formed a heteroduplex with the complementary strand of the recipient, LE markers undergo an excision process but HE markers do not. Ephrussi-Taylor and Gray (1966) suggested that an excision enzyme recognized some feature of the LE site in the heteroduplex, such as an absence of pairing between one base of each complement, and excised one strand in the region recognized.

(2) The excision enzyme chooses to excise the donor strand nine times out of 10, but chooses the recipient strand for excision the remaining time.

(3) The excision is extensive, such that with two closely linked mutant sites it often extends to the second site.

The excision of the donor strand nine times out of 10 with an LE marker accounted for the 10-fold higher integration efficiency of HE mutants compared with LE. The extensive excision explained why, with a double mutant (*cis* relationship) consisting of closely linked HE and LE mutants, the integration efficiency of the HE mutant was reduced. The falling off of this effect with increasing separation of the mutant sites was accounted for by the reduced likelihood that the excision, initiated at the LE site, would reach the HE site. The absence of excision initiated at HE sites provided an explanation for their delayed transmission (Fig. 2). Ephrussi-Taylor and Gray (1966) supposed that the initial heteroduplex arose during replication of the recipient DNA. With an HE marker, if the recipient is wild type, a heteroduplex will then segregate from wild type at the first cell division after DNA uptake, and mutant and wild type will segregate at the second division. The donor marker will not be transmitted to both progeny cells until the third division (Fig. 2(a)). On the other hand, an LE marker in the donor will either be excised from the heteroduplex (nine times out of 10) or become a homoduplex through excision of the wild-type strand of recipient origin. Thus, mutant and wild type will segregate at the first cell division after uptake, and the donor marker will be transmitted to both progeny cells at the second division (Fig. 2(b)). These predictions of the hypothesis were in agreement with observation.

Ephrussi-Taylor and Gray's hypothesis also explained the relationship between integration efficiency and recombination frequency in two-point *trans* crosses. Lacks and Hotchkiss (1960a) had shown that the frequency of wild-type recombinants was grossly influenced by the efficiency with which a mutant site in the

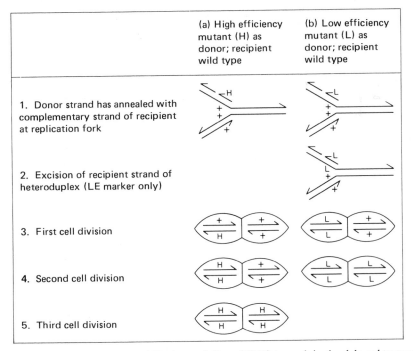

Fig. 2. Hypothesis of Ephrussi-Taylor and Gray (1966) to explain the delayed transmission of mutants with high integration efficiency compared with those with low integration efficiency.

recipient was replaced by the wild-type sequence at this site in the donor. Sicard and Ephrussi-Taylor (1965) mapped their *amiA* mutants using the recombination frequencies in two-point crosses. They found that whenever an HE mutant was crossed as recipient with an LE mutant as donor it gave wild-type recombinants 10 times as frequently as the reciprocal cross. A possible source of wild-type recombinants is from a heteroduplex that extends over both mutant sites, followed by excision of one or other of the mutant strands, the excision not extending to the second site (Fig. 3(i), (ii)). But a heteroduplex extending over both sites will make little or no contribution to wild-type recombinants because their occurrence depends on excision that does not extend to the second site. With HE mutants there will be no excision, and with LE mutants, although excision is postulated, it will usually include the second site.

The primary source of wild-type recombinants is believed to be from a heteroduplex that includes only the recipient mutant site, the donor site being excluded (Fig. 3(iii), (iv)). Lacks and Hotchkiss (1960*a*) had pointed out, with reference to crosses between amylomaltase-deficient mutants, that recombination to give maltose-positive cells would occur only when the genetic deficiency of the host was replaced by donor DNA that did not at the same time introduce its own deficiency. The frequency of these recombinants is expected to depend on the site separation. This is because the closer the sites the more often the heteroduplex will be expected to extend to both of them, when, as indicated above, it will rarely or never give wild-type recombinants. A relationship between recombination frequency and site separation was evident from the ability of Lacks and Hotchkiss (1960*a*), Lacks (1966) and Sicard and Ephrussi-Taylor (1965) to construct consistent linear maps of the *mal* and *amiA* mutant sites using recombination frequencies.

Sicard and Ephrussi-Taylor (1965) pointed out that in two-point *trans* crosses the integration efficiency of the mutant in the recipient was more important, in determining recombination frequency, than the distance between the mutant sites. The occurrence of wild-type recombinants primarily as a result of a heteroduplex that includes only the recipient mutant site explained why recombination frequencies depend in this way on the integration efficiency of the mutant in the recipient and not on that in the donor. Sicard and Ephrussi-Taylor (1965) had shown with two-point *trans* crosses of *amiA* mutants that the integration efficiency of the mutant in the donor DNA was without appreciable influence on the recombination frequency. This lack of influence was described by Ephrussi-Taylor and Gray (1966) as having been a major obstacle to explaining efficiency, because it ruled out the simplest explanation, which was faulty pairing of nucleotide chains near the site of an LE mutant. This explanation predicts that an LE marker will have the same effect whether in the recipient or the donor. The irrelevance, as observed, of the donor marker integration efficiency for recombination frequency, can be seen in Fig. 3, where the primary sources of wild-type recombinants, according to Ephrussi-Taylor and Gray's hypothesis, are not influenced by the efficiency of the donor marker. Lacks (1966) did find, however, that when *mal* mutants were closely linked (less than 4% of the total length of the

Event	Genotype	Mutant m_1 in recipient			
		HE		LE	
		Mutant m_2 in donor			
		HE	LE	HE	LE
(i) Heteroduplex at both sites, with excision of donor at m_2 only	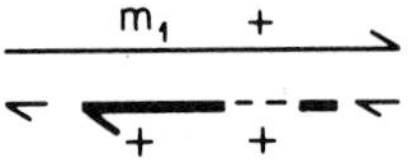 m_1 + / + +	Absent[1]	Rare[2]	Absent[1]	Rare[2] or absent[3]
(ii) Heteroduplex at both sites, with excision of recipient at m_1 only	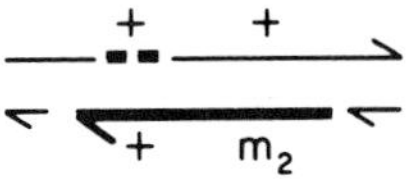 + + / + m_2	Absent[1]		Rare[2]	Rare[2] or absent[3]
(iii) Heteroduplex at m_1 only, with no excision	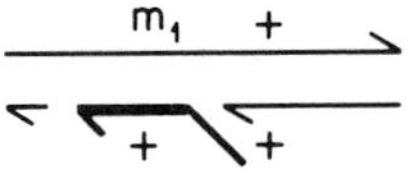 m_1 + / + +	Primary source of wild-type recombinants		Absent[3]	
(iv) Heteroduplex at m_1 only, with excision of recipient	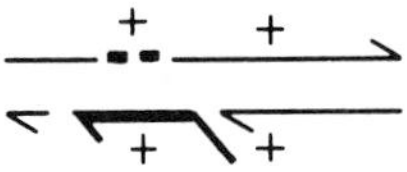 + + / + +	Absent[1]		Primary source of wild-type recombinants	

Fig. 3. Explanation, according to Ephrussi-Taylor and Gray (1966), for the difference in recombination frequency between reciprocal crosses of a high and a low efficiency marker, and for the dependence of recombination frequencies on the integration efficiency of the marker in the recipient, but not on that of the donor marker. Strands with donor nucleotide sequence are shown with a thick line, and newly synthesized strands with a broken line. Cross:

$$\begin{array}{lcc} \text{Recipient} & m_1 & + \\ \text{Donor} & + & m_2 \end{array}$$

Events may be absent (marked 1) because they depend on excision at an HE site, where it is not expected. Events may be rare (marked 2) because they depend on excision not extending to the second site, which it is expected often to do; excision at both sites would not give a wild-type recombinant. Events may be absent (marked 3) because they depend on lack of excision at an LE site, where it is expected to occur.

gene apart) the integration efficiency of the mutant in the donor affected the recombination frequency. The situation described in Fig. 3(i) and (ii), with a heteroduplex extending to both sites, might provide an explanation.

The exclusion of an HE marker by an LE marker occurs in both directions along the DNA. Gray and Ephrussi-Taylor (1967) found in two-point *cis* crosses that *amiA* LE mutant no. 39 reduced the integration efficiency of HE mutant no. 36, which mapped to its left, and also, in a separate cross, that of HE mutant no. 33 situated to its right. In both crosses the double mutant was in the donor DNA. It appeared as if excision occurred in both directions from the LE site (cf. Chapter VII, Section 10). The exclusion of an HE mutant by an LE was greater if the HE was to the left of the LE than if it was to the right. Two LE mutations in *cis* showed an integration efficiency no lower than either alone, as though one LE mutation were sufficient to trigger excision.

By combined physical and genetical studies, Gurney and Fox (1968) obtained detailed information about the molecular events in pneumococcal transformation. As in the earlier experiments of Fox and Allen (1964) (see Section 2), the donor DNA was labelled with the heavy isotopes ^{2}H and ^{15}N, and the physical nature of the DNA of the transformed cells established by extracting it after various time intervals, breaking the molecules to particular size ranges by shearing in a homogenizer at the appropriate speed, and then fractionating in a caesium chloride density gradient. The genetic markers were an erythromycin (*ery*)-resistance mutant, r2, in the donor DNA and a streptomycin (*str*)-resistance mutant, r41, in the recipient, or the converse. The mutants are unlinked, and both were integrated with high efficiency. The genotype of the DNA extracted from the transformed cells was determined by making a second transformation using cells sensitive to both erythromycin and streptomycin as recipient.

With DNA extracted immediately after the initial transformation, both donor and recipient markers were found at the hybrid density (Fig. 4(a), (a')). If a particular strand only of the DNA can be incorporated into the recipient DNA molecule, no recipient activity should appear at the hybrid density. This is because, to produce such a hybrid, the wild-type allele in the donor of the genetic marker in the recipient molecule must have been inserted into the recipient genome (Fig. 4(a')). With DNA extracted immediately after transformation and therefore, it is presumed, before any DNA replication has taken place, the recipient marker could appear after the second transformation only if the recipient strand in the first transformation could act as donor in the second. Gurney and Fox (1968) concluded that this could happen, or in other words, either strand could participate in transformation.

With DNA extracted at a later time after the initial transformation, one or more replications of the DNA would have occurred. It was found that the donor marker did not begin to appear in DNA of light density until the second replication. By the end of this replication the donor activity was found to be equally divided between DNA of hybrid and of light density (Fig. 4(c)). These are the results expected if the physical hybrid produced in the initial transformation is genetically heterozygous, that is, the heavy strand carrying the donor mutant has annealed with the complementary recipient strand, which would be of light density and would carry the wild-type allele of the donor marker. At the first replication these strands would separate again and new strands of light density would be synthesized, one using the donor heavy strand as template and the other the recipient light one (Fig. 4(b)). So the donor marker would remain in molecules of hybrid density until the second replication.

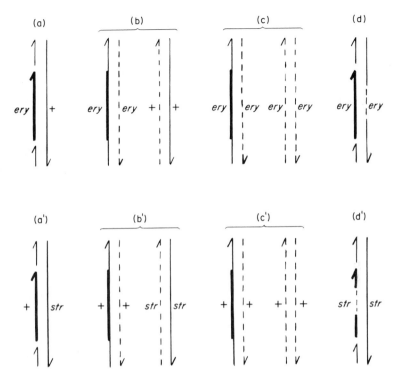

Fig. 4. Diagram to illustrate Gurney and Fox's (1968) results with pneumococcal transformation. The donor material, which carried heavy atoms, is shown with a heavy line. Newly synthesized strands are shown with a broken line. *ery* indicates erythromycin-resistance marker, r2, carried by donor; *str* indicates streptomycin-resistance marker, r41, carried by recipient. (a), (a') Initial heteroduplexes. (b), (b') Products of first replication. (c), (c') Products of second replication of molecule carrying donor marker. (d), (d') Initial heteroduplexes in which mismatch repair has given homozygosity for the markers.

Gurney and Fox (1968) found that the recipient activity disappeared from the hybrid density during the first replication (Fig. 4(b')). This result is also that expected if the initial physical hybrid is a genetical heteroduplex. They could not rule out, however, the possibility that a small amount of the recipient activity remained near the hybrid density during the first replication, nor the possibility that a small amount of donor activity appeared in DNA of light density after the first replication. These are the results that would be expected if, after formation of the initial heteroduplex, there was local destruction of one of the strands, followed by repair using the other strand as template (Fig. 4(d), (d')). Gurney and Fox pointed out that such destruction, if it occurred, might take place randomly or it might be specifically induced by the mismatch inherent in a genetically heterozygous duplex (cf. Chapter VII, Section 7). Since their markers were both integrated with high efficiency, their experiments provided no test of Ephrussi-Taylor and Gray's hypothesis that low efficiency integration is a consequence of mismatch repair.

Further evidence for genetic heterozygosity in pneumococcal transformation was obtained by Guerrini and Fox (1968*a*) using mutants *a* and *d* conferring resistance to sulphanilamide. These markers are linked, the double mutant showing joint transformation of both markers with a frequency of about 50%. The wild-type allele of *d* confers resistance to *p*-nitrobenzoic acid, the *d* mutant being sensitive to this substance. There is no comparable effect with the *a* mutant, which does not influence the behaviour of *d*. Most of the possible combinations of genotypes that might arise in a transformed cell can therefore be distinguished by growth tests with the resulting colony. Ultrasound was used to separate the individual cells of bacterial chains. The survivors of heat treatment also yielded colonies derived from single cells. Transformed colonies were recognized by plating in the presence of a low concentration of sulphanilamide.

A total of 227 transformed colonies, when both *a* and *d* were present in the donor DNA, were analysed for genotype, and a further 139 with *d* only in the donor. The results are given in Table 2. It is evident that almost all the colonies transformed for *d* also carried the wild-type allele of *d*, implying a heteroduplex in the initial transformed cell. The results for *a* are similar, though the evidence is not quite so complete. Furthermore, the heterozygosity was lost if the transformed cells were allowed to replicate before being isolated, as expected if the heterozygosity arises from a heteroduplex. Guerrini and Fox (1968*a*) allowed the transformed population to undergo one doubling by growing it for 60 min at 37 °C after the exposure to the donor DNA and before sonication and plating. They found that about 90% of the colonies containing transformants were pure, unlike the earlier result.

Guerrini and Fox (1968*a*) concluded that the product of transformation is genetically heterozygous and that this heterozygosity is rarely, if ever, converted to homozygosity prior to DNA replication. Both the *a* and *d* mutants, however, were integrated with high efficiency (cf. Tiraby and Fox, 1973), so these results,

Table 2.   Results obtained by Guerrini and Fox (1968*a*) for the numbers and genotypes of colonies derived from single cells of the pneumococcus transformed to sulphanilamide resistance. Each of the mutants *a* and *d* conferred such resistance. They could be distinguished by their differing sensitivity to *p*-nitrobenzoic acid. The solidus separates the genotypes of different cells of the colony.

Genotype of donor	Genotype of transformed colonies		
	Heterozygous $d/+$	Homozygous d	Homozygous d^+
a d	$a\,d\,/++$ 76 $a\,d\,/\,a+$ 0 $+d\,/++$ 71 $+d\,/\,a+$ 0	$\left.\begin{array}{l} a\,d\,/\,a\,d \\ a\,d\,/+d \\ +d\,/+d \end{array}\right\}$ 1 1	$\left.\begin{array}{l} a+/\,a+ \\ a+/++ \end{array}\right\}$ 78
$+d$	$+d\,/++$ 137	$+d\,/+d$ 2	—

like the earlier ones of Gurney and Fox, provide no test of Ephrussi-Taylor and Gray's hypothesis.

Guerrini and Fox (1968b) also investigated the effects of DNA repair on transformation. The repair followed treatment with ultraviolet light or mitomycin C. The genetic markers were the a and d sulphanilamide-resistant mutants, as before, and also a streptomycin-resistant mutant, S, linked to them. Wild-type bacteria were exposed to one of the damaging agents immediately after transformation with DNA carrying the three markers. It was found that relatively fewer cells carrying the donor markers survived compared with the untransformed recipient cells, and the larger the number of donor markers present the fewer survived. This is the result expected if repair of the damage to the DNA involves excision of a stretch of one strand, followed by resynthesis using the other strand as template. Such excision would be expected sometimes to remove the donor marker, if it was present in only one strand of the DNA. There was no corresponding effect if the cells were allowed to grow for 1 h, implying 1–1.5 doublings of the DNA, before treatment with the damaging agent. The donor markers, as a result of the replication, would now be in both strands, and excision of one strand would not eliminate them.

Confirmation that the higher sensitivity of the transformants to agents which damage DNA resulted from their elimination by excision was obtained by exposing the recipient bacteria to the damaging agent before transformation. No selective loss of transformed cells was found, as expected since there should be no damage in the donor DNA and hence no excision of it. Furthermore, with increasing dose of mitomycin C applied before transformation, an increasing proportion of the transformed cells were found to be homozygous for a donor marker. This is the result expected if excision of the recipient strand, in the repair of damage, leads to the incorporation of the donor nucleotide sequence into both strands. With two donor markers, a and d, cells homozygous for a and heterozygous d/+ were frequent, following treatment with either damaging agent. This was contrary to the results with untreated cells (Table 2). This genotype was interpreted as the result of excision and resynthesis of the recipient strand at the site of a, but the excision not extending as far as the site of d.

The conclusion that repair of damage leads to homozygosity for donor markers, or their elimination, depending on which strand is excised, implies that there is no such excision in the absence of damage repair. This supports the earlier conclusion that the a and d markers do not show mismatch repair.

(b) The *hex* system in the pneumococcus

A step forward in understanding integration efficiency was taken with the discovery of pneumococcal strains lacking the ability to give low efficiency transformation of point mutants. Lacks (1970) reported that Guild and Fox had found that a novobiocin-resistant mutant (*nov*) that transforms wild-type pneumococcus strain R6 with low efficiency will transform strain Rx with high efficiency. LE amylomaltase-deficient mutants on R6 were found likewise to become HE mutants on Rx. The genotype of Rx responsible for this behaviour was called *hex* (high efficiency on Rx).

Lacks (1970) isolated *hex* mutants by treating the normal strain (R6) with the mutagen 1-methyl-3-nitro-1-nitrosoguanidine (MNG) and testing for transformation with normally low efficiency markers. The *hex* mutants showed transformation of all single-site markers with high efficiency, but multisite markers such as deletions continued to be integrated with low efficiency. He pointed out that the occurrence of *hex* mutants suggests that low integration efficiency of single-site mutants might be due to a positive gene-determined factor, such as an enzyme, as postulated by Ephrussi-Taylor and Gray (1966). Lacks found that the ultraviolet resistance of *hex⁻* strains was about normal. This showed that the LE mechanism did not depend on the agents of scission and excision functioning in the repair of ultraviolet-induced damage. Tiraby and Sicard (1973*a*) described a mutant, 401, with properties similar to the *hex* mutants. Furthermore, they found that mutants which were integrated with low efficiency on wild type and high efficiency on 401 showed the late transmission to daughter cells on 401 characteristic of HE markers (see above). This was evidence that insertion of a marker in strain 401 by transformation results in alteration of only one of the strands of the recipient genome, whereas in wild type the LE markers result in alteration of both recipient strands.

They also found that spontaneous aminopterin-resistant (*amiA*) mutants were about three times as frequent in strain 401 as in wild type (Tiraby and Sicard, 1973*b*). They suggested that the discriminating process which acts during transformation in the wild type to give the low integration efficiency of some mutants may also act to eliminate some mutations, if initially they are present as a heteroduplex. This idea was supported by the observation that the excess of mutants with strain 401 were those of low efficiency when tested on wild type. Similar results were obtained by Tiraby and Fox (1973) with two other *hex* mutants, namely, that found in strain Rx and an MNG-induced mutant, *hex-1*, obtained by Lacks (1970): see above. They concluded that the discrimination between LE and HE markers in *hex⁺* strains pointed to the existence of one or more nucleases capable of distinguishing different mismatched base pairs. Excision of one strand at the mismatch, followed by repair synthesis, would make the surviving LE transformants homozygous, as found by Ephrussi-Taylor and Gray (1966). To explain the low integration efficiency, Ephrussi-Taylor and Gray had postulated preferential attack by the enzyme on the donor strand, but Tiraby and Fox pointed out that another possibility was that an attack on the recipient strand was frequently lethal (see below). In *hex⁻* strains the enzyme postulated to recognize some mismatches appears to be deficient, with the result that LE markers are integrated with high efficiency and spontaneous mutations are much more frequent, the increase being in LE mutants (when tested on *hex⁺*). This is the result expected if the *hex* function often eliminates mutational heterozygotes within the duplex DNA that resemble those formed by transformation with LE markers.

Tiraby and Fox (1974*a*) investigated the base mismatches to which the *hex* function responded by treating the donor DNA with the mutagens hydroxylamine and nitrous acid, which are known to induce transitions: hydroxylamine induces a transition from $G \cdot C$ to $A \cdot T$, and nitrous acid induces both this transition and the reverse change. Mutants showing resistance to fusidic acid were selected. The mutants induced with either mutagen fell almost exclusively in the LE class. This result was expected in view of the comparable discoveries by Ephrussi-Taylor, Sicard and Kamen (1965) with aminopterin-resistant mutants and by Lacks (1966) with amylomaltase-deficient mutants (see above). Tiraby and Fox (1974a) concluded that the mispaired bases $A \cdot C$ and $G \cdot T$ were recognized by the *hex* function, and that at least one class of transversion mispairing ($G \cdot G$ and $C \cdot C$; or $A \cdot A$ and $T \cdot T$; or $G \cdot A$ and $T \cdot C$) was not recognized, since it was known that HE markers can result from single base substitutions. A more surprising result was the discovery that mutants arose with the same frequency, following the treatment of the donor DNA with the mutagen, whether the recipient was *hex*+ or *hex*−. This implied that the heteroduplex comprising the incipient mutant in the donor and the untreated recipient strand was not sensitive to *hex* action.

Information about the time of action of the mechanism determined by the *hex* function was obtained by Shoemaker and Guild (1974) from kinetic studies. The donor DNA contained a novobiocin-resistance mutant integrated with low efficiency and a streptomycin-resistance mutant integrated with high efficiency. Recovery of the LE and HE donor markers from the eclipse phase that occurs immediately after the donor DNA has entered the recipient cells (see Section 2) was monitored by sampling at different times, and using these lysates in a further transformation experiment. It was found that in *hex*+ cells the activity of the LE marker rose substantially after the eclipse, at 37 °C reaching a maximum after 3–4 min. This rise was relative to an erythromycin-resistance marker in the recipient, and was taken to indicate the formation of heteroduplex structures, since recovery from eclipse is known closely to parallel the transition of single-stranded donor DNA segments to donor–recipient heteroduplex structures: see discussion of the work of Lacks (1962) and others in Section 2. The activity of the LE marker was found by Shoemaker and Guild (1974) then to fall nearly twofold until becoming constant, relative to the marker in the recipient, at 10 min from the start of the rise. At 30 °C the reactions were 2–2.5 times slower. By contrast with these results for LE markers, they found that HE markers showed no second decline in activity after their recovery from eclipse. This recovery took about 5 min at 37 °C, so LE markers continued to decline in activity for a further 5 min after HE markers had fully recovered from eclipse. In *hex*− cells all the markers showed recovery similar to the HE markers in *hex*+ cells.

The decline in LE marker activity in *hex*+ cells 3–4 min after the start of the rise was attributed to the excision of the marker from the donor strand at the site of the mismatch in the heteroduplex. Shoemaker and Guild (1974) concluded that *hex* action was relatively slow and occurred after the main steps in heteroduplex formation had taken place.

According to Ephrussi-Taylor and Gray (1966) donor excision occurs in 90% of the LE mismatches, while in the remaining 10% the recipient allele is excised, thus giving rise, following repair of the recipient strand, to the LE marker in both strands. By contrast, the mismatches arising in hybrid DNA from HE markers remain unrepaired and so the marker is in the donor strand only. Shoemaker and Guild (1974) therefore predicted that in experiments such as theirs, where the donor DNA contained both an LE and HE marker, the ratio (LE/HE) of their activities in the lysate when these had become constant (10 min at 37 °C after the start of the rise) should be twice that observed by plating the recipient cells immediately after the donor DNA has entered them. These cells will yield a transformed clone if the donor marker is integrated, irrespective of whether the integration involves one strand or both. The observed values for

$$\frac{\text{final activity (LE/HE)}}{\text{initial transformants (LE/HE)}}$$

in four experiments with the novobiocin- and streptomycin-resistance markers were all less than the 2.0 predicted: they ranged from 1.1 to 1.5. Shoemaker and Guild (1974) concluded that an appreciable fraction of the LE (novobiocin-resistance) markers that were not eliminated by donor excision also failed to undergo recipient excision, that is, they escaped *hex* action. A likely explanation for such escape was that the presumed excision–repair system was easily saturated with substrate, with the result that some LE mismatches were not repaired before replication supervened.

They tested this hypothesis by adding DNA that would increase the number of mismatched sites, and studying its effect on integration efficiency and on sensitivity to ultraviolet light (Guild and Shoemaker, 1974). They found that uptake of pneumococcal DNA of a different strain, along with the donor DNA, increased six- to sevenfold the efficiency of integration of the LE novobiocin-resistance marker. This is the result expected if the added non-isogenic pneumococcal DNA is providing extra mismatched sites which compete with the LE marker site for the *hex* system. In agreement with this conclusion, *B. subtilis* DNA, which enters pneumococcal cells but is non-homologous with pneumococcal DNA and so would not form a heteroduplex, had no effect on the integration efficiency of the LE marker. Completely homologous, that is, isogenic, pneumococcal DNA also had no effect. This result was also predicted, as such DNA would not increase the number of mismatched sites. Guild and Shoemaker (1974) concluded that intracellular competition for a part of the *hex* system was the likely reason why heterozygosity for unlinked genes leads to escape of LE markers from the system. If at least one component is present in only a small amount per cell, this would also explain why *hex* action continues for 10 min at 37 °C, as already described.

As regards number of transformants, LE markers are known to be much more sensitive than HE markers to ultraviolet irradiation of the donor DNA carrying them. Lacks (1970) showed that this effect was evident only in *hex*[+] recipient cells. In a *hex*[−] recipient, LE markers are not only integrated with high efficiency but also lose their ultraviolet sensitivity. In both these respects they become indistinguishable from HE markers. Guild and Shoemaker (1974) discovered that competitor DNA, that is, non-isogenic pneumococcal DNA, protects the LE novobiocin marker from ultraviolet light in a rather dramatic way. They suggested that the sensitivity of LE markers to ultraviolet light was because *hex* attack on the recipient strand in the heteroduplex in conjunction with attack by ultraviolet-endonuclease on pyrimidine dimers in the donor strand leads to breakage of the duplex and hence loss of the potential transformant. The marker-specific rescue from ultraviolet inactivation by non-isogenic but homologous DNA would then be the consequence of deflecting the *hex* attack elsewhere.

In order to explain the low efficiency of integration of certain mutants (transitions, at least) in *hex*$^+$ strains, Tiraby and Fox (1974*b*) tested two hypotheses:

(1) *Survival of LE transformants requires replication to occur immediately after the mismatch has arisen, before it is recognized by an endonuclease that cuts both strands.*

They postulated that the mismatch formed by LE markers was recognized and that formed by HE markers was not recognized by an endonuclease that cut both strands of the DNA in proximity and so was lethal for the potential transformant. If an LE mismatched site happened to be just ahead of a replication fork, however, the mismatch might be so transitory that replication supervened before the endonuclease could act. Such replication would then not only allow the transformant to survive but would explain how the marker nucleotide sequence came to be in both strands of the duplex.

Tiraby and Fox (1974*b*) tested this hypothesis in two ways. First, they used 6-(*p*-hydroxyphenylazo)uracil to inhibit DNA replication. Incorporation of ^{3}H-labelled thymidine ceased almost immediately, but cell viability and the yield of transformants for individual LE markers were unaffected by the inhibitor. Moreover, the frequency of double transformants for two unlinked LE mutants (three different pairs were tested) was unaffected. This was a sensitive test because in order to explain simultaneous transformation by two widely spaced LE markers, it would be necessary to invoke two replication forks, the frequency of which would be expected sharply to decline in the presence of an inhibitor of DNA synthesis.

For the second test of the replication hypothesis, recipient cells (*hex*$^+$) were transferred to a medium containing nutrients labelled with the heavy isotopes ^{15}N and ^{13}C, and after incubation for 10 min donor DNA carrying an LE (novobiocin-resistance) marker and an HE (streptomycin-resistance) marker, but no heavy isotopes, was added. After 30 min exposure, any remaining DNA external to the cells was broken down by deoxyribonuclease treatment. After a further 5, 30 and 60 min incubation (corresponding to 5%, 50% and 75% of the recipient DNA having undergone replication in the heavy medium) the DNA was extracted and centrifuged to equilibrium in a caesium chloride gradient. The transforming activities of extracts were determined by a further transformation experiment. The hypothesis postulates that LE markers successfully bring about transformation only in the neighbourhood of replication forks. The newly synthesized DNA will have one heavy and one light strand, and so this DNA (recognized by its hybrid density in the gradient) should be enriched in LE marker activity. This is relative to the HE marker activity, which will show no relation to replication. Conversely, the DNA of light density, and therefore unreplicated, should show reduced LE marker activity. These predictions of the hypothesis were not confirmed, however, by the experimental results, which showed that DNA carrying the LE marker activity did not become of hybrid density any more rapidly than did that carrying the HE marker activity.

From these two experiments, Tiraby and Fox (1974b) concluded that the replication fork was not involved in the survival of LE transformants.

(2) *Survival of LE transformants requires exonucleolytic breakdown of the recipient strand, not the donor one, at the site of mismatch, but this breakdown must not extend to the end of the inserted fragment (unless covalent joining has already occurred).*

Tiraby and Fox (1974b) assumed that the mismatched bases in the heteroduplex at the site of the LE mutant were recognized by an endonuclease that cut one strand, and that there was then exonucleolytic breakdown of that strand, before repair synthesis took place. If the donor strand was cut, the newly introduced marker would be lost (Fig. 5(a)). If the recipient strand was cut, the transformant would survive (Fig. 5(b)), provided that the excision of the recipient strand did not extend as far as the end of the donor fragment. Such extension would be lethal because both strands of the duplex would be broken at the same position (Fig. 5(c)).

They tested this hypothesis by carrying out transformation using donor DNA molecules of various sizes. These were obtained by shearing the molecules extracted from the donor strain by stirring them at 40 000 rev min^{-1} in a homogenizer for various periods of time (30, 90 and 210 min). It was found that in a discriminating (*hex$^+$*) strain, LE markers were much more sensitive to shear than HE, as predicted by the hypothesis. In other words, the chances of an LE transformant surviving were greatly reduced as the donor fragment carrying it was shortened.

Tiraby and Fox (1974b) concluded that low integration efficiency arose in the way that Ephrussi-Taylor and Gray had suggested, as a consequence of mismatch repair at the mutant site in the heteroduplex often leading to loss of the donor nucleotide sequence (Fig. 5(a)). But Tiraby and Fox also concluded that there was no need to suppose, as Ephrussi-Taylor and Gray had done, that the donor strand was excised 10 times more often than the recipient one. The need for this bias was removed, with the evidence supporting the alternative hypothesis that breakdown of the recipient strand was often lethal, through excision extending in the recipient strand of the heteroduplex to the point where the donor segment ended (Fig. 5(c)).

The idea that excision of the recipient strand was often lethal was questioned by Guild and Shoemaker (1976). They repeated Tiraby and Fox's experiment, shearing the donor DNA to molecular weights ranging from over 10^7 to 3.6×10^5 daltons, and found no effect on the relative integration efficiencies of LE and HE markers in a *hex$^+$* strain. Guild and Shoemaker concluded that in order to

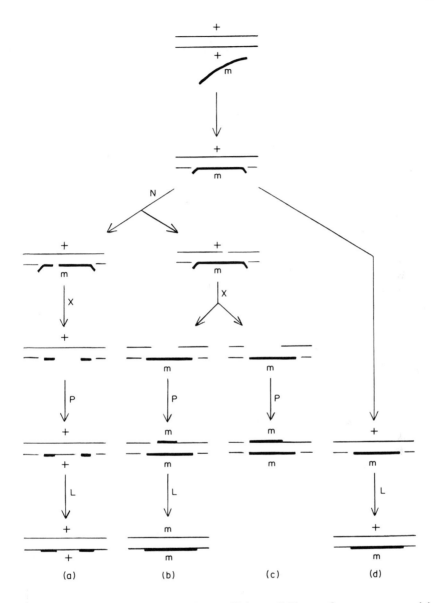

Fig. 5. Mechanism of elimination of low efficiency (LE) transformants proposed by Tiraby and Fox (1974*b*). Strands with donor nucleotide sequence are shown with thick lines. The donor carries a mutant, m, the recipient having the wild-type allele, +. L, ligase; N, endonuclease; P, polymerase; X, exonuclease. (a) Donor strand cut at site of LE mismatch followed by exonucleolytic breakdown and resynthesis, leading to loss of donor mutant. (b), (c) Recipient strand cut at site of LE mismatch, followed by exonucleolytic breakdown and resynthesis, leading to (b) a double-strand transformant, or (c) a lethal two-strand break as a result of the exonucleolytic breakdown extending to a point in the recipient strand opposite the end of the donor segment (not yet covalently joined). (d) Lack of recognition of HE mismatch leads to single-strand transformant.

maintain the lethal event hypothesis the excision must often be of greater extent than the size range studied, implying at least 7000 nucleotides excised. They preferred Ephrussi-Taylor and Gray's hypothesis of preferential excision of donor strands, and suggested that the nuclease, after recognizing an LE mismatch, migrates along the DNA and if it finds a free end (such as might still be present at the ends of the donor segment) excises that strand. This hypothesis is discussed in Chapter VIII.

Further experiments of the same kind were made by Lefevre, Claverys and Sicard (1979), using mechanical shearing, and also restriction endonuclease treatment, to reduce the size of the donor DNA molecules. They found that transformation ability of LE markers decreased with DNA size to the same extent as that of HE markers, even for very short DNA fragments, in agreement with Guild and Shoemaker (1976). Moreover, Lefevre *et al.* (1979) searched for *hex*-mediated duplex breaks using as donor homologous but non-isogenic DNA in order to saturate the *hex* system with the numerous mismatches that this would produce in the heteroduplex segments (see above). They followed the growth of hex^+ recipient cultures derived from single cells and found no difference in survival between those transformed with isogenic and with non-isogenic DNA. They concluded that the lethal events postulated by Tiraby and Fox were not taking place.

A new approach to an understanding of integration efficiency was made by Roger (1977), by using artificially prepared heteroduplexes as donor DNA in transformation. Her technique for resolving the heavy (H) and light (L) strands is described in Section 4 below. Heavy or light refer to the buoyant density in a caesium chloride density gradient. Complementary heteroduplexes, that is, H wild type/L mutant, and L wild type/H mutant, were prepared for a number of HE mutants. For most of the mutants she found a five- to sevenfold difference in integration efficiency between the two heteroduplexes when the recipient strain was hex^+, with some mutants showing the low efficiency when the H strand was mutant and others when the L strand was the mutant one. Only small differences or none were found when the recipient was hex^-. Sulphanilamide-resistance mutant *d* was known to integrate with unusually high efficiency (about twice that of normal HE mutants) and Roger found that the two heteroduplexes of this mutant and wild type behaved alike on a hex^+ recipient. She concluded that this very high efficiency (VHE) mutant was unaffected by the *hex* system but that, contrary to previous ideas, one of the two kinds of mismatch of ordinary HE mutants was sensitive to *hex* action. In agreement with this, HE mutant *a* at the sulphanilamide-resistance (*sul*) locus, which showed low efficiency integration when in the L strand, affected *d* similarly when the donor heteroduplex comprised both mutants in a *cis* configuration, the integration efficiency of the VHE mutant being nearly halved under these circumstances. Evidently excision, triggered by the mismatch when mutant *a* is in the L strand of a donor–recipient heteroduplex, is extensive and often reaches the site of mutant *d*. This influence of an HE mutant on a closely linked VHE mutant was comparable to the effect of an LE mutant on a closely linked HE mutant reported by Lacks (1966) and others and described above.

Roger used the *a* and *d* mutants at the *sul* locus to confirm that the mismatch excision was occurring in the donor–recipient heteroduplex formed during transformation and not in the artificial heteroduplex used as donor. This distinction was made possible by using various donor heteroduplex genotypes (involving *a*, *d* or both) to transform various recipient genotypes (wild type, *a* or *d*). The relative integration efficiences when L and H strands were transposed were found to be constant for each donor–recipient heteroduplex genotype, irrespective of which donor heteroduplex gave rise to it.

Roger's investigation of the transformation behaviour of individual DNA strands was extended to LE mutants by Claverys, Roger and Sicard (1980). As expected, they found that the two heteroduplexes of such mutants with wild type behaved alike on a *hex*+ recipient. Evidently excision occurs with both kinds of mismatch produced by LE mutants. Claverys *et al.* confirmed with 11 VHE mutants at the aminopterin-resistance (*amiA*) locus that there was no strand bias in integration, whereas HE mutants showed a marked strand preference. Moreover, this bias was allele specific, some HE mutants at both the *amiA* and the rifampicin-resistance (*rif*) loci showing low efficiency integration when the mutant was in the H strand, and others when in the L strand (Table 3). No such

Table 3. Data on pneumococcal transformation obtained by Claverys, Roger and Sicard (1980) using artificial heteroduplexes of mutant and wild type as donor (H+/Lm and L+/Hm) and a *hex*+ wild-type recipient. The integration efficiency of the mutants is indicated by LE, HE and VHE, standing for low, high and very high, respectively. The mutants were in the genes *amiA, opt* and *rif* (resistance to aminopterin, optochin and rifampicin, respectively). L and H indicate the light and heavy strands, respectively, of the DNA; + and m stand for wild type and mutant, respectively.

Class of mutant	Mutant (gene in italics)	Approximate relative strand efficiency Lm/Hm	Mismatches inferred to be recognized and excised in donor–recipient heteroduplex	Relative rates of expression of donor mutant character in transformants
VHE	*amiA* 1 *amiA* 158	1	Neither	H earlier than L
HE	*amiA* 144 *rif* 10	5	L+ · Hm	H earlier than L
HE	*amiA* 22 *amiA* 150 *rif* E	0.2	H+ · Lm	H and L equal
LE	*amiA* 9 *opt* 2 *rif* 23	1	Both	H and L equal

difference was found with a *hex*⁻ recipient. The allele specificity in strand preference for HE markers provides strong support for the idea that the *hex* function recognizes and excises specific mismatched base pairs.

Claverys *et al.* (1980) carried out a further experiment. At various times after exposure to the donor DNA the recipient culture was sampled and scored for the donor character. This enabled a comparison to be made of the time of expression of each donor mutant character when introduced in the H strand and when introduced in the L strand. With a *hex*⁻ recipient all markers irrespective of gene or integration efficiency were expressed earlier when in the H strand. This result would be accounted for if the H strand is that which is transcribed. With a *hex*⁺ recipient, VHE mutants were expressed earlier when introduced in H than in L, as expected if there is no mismatch correction; LE mutants showed the same expression rate for both strands, as expected if mismatch repair regularly transfers the mutation to both; and HE mutants were diverse, some behaving like the VHE mutants and others like the LE mutants. Significantly, the former (with earlier expression in H than in L) were the HE mutants that showed high efficiency integration when introduced in L (Table 3), implying no mismatch repair when in this strand, and hence no opportunity for transfer to the transcribed H strand until replication occurred. Conversely, the HE mutants that showed the same expression rate for both strands were those which showed low efficiency integration when introduced in L, with the implication that the mutant nucleotide sequence was transferred to H by mismatch repair. It was evident that correction by the *hex* function of one of the two kinds of mismatch will explain the results obtained for HE mutants both as regards integration efficiency and rate of gene expression. The nature of these mismatches has not yet been established.

(c) Other species

Marker effects, such as have been studied so extensively in *D. pneumoniae*, have not been found in *B. subtilis* or *H. influenzae*. The presumption, therefore, is that there is no mismatch repair in transformation in these bacteria. Support for this conclusion for *B. subtilis* was obtained by Darlington and Bodmer (1968). Competent cells of a strain carrying mutants at four closely linked loci were exposed to wild-type DNA as donor at low concentrations, such that transformation brought about by more than one molecule was negligibly frequent. The markers in the recipient were at the *aro-2*, *try-2*, *his-2* and *tyr-1* loci, which map in the sequence given. Transformants for the two outside markers, that is, wild type for *aro-2* and *tyr-1*, were selected. It was found that 85% were wild type also for both of the intervening markers. It was evident that excision and repair of these *try-2* and *his-2* mutants could not be frequent.

4.   THE CLUSTERING OF SINGLE-STRAND INSERTIONS

From three-point transformation crosses with aminopterin-resistant (*amiA*) mutants of the pneumococcus, Ephrussi-Taylor and Gray (1966) found that the recombination frequencies which they observed between very closely linked markers were too high to be accounted for by the insertion of one stretch of nucleotide chain per donor DNA molecule. They estimated that about 10 stretches were inserted, that is, 20 alternations of donor and recipient parentage, with a random distribution of exchange points.

Lacks, Greenberg and Neuberger (1974) had discovered mutants of the pneumococcus that were able to bind DNA to the cell surface but unable to take it into the cell: see Section 2. Lacks and Greenberg (1976) set out to examine the structure of this DNA bound to the mutant cells. For this purpose they used the DNA of phage T7 of *E. coli*. They compared the sedimentation behaviour in sucrose gradients of DNA that had been bound to the cells of the pneumococcal mutant with that of the control, that is, unbound DNA. The sedimentation rate in neutral gradients did not change on binding. This showed that no duplex cuts occurred. Sedimentation in alkaline gradients was much slower after binding, however, with the implication that single-strand breaks had occurred. Lacks *et al.* (1974) suggested that the development of susceptibility to transformation, that is, the condition called competence, might correspond to the formation on the cell surface of the mechanism for binding, and Lacks and Greenberg (1976) suggested that the binding involved the formation of the single-strand breaks. The binding reaction requires energy, so they suggested that the DNA binds to a surface protein that nicks the DNA in an energy-requiring reaction. The binding is not readily reversible, so they proposed that the nicking protein becomes bound to a newly formed end of the broken strand.

When they used a mutant that was only partially blocked in entry of DNA into the cells, they found that duplex cuts occurred, causing fragmentation of the molecule. Such double-strand cleavage of the bound DNA had been discovered by Morrison and Guild (1973), and it now appeared that this duplex breakage was related to entry rather than to binding. Lacks and Greenberg (1976) suggested that a deoxyribonuclease localized in the membrane adjacent to the binding protein might initiate entry by cutting the strand opposite the nick in the bound strand. The latter would then enter the cell, while the DNase acted progressively on the strand it cut, releasing oligonucleotides. Their hypothesis is illustrated in Fig. 6. It has important genetical consequences, because it envisages one DNA molecule binding at several sites, followed by entry of single strands derived from different, that is, non-homologous, parts of the molecule, with the strand chosen for entry picked at random. In adjacent segments the incoming strands might thus be of opposite polarity, that is, have come from different strands of the original duplex. Multiple alternations of donor and recipient parentage in the transformed molecule, as postulated by Ephrussi-Taylor and Gray (1966), might thus be accounted for, following insertion of the single strands of donor origin into the DNA of the recipient.

Roger (1972) had prepared heteroduplex molecules of pneumococcal DNA *in vitro*. She separated the strands by denaturation with sodium hydroxide, and then isolated the heavy (H) and light (L) strands either by salt gradient elution from methylated albumin–kieselguhr columns or by interaction with poly(U,G). To

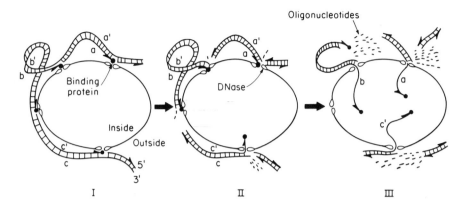

Fig. 6. Molecular model for DNA binding and entry, I, Formation of single-strand breaks on attachment of DNA to binding protein. II, Formation of double-strand breaks on initiation of DNA entry. III, DNA entry catalysed by membrane DNase that degrades one strand of DNA to oligonucleotides as it draws the complementary strand segment into the cell. Different parts of one strand of the original molecule are labelled a, b and c, and of the complementary strand a′, b′ and c′. (Reproduced with permission from S. Lacks and B. Greenberg, *J. molec. Biol.*, **101**, 273 (1976). Copyright by Academic Press Inc. (London) Ltd.)

obtain heteroduplex molecules, equal amounts of H and L strands derived from DNA carrying different genetic markers were mixed and annealed at 65 °C. The duplex DNA molecules were then used as donors to bring about transformation of recipient cells. When the heteroduplex molecules carried two widely spaced genetic markers in the *trans* configuration, that is, one in H and the other in L, doubly transformed progeny, that is, carrying both markers, were as frequent as in homoduplex controls. Lacks and Greenberg (1976) suggested that this co-transformation might be the consequence of strands of opposite polarity from adjacent segments of the donor molecule entering the cell and becoming integrated into the recipient genome, followed by mismatch correction to the donor genotype for at least one of the mutants.

Multiple insertions per donor DNA molecule have also been inferred for transformation in *B. subtilis* by Dubnau and Cirigliano (1972) using a different approach. They studied the efficiency of the transformation process by using labelled donor DNA and, after transformation, obtaining a lysate at various times, subjecting the DNA to sucrose gradient sedimentation and measuring the radioactivity at the various densities. From their results they concluded that an appreciable part of the recipient genome may be replaced by donor DNA. Earlier studies (Dubnau and Davidoff-Abelson. 1971), some aspects of which have already been discussed (Section 2), had indicated an average molecular weight of 750 000 daltons for an individual simplex donor segment inserted into the recipient genome. Dubnau and Cirigliano (1972), however, found that the average molecular weight of the transforming DNA entering the cell was much higher, namely, about 8×10^7 daltons. Knowing the high efficiency of integration, they concluded that each donor molecule must give rise to a clustered series of single-strand insertions.

As already pointed out in Section 3, Darlington and Bodmer (1968) found that closely linked markers in *B. subtilis* tend to insert as a single piece. Dubnau and Cirigliano (1972) assumed that the mutants used by Darlington and Bodmer had sites within the length of an individual donor insert.

Direct evidence for the occurrence of multiple insertions of segments of a single donor DNA molecule in *B. subtilis* was obtained by Fornili and Fox (1977). They took advantage of the fact that the alkalinity required to denature DNA is reduced as a result of replacement of thymine by bromouracil. They used such DNA as donor, the recipient having normal DNA, isolated the DNA from transformed cells and denatured it at a pH (10.8) that caused separation of strands containing bromouracil but not of those without it. The molecules were examined under the electron microscope and a series of loops ('bubbles') were seen, where the strands had separated. These evidently corresponded to regions of donor DNA insertion, as no such loops were found in control experiments at this pH using normal DNA, or DNA in which one or both strands contained bromouracil. The loops were found often to occur in clusters with an average overall length of loops and short intervening regions of 3.5 μm. Each cluster of denaturation loops was believed to represent one single-strand segment of donor DNA integrated into the recipient molecule. This length would be equivalent to about 10 000 nucleotides.

A striking feature of Fornili and Fox's electron micrographs was the abundance of molecules with two or more well separated loops or loop clusters. These did not appear to be the result of independent interactions with more than one donor molecule, because calculations showed that the high frequency observed of two or more loop clusters within the 10–30 μm lengths of DNA examined would not be expected. They concluded that the multiple loop clusters were the products of coupled integration events. Such associated events would be expected if transformation in *B. subtilis* involves one donor DNA molecule binding to the recipient bacterium at several sites, followed by multiple entry of single strands from different parts of the molecule, as believed to occur in the pneumococcus from the studies by Lacks and Greenberg (1976) described above.

Clustering of single-strand insertions may not occur in *H. influenzae* because the uptake of DNA by competent cells is differently organized. In the pneumococcus and *B. subtilis* the transport systems are non-specific and DNA from any source may be taken up. In *H. influenzae*, on the other hand, uptake is highly specific and only *Haemophilus* DNA enters the cell. Sisco and Smith (1979) showed that the specificity in *H. parainfluenzae* depends on base sequence. They cloned an 8100-base sequence in a plasmid and determined the number and location of uptake sites by testing the fragments produced by restriction endonucleases, using ^{32}P at the 5' end of each fragment as a radioactive label. Only two fragments retained the ability to be absorbed. Sisco and Smith thus estimated that there were about 600 copies of the uptake site sequence in the whole genome. They concluded that a receptor that recognizes the uptake sequence must exist on the surface of competent cells.

5. CONCLUSIONS

A substantial body of evidence supports the conclusion that genetic transformation in *Diplococcus pneumoniae, Bacillus subtilis* and *Haemophilus influenzae* involves the insertion of a single strand of the donor DNA into the recipient genome in place of the corresponding segment of its DNA.

The resulting heteroduplex is subject to mismatch repair in some strains of the pneumococcus, but not in other strains, nor in the other two species of bacteria. The mismatch repair system recognizes some kinds of base mismatch but not others. Following recognition, one of the mispaired nucleotides is excised. If both of the complementary mismatches that can be produced by a substitution mutant are recognized, the mutant is integrated with low efficiency (LE); if only one is recognized, the mutant is integrated with high efficiency (HE); and if neither is recognized, with very high efficiency (VHE). The excision in the repair process is extensive, accounting for the influence of LE on HE mutants, and of HE on VHE, when closely linked. Excision appears to occur in the donor strand about 10 times more often than in the recipient one. The reason for this bias is not known. A component of the repair system is in relatively short supply, with the result that the system is easily saturated by substrate.

At least in *D. pneumoniae* and *B. subtilis*, multiple insertions take place of single strands of either polarity from different parts of one donor DNA molecule, apparently as a consequence of multiple binding sites on the cell surface, to which different parts of the same molecule become attached.

III.   Recombination in single-stranded DNA phages of *Escherichia coli*

1. INTRODUCTION

There are two groups of single-stranded DNA phages of *Escherichia coli* which are serologically unrelated and differ in many characters. They are icosahedral and filamentous in shape, respectively.

2. ICOSAHEDRAL PHAGES S13 AND φX174

Phage S13 is one of about 30 phages studied at the Lister Institute, London, by Burnet (1927), who gave it its identification number. He derived it from a lysogenic strain of *Bacillus enteritidis* (Gaertner), that is, *Salmonella enteritidis*. According to Buchanan, Holt and Lessel (1966) this bacterium was isolated by Gaertner in 1888 from faeces in an epidemic of meat poisoning at Frankenhausen, Germany, so this may have been the original source of the virus. The prefix S to the identification number was added by Burnet and McKie (1930*a,b*) to indicate that it was a *Salmonella* phage.

Phage φX174 was isolated by Sertic and Boulgakov (1935*a,b*) from a Paris sewer. Their work was conducted from d'Herelle's laboratory. His original discovery of bacteriophage had been from the same source. Sertic and Boulgakov isolated some 75 phage strains and gave them serial numbers from 110 onwards. They found that the phages belonged to 14 antigenic types, which they denoted by Roman numerals preceding the isolation number, 174 and one other being of antigenic type X. They added a further prefix: φ (phi) for those phages, including X, that were virulent on several species of bacteria and τ (tau) if they multiplied only on *Salmonella typhi*.

S13 and φX174 have been shown to be closely related physically and chemically. They both have virions which are icosahedral in shape and 25 nm in diameter. Both are released from the host cell in the normal way by breakdown of its membrane (lysis). φX174 is inactivated by anti-S13 serum, though more slowly than S13 (Zahler, 1958).

35

3. FILAMENTOUS PHAGES f1, fd AND M13

Loeb (1960) at the Rockefeller Institute. New York, isolated from sewer water seven phages which he called f1, f2, etc. They were selected for their ability to multiply on male strains (F+ and Hfr) of *E. coli* but not on the female (F−). Of these phages, f1 was found subsequently to be a DNA phage (Zinder *et al.*, 1963) and f2 an RNA phage.

Marvin and Hoffmann-Berling (1963) at Tübingen and Heidelberg isolated from sewage two small phages which they called fd and fr, since their nucleic acid was DNA and RNA respectively. These phages proved it to be quite similar to f1 and f2 respectively.

Hofschneider (1963) at the Max-Planck-Institut für Biochemie, Munich, isolated several phages, prefixing their serial numbers by M for Munich. M13 was found to be similar to f1 and fd.

Salivar, Tzagoloff and Pratt (1964) showed that these three viruses were closely related physically and chemically, though slight differences were detected serologically. They all have a virion shaped like a flexible rod, as first discovered by Marvin and Hoffmann-Berling (1963) with fd. It measures about 800 nm x 5 nm. They are all restricted to male strains of *E. coli*. Furthermore, the host cells are not lysed but continue to grow and divide as progeny virions are extruded through the cell membrane.

4. SIMILARITIES BETWEEN ICOSAHEDRAL AND FILAMENTOUS PHAGES

Despite their differences in structure and life-cycle, the icosahedral and filamentous phages have many features in common:

(a) DNA

They have as their genome a closed circular simplex DNA molecule of molecular weight about 1.7×10^6 daltons, equivalent to about 5500 nucleotides. The genome comprises about nine genes.

(b) Replication of DNA

(1) The strand of DNA present in the mature virus particle is one particular one and is called the plus strand. On entering the host cell, a complementary minus strand is synthesized by the host DNA polymerase III, a polyribonucleotide having first been synthesized to act as a primer. The circular duplex molecule formed is called the parental replicative form (RF). It has a contour length on electron micrographs of about 1.7 μm.

(2) Further circular duplex molecules, called progeny RF, are then formed by replication of the parental RF by rolling circle. To initiate this process a specific endonuclease, the product of a phage gene (called *A* in the icosahedral and *2* in the filamentous phages), cuts the plus strand at a specific site in this gene.

(3) At a later stage, circular plus strands for inclusion in the progeny particles are synthesized using the minus strands of the progeny RF as templates. The specific endonuclease is believed to function here also, cutting off unit lengths at the specific site. The switch to single-strand synthesis requires a DNA-binding protein, the product of a phage gene (called B in the icosahedral and 5 in the filamentous phages).

(c) Transcription

The plus strand, that is, the strand present in the mature virus particle, is antisense, so messenger RNA has to be synthesized using the minus strand as template. There is evidence that transcription occurs from the minus strands of the progeny RF.

5. RECOMBINATION IN THE ICOSAHEDRAL PHAGES

(a) Introduction

Zahler (1958) looked for recombination between S13 mutants without success and concluded that there was less than one recombinant per 1000 parental types. By increasing the frequency of recombinants by exposure of the parents to ultraviolet light, Tessman and Tessman (1959) observed recombinants with a frequency of about 0.1%. The mutants used affected plaque type or host range. Recombination in ϕX174 was first demonstrated by Pfeifer (1961a,b) using host-range mutants and revertants of them. Tessman and Shleser (1963) showed that recombination could occur between S13 and ϕX174. By mixed infection of *E. coli*, plaque-type mutants of S13 were crossed with wild-type ϕX174, both with and without pretreatment with ultraviolet light. Recombination was detected because wild-type S13, unlike the parental phage strains, will form plaques on a resistant host strain.

Baker and Tessman (1967) showed that the S13 linkage map is circular. The seven genes known at that time were mapped. Double recombination, that is, high negative interference, was found to be the rule, as expected with a small circular DNA molecule as genome. Benbow *et al.* (1971) made comparable discoveries for ϕX174. They used temperature-sensitive and nonsense mutations in eight genes, and found that the total length of the linkage map was 24×10^{-4} wild-type recombinants per progeny phage. The map was revised by Benbow *et al.* (1974a) to include nine genes, and subsequently the astonishing discovery was made that some of these overlap, using different reading frames (Barrell, Air and Hutchison, 1976; Weisbeck *et al.*, 1977): see Fig. 7(a).

(b) Primary and secondary mechanisms

An important discovery concerning recombination in the icosahedral phages was made by Tessman (1966). He studied the effect of a *recA* mutant of *E. coli* (see

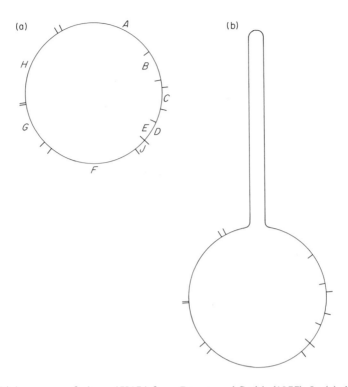

Fig. 7. Linkage map of phage φX174 from Brown and Smith (1977). In (a) the letters *A–J* show the position and size of the nine genes. The replication origin is within gene *A* and is at 12 o'clock on the map. In (b) the hot spot for recombination at the replication origin is indicated. This diagram is based on that of Benbow *et al.* (1971) with distances proportional to recombination frequencies. The panhandle shows the discrepancy at the replication origin between the genetic map and the physical map.

Chapter VI, Section 4) isolated by Clark and Margulies (1965). It was known from their work to reduce host recombination more than 1000-fold. Tessman (1966) found that recombination of mutants in genes *F* and *H* (called complementation groups I and IIIb at that time) of S13 was reduced 20–40 times in a *recA⁻* host, compared with *recA⁺*, and in gene *B* (complementation group II) four to eight times. These much smaller reductions than in the host were taken to mean that S13 has a secondary recombination mechanism that is independent of the host *recA⁺* gene product. He extended this conclusion (Tessman, 1968) by studying the effect of ultraviolet irradiation on S13 recombination. He found that ultraviolet treatment of the virions before infection caused a large increase in S13 recombination on a *recA⁺* host, but had no effect with *recA⁻*. This selective stimulation of the *recA*-sensitive process supported the idea that there were two recombination mechanisms in S13. Tessman (1968) found that S13 recombination was unaffected in a *recB* mutant of the host.

Baker, Doniger and Tessman (1971) investigated in S13 the stages of DNA replication during which the two recombination mechanisms were effective. Their experiments were made possible by the existence of mutants of gene A of the phage. As already indicated, the product of this gene is required for replication of the parental replicative form (RF), but not for its formation. In A^- mutants, therefore, DNA replication is blocked at this stage. Baker et al. (1971) found that S13 recombination by the primary mechanism occurred with the same frequency in A^+ and A^- phage, whereas recombination by the secondary mechanism was abolished in A^- mutants. The mechanisms were distinguished by their response to a $recA^-$ host. The authors concluded that the primary mechanism functions entirely (or almost so) with parental RF, and that in the secondary mechanism the gene A product may be needed to provide progeny RF as substrate, or it may function directly in the pathway. They pointed out that if the progeny RF is the principal substrate for the secondary mechanism, this would explain the lack of response to ultraviolet light, since the ultraviolet-induced lesions would be confined to the parental DNA.

Comparable discoveries were made for ϕX174 by Benbow et al. (1974a). They showed that the primary recombination mechanism operates early in the infection process, requires the host $recA$ protein and involves two parental RF molecules. None of the nine ϕX174 gene products was essential, suggesting that the mechanism shares many features with the host $recA$ pathway, rather than being specific to ϕX174. An alternative, less efficient recombination pathway was observed in the absence of the host $recA$ gene product and this mechanism required an active viral gene A protein. Recombination frequencies in a $recA^-$ $recB^-$ host were found to be five to 10 times lower than in $recA^-$ $recB^+$, suggesting that the host $recB$ gene product is involved in the secondary recombination mechanism in ϕX174.

(c) Multiple-length molecules

Rush et al. (1967) discovered from electron micrographs that about 3% of the replicative form DNA molecules of ϕX174 consisted of multiple-length rings, mostly dimers. These were evidently closed circular molecules because the procedure for obtaining RF preparations depended on the resistance of such molecules to denaturation by alkali.

Rush and Warner (1968) made mixed infections of E. coli with temperature-sensitive mutants in genes F and H, respectively, of S13. They purified the RF, separated the monomer and dimer fractions, and then infected spheroplasts with each. Assay of the spheroplast lysates revealed that 0.03% and 0.45%, respectively, of the liberated phage were wild type. They concluded that the dimers may be intermediates in recombination.

Gordon, Rush and Warner (1970) examined the ϕX174 replicative form dimers under the electron microscope and found they were of three kinds (see Fig. 8), circles, figures of eight, and interlocked monomers, in the approximate ratio of

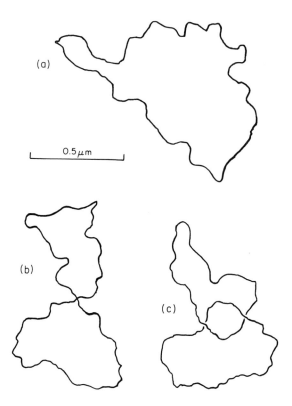

0.5 μm

Fig. 8. Phage φX174 replicative form dimers, as seen under the electron microscope by Gordon *et al.* (1970). The dimers consist of duplex DNA and are of three kinds: (a) a circle; (b) a figure of eight; (c) two interlocked monomers.

7:2:1. They also looked at a gene *A* mutant of S13, replication of the parental RF therefore being blocked, and found that the circular dimers were reduced to about 10% of the frequency observed in wild-type φX174. On the other hand, the combined frequency of the figures of eight and the interlocks (which were not recorded separately) was not reduced significantly (it was about 70% of the wild-type frequency). Gordon *et al.* (1970) concluded that many of the circular dimers in the wild-type virus resulted from replication, but that such dimers could also arise without replication.

Benbow, Eisenberg and Sinsheimer (1972) made mixed infections of *E. coli* with an amber mutant and a deletion mutant of φX174. Both mutants affected gene *E*. This gene controls the host lysis function and is the only gene that is not essential for the production of infective viral particles. Hence it is in this gene alone that deletions can survive. The virus particles in gene *E* mutants remain trapped in the host cell. The DNA of the deletion mutant had a contour length of 1.55 μm compared with the normal length of 1.70 μm, as found in the amber mutant. The RF, following the mixed infection, was examined under the electron microscope. The circular dimers (or higher polymers) were found to have contour lengths that were exact multiples of one or other parent, but about half of the dimers that consisted of interlocked monomers or figures of eight (not distinguished) were found to have one molecule of each parental genotype. Benbow *et al.* (1972) concluded that many of the non-circular dimers arise by recombination. Moreover, they confirmed for φX174 the finding of Gordon *et al.* (1970) for S13 that non-circular dimers predominate over circular when replication of the parental RF is blocked by a gene *A* mutant. Evidently most of the circular dimers arise by replication of a monomer and not by recombination. Further evidence for this was obtained by studying the multiple-length molecules formed in a *recA*⁻ host, the primary recombination mechanism then being blocked: a majority of the dimers (or higher polymers) were circular.

Doniger, Warner and Tessman (1973) achieved a partial separation of the circular and non-circular DNA dimers in RF preparations of S13 by density-gradient centrifugation. From electron micrographs they estimated that in a fraction containing mainly circular dimers not more than 10% were non-circular ones, that is, figures of eight or interlocks. *E. coli* was infected with a mixture of two S13 double mutants, one with mutations in genes *A* and *F* and the other in genes *E* and *H*. The mutants were all conditional, that is, amber (genes *A* and *E*) or temperature-sensitive (genes *F* and *H*). The dimer fraction of the RF containing predominantly circular molecules was used to infect spheroplasts, and the progeny types found in one-drop samples were investigated, if there was evidence for recombination. This evidence was obtained by testing for the presence of *am*⁺ phage, that is, lacking either amber mutant. Twenty-one of the one-drop samples, representing about 1% of them, yielded *am*⁺ recombinants. From the proportion of samples containing bursts, it was estimated that about 81% of the 21 were likely to contain single bursts, the remainder having more than one. Yet all 21 were found to have *am*⁻ phage particles as well as the *am*⁺, implying that the original dimers were heterozygous. Doniger *et al.* (1973) concluded that these dimers must result from recombination, since a dimer arising by replication of a recombinant monomer would be homozygous.

In order to establish whether the recombinant dimers belonged to the primary or the secondary pathway for recombination, Doniger *et al.* (1973) measured the frequency of recombinants in a *recA*⁻ host. They found that the frequency was 10 times lower than in *recA*⁺, in both the monomer and dimer fractions of the RF. They concluded that formation of recombinant dimers requires the *recA* function and is therefore part of the primary mechanism. Using a *recA*⁺ host, Doniger *et al.* compared the frequency of recombinants in the circular and non-circular dimers with that in the monomers and found that both kinds of dimers showed three times as many recombinants as monomers. This ratio is lower than that found by Rush and Warner (1968).

Benbow, Zuccarelli and Sinsheimer (1975) infected *E. coli* with a mixture of two amber mutants of φX174, in genes *A* and *E* respectively. Parental RF molecules were isolated, purified and fractionated in a propidium bromide–caesium chloride gradient. The dimer composition of each fraction was determined by electron microscopy, and the frequency of recombinants in each fraction was assayed by infecting spheroplasts and recording the number of wild-type recombinants in the progeny phage. To minimize recombination during the assay, a $recA^- recB^- uvrA^-$ host strain was used. (The $uvrA$ gene is discussed in Chapter VI, Section 9). The results are shown in Table 4. The multiple-length molecules showed only a twofold increase, at most, in recombinants compared with a direct burst of phage. The recombination frequencies in relation to dimer composition suggest that the figure of eight molecules are the primary source of recombinants amongst the three kinds of dimer. From electron micrographs, Benbow et al. (1975) found that figure of eight molecules were at least 10 times more frequent in $recA^+$ strains than in $recA^-$. Furthermore, in a cell-free system involving cells broken up by sonication, figure of eight structures were formed much more often in sonicates of $recA^+$ than of $recA^-$ cells. These results are in keeping with the idea that figure of eight molecules play a part in the primary recombination pathway. On the other hand, measurements of the contour lengths of these molecules on electron micrographs, following mixed infection with the amber and deletion mutants of gene E used in the earlier study (Benbow et al., 1972), revealed that over 60% had two parental genotypes. Evidently only a minority of the figures of eight arise by recombination.

Improved shadowing techniques in the preparation of electron micrographs were used by Thompson et al. (1975) to distinguish figure of eight molecules from twisted circular dimers and interlocked monomers in the dimer fraction of S13 and ϕX174 RF DNA. It was confirmed, also, that the node in the figures of eight is at the mid-point. Moreover, when a double-strand scission was introduced into

Table 4. Data of Benbow, Zuccarelli and Sinsheimer (1975) for recombination frequencies following mixed infection of E. coli by ϕX174 phage or DNA with amber mutants in genes A and E, respectively. The dimer composition, determined from electron micrographs, is given for the DNA from different regions of the density gradient used to infect the host. RFI is the replicative form DNA with covalently closed supercoiled duplex circles.

Source	Dimer composition (%)			Recombination frequency (wild type/total phage) $\times 10^4$
	Circle	Figure of eight	Interlock	
Direct burst of phage	—	—	—	$\begin{cases} 8.3 \pm 0.9 \\ 9.7 \pm 1.2 \end{cases}$
RFI	—	—	—	15.5 ± 1.7
Total multiple-length DNA molecules	—	—	—	29.6 ± 2.3
Upper band	61	12	27	27.2 ± 2.1
Middle band	<10	20	70	40.3 ± 2.1
Interband	44	32	24	90.7 ± 10.3

the dimers by restriction enzymes from *Haemophilus influenzae* (a mixture of the *Hin*d II and III enzymes), the figures of eight were converted to α-shaped molecules. Another technique was to make preparations for electron microscopy in the presence of ethidium bromide, which intercalates in the DNA helix and results in preparations in which circular molecules, if nicked, are much less tangled. Counts obtained using these various treatments agreed with one another, and indicated that about 5% of the dimers were figures of eight, 25% being interlocked monomers, and the remainder circles. This is a lower frequency of figures of eight than previous estimates.

(d) Data from bursts of single cells

In their study of progeny types from a four-point cross in phage S13, Doniger, Warner and Tessman (1973) obtained 21 samples yielding recombinants, of which four-fifths were believed to represent bursts from single cells (see above). Eleven of the 21 samples were found to contain one recombinant and one parental genotype, six contained two or three non-reciprocal recombinants and usually one parental genotype, three contained reciprocal recombinants only, and one comprised both reciprocal and non-reciprocal recombinants as well as a parental genotype. Doniger *et al.* (1973) pointed out the similarity between these results and those of Boon and Zinder (1971) for phage f1, notably in the high frequency of bursts containing only one parental and one recombinant genotype (see Section 6(c) below).

Benbow *et al.* (1974*a*) crossed an amber temperature-sensitive double mutant of φX174 with an opal mutant. The amber and opal mutations were in gene *F* and the temperature-sensitive mutant was in gene *G*, the amber mutation being located between the other two on the genetic map. The cross was made on a non-permissive (Su^-) host, with the result that only those cells in which a wild-type recombinant has been generated will produce a burst. Nine single bursts were examined and each yielded phage of predominantly one parental and one recombinant genotype. The reciprocal recombinant genotype for the two *F* mutants, that is, the amber–opal double mutant, would have been detected if it had been generated, but none were found in any of the nine bursts. The similarity to the data of Doniger *et al.* (1973) for S13 and of Boon and Zinder (1971) for f1 is striking.

(e) Effect of single-strand breaks

In a study of complementation groups in phage S13, Ethel Tessman (1965) discovered that a relatively large amount of recombination occurred in crosses between mutants in gene *A* (called at that time complementation group IV) compared with crosses between mutants within other genes of the phage. Similarly, Benbow *et al.* (1971) found a high frequency of recombination between mutants in a particular region of gene *A* of φX174 (Fig. 7(b)). Because recombination events in these phages are paired – the high negative interference expected with a

44

small circular genome – mutations in the genes (*B* and *H*) on either side of *A* are tightly linked. Benbow *et al.* (1971) suggested that the high frequency of recombination in gene *A* may arise because one strand of the DNA is frequently nicked here during replication (Fig. 9(a),(b)). In a *recA⁻* host, the recombination anomaly in gene *A* was not found (Benbow *et al.*, 1974a).

Benbow, Zuccarelli and Sinsheimer (1974b) found a 50-fold stimulation of recombination in φX174 by ultraviolet irradiation or thymine starvation in a *recA⁺* host, but no increase in a *recA⁻* host. This was comparable to the effect of ultraviolet light on recombination in S13 described by Tessman (1968) (see above). In a *recA⁺* host Benbow *et al.* (1974b) found that ultraviolet irradiation of one of the two parental phages stimulated wild-type recombinant formation almost as effectively as irradiation of both parents. There was no stimulation of

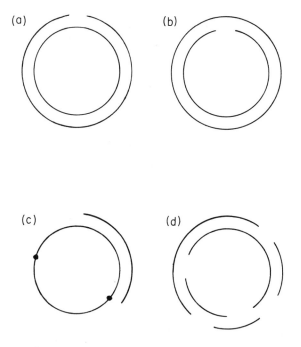

Fig. 9. Structures that promote recombination in phage φX174, from the work of Benbow *et al.* (1971, 1974b). In each diagram the inner circle shows the plus (virion) strand and the outer the complementary minus strand. The replication origin is at 12 o'clock. (a) The structure presumed to exist when the parental replicative form DNA has been synthesized, but the gap or nick at the replication origin and terminus has not yet been closed. (b) At a later stage, this closure has occurred and the plus strand has been nicked at the origin by the gene *A* product. (c) The structure inferred to result from replication following irradiation of the virions with ultraviolet light. The dots represent pyrimidine dimers. (d) The structure believed to result from thymine starvation during formation of the parental RF. The common feature in (a)–(d) is the presence of single-strand breaks, that is, nicks or gaps.

recombination by ultraviolet irradiation in the high recombination region of gene *A*. This suggests that the ultraviolet irradiation effect and the gene *A* effect involve the triggering of the same mechanism, the ultraviolet light in consequence being unable to increase recombination further.

Benbow *et al.* (1974*b*) examined by various techniques the structure of the RF produced on infecting *E. coli* with ultraviolet-irradiated ϕX174. The normal RF is of two kinds: covalently closed supercoiled duplex circles called RFI, and relaxed duplex circles called RFII; these have one or more nicks or gaps. On a neutral sucrose velocity sedimentation gradient, the RFI and RFII sediment at different positions. When replication of the parental RF was blocked by treatment with chloramphenicol or in other ways, 65% of the DNA sedimented as RFI and the remainder as RFII. With increasing ultraviolet radiation dose to the parent phage the proportion of the parental RF sedimenting at the RFII position increased. In addition, many ultraviolet-damaged molecules sedimented anomalously. Similar results were obtained whether the host was *recA*$^+$ or *recA*$^-$, except that with *recA*$^+$ 20% of the DNA sedimented at a position indicative of multiple-length molecules. In an alkaline sucrose velocity gradient, molecules other than RFI are denatured. With ultraviolet-irradiated phage, apart from some RFI molecules, the viral (plus) strands (which had been labelled with [^{14}C]thymidine) sedimented primarily as full length circles, while the complementary (minus) strands (which had been labelled with [^{3}H]thymidine) sedimented as linear molecules of various sizes but shorter than full length. These results, both for the neutral and the alkaline gradients, confirm those of Francke and Ray (1971) for a *recA*$^+$ host. Sedimentation of the doubly labelled RF on a neutral gradient showed less ^{3}H than ^{14}C in the RFII when the parent phage had been treated with ultraviolet light. This result indicates that the minus strands were incomplete. All the sedimentation data fit the hypothesis that, following ultraviolet irradiation of the phage, the irradiated plus strand remains as an intact circle, but synthesis of the complementary minus strand proceeds from the replication origin but stops at the first pyrimidine dimer or other ultraviolet-induced lesion which it encounters in the plus strand template (Fig. 9(c)).

Electron microscopy of parental RF molecules spread in an aqueous film of cytochrome *c* revealed one duplex and one simplex region within each circular molecule, following ultraviolet irradiation of the ϕX174 particles. The length of the duplex region was inversely related to the ultraviolet dose. These findings are in keeping with the conclusions from the sedimentation data.

The experiments on the effect of thymine starvation were carried out by infecting a thymine-requiring (*thy*$^-$) host strain with the phage and growing the cells for 20 min in a medium lacking thymine. The data for the parental RF from a neutral sucrose gradient, using thymine-starved cells, were found to be similar to those for ultraviolet-irradiated phage, but the alkaline gradient data showed ^{3}H- and ^{14}C-labelled DNA principally as linear molecules of up to full length. Under the electron microscope the molecules were seen to be predominantly circular duplex RFII (or at least of RFII appearance) of normal length, but 10–20% were linear, also of full length. Simplex regions of more than 100 nucleotides would have been detected and were evidently absent. Benbow *et al.* (1974*b*) concluded that, following thymine starvation, the parental RF contains random single-strand breaks in either or both strands (Fig. 9(d)).

The common feature promoting recombination with ultraviolet treatment, with thymine starvation, and in gene *A* was believed to be the presence of single-strand breaks (Fig. 9). It was noteworthy that these breaks were present in *recA*⁻ cells as well as *recA*⁺, following ultraviolet treatment of the phage. Evidently the *recA* gene product promotes a later stage in recombination.

(f) Branch migration

Kim, Sharp and Davidson (1972) discovered a deletion mutant of φX174 of spontaneous origin. From electron micrographs it was found to lack about 9% of its DNA. About 2% of the molecules were dimers and 0.2% higher multiples. Heteroduplexes were obtained by renaturation between wild-type circular plus strands and the deletion mutant dimers. Only linear molecules can renature with a circular one. The linear dimers were sometimes associated with one circular plus strand, giving a duplex circle with attached single strands, and sometimes with two circular plus strands giving duplex figures of eight. The position of the deletion in the mutant molecule was revealed in the heteroduplex, because the wild-type segment missing in the deletion formed a simplex loop. The relative positions of the two loops formed in this way in the figure of eight molecules (Fig. 10(d)) showed that the dimers were tandem (head-to-tail) duplications of monomers. In the heteroduplexes containing only one wild-type circle, the simplex loop at the position of the deletion may or may not correspond in position to that of the attached single strands (Fig. 10(a),(b)). The frequencies of these symmetrical and asymmetrical tailed circles were in the ratio 2:1. The corresponding structures with two wild-type circles have the two simplex loops together at the node of the figure of eight (Fig. 10(c)), or away from the node (Fig. 10(d)), respectively. The ratio of occurrence of these symmetrical and asymmetrical figures of eight was 50:1. The very different relative frequencies of these alternatives in the one-circle and two-circle heteroduplexes led Kim *et al.* (1972) to conclude that branch migration was occurring (see Chapter IV, Section 8), and that the relative frequencies represented the equilibrium proportions of the alternatives. Single-strand branch migration would be involved in converting a symmetrical tailed circle to an asymmetrical one or vice versa, and double-strand branch migration in interchanging symmetrical and asymmetrical figures of eight. The one-circle and two-circle heteroduplexes might have different equilibrium ratios of symmetrical and asymmetrical configurations because of their structural differences.

Thompson, Camien and Warner (1976) measured the rate of branch migration in duplex DNA. To do this they took figure of eight molecules of RF DNA of G4, a phage allied to φX174, and opened the loops by *Eco*RI restriction endonuclease action to give X-shaped molecules. Branch migration allows the point of strand exchange in the X-shaped dimer to move along the molecule in either direction. If by chance the exchange point reaches either end, two linear monomers will be

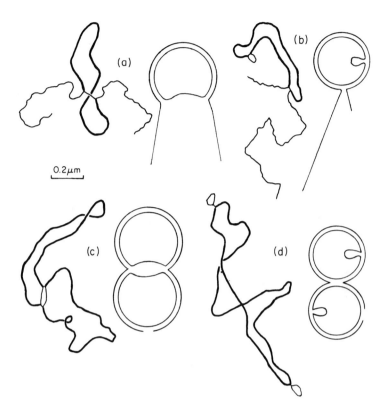

Fig. 10. Structures observed on electron micrographs by Kim *et al.* (1972) when circular plus strands of φX174 DNA anneal with linear dimers. These dimers, originally circular, were derived from a deletion mutant and were tandem (head-to-tail) duplications of monomers. On the electron micrographs duplex regions of DNA appear thicker and straighter than simplex. The linear dimer may associate with one ((a), (b)) or two ((c), (d)) virion strands. For each configuration, an example of its appearance on the micrograph is shown on the left and a diagrammatic representation on the right. (a) Symmetrical tailed circle. (b) Asymmetrical tailed circle. (c) Symmetrical figure of eight. (d) Asymmetrical figure of eight. The scale refers to the electron micrographs and not to the diagrams.

formed. The rate of disappearance of X-forms was measured by counting them, and the linear monomers, on electron micrographs prepared at various times after the *Eco*RI digestion. By treating branch migration as a random walk, they were able to calculate the time taken for each step, that is, movement of the junction by one base pair. A value of about 6000 base pairs per second at 37 °C was obtained. They calculated that there was a probability of 0.32 that the junction would be found more than 300 nucleotide pairs in either direction from its point of origin in 1 min, or more than 850 pairs in 8 min. This rate of branch migration appeared to be just rapid enough to account for recombination in G4 and similar phages, a few minutes being the maximum time available.

(g) Mismatch repair

Experiments were carried out by Weisbeek and van de Pol (1970) with φX174 DNA which pointed to the occurrence of a correcting mechanism when mispaired bases occur in heteroduplex DNA. The heteroduplexes were obtained by mixing circular simplex plus strands carrying an amber mutant with minus strand fragments from a temperature-sensitive mutant, under renaturing conditions. The high frequency of wild-type progeny pointed to mismatch correction in the heteroduplex RF.

Baas and Jansz (1972*a*) annealed wild-type minus strands of φX174 DNA with plus strands from various conditional lethal mutants and infected spheroplasts with the heteroduplex molecules. The bursts of progeny phage from single cells were usually all of one genotype, implying that the mismatch in the heteroduplex had been repaired before replication. Either genotype appeared equally often. In a small fraction of the spheroplasts both genotypes were produced, and the frequency of these mixed bursts depended on the position of the mutation on the linkage map. Baas and Jansz (1972*b*) found that a mutation in gene *A* near the gene *B* end, and mutations in genes *B*, *D* and *E*, all gave mixed bursts with frequencies of 26–32%, a mutation in gene *F* with frequency 17%, *G* 10%, *H* 7%, and in *A* near the gene *H* end 4%. They concluded that the replication origin is in gene *A*, that replication proceeds in alphabetical order through these genes on the map, that is, clockwise as drawn (Fig. 7), and that mispaired nucleotides near the origin have a better chance of escaping repair than those replicated later. It is of interest, as Benbow *et al.* (1974*b*) pointed out, that Baas and Jansz's data indicate that the replication origin is at the high recombination region in gene *A*. Confirmation of this position for the replication origin and of the clockwise direction has since been obtained from several sources (Johnson and Sinsheimer, 1974; Godson, 1974).

(h) A single-strand aggression model

On the basis of the experimental results described above, Benbow, Zuccarelli and Sinsheimer (1974*b*, 1975) proposed a model for recombination in φX174 based

---

Fig. 11. The single-strand aggression model for recombination in φX174 proposed by Benbow *et al.* (1974*b*, 1975). The numbers indicate mutants and the plus signs their wild-type alleles. The DNA molecules are really circular. (a) One strand of one molecule is nicked. (b) In the presence of the *recA*⁺ gene product, single-strand aggression occurs, leading to annealing with the complementary strand of the other molecule to give a figure of eight structure. The displaced strand of the invaded molecule is broken. (c) The latter strand also forms a heteroduplex, or (c′), following Meselson and Radding (1975), it is broken down by an exonuclease (shown as a dot), the gap in the aggressing molecule being filled by DNA synthesis (shown as a broken line). (d), (d′) The heteroduplex regions are extended by branch migration, with (d′) further polymerase and exonuclease activity. (e), (e′), (e″) Mismatch repair by excision and resynthesis takes place, and gaps are closed and sealed. The disappearance of the strand exchange can be attributed either to DNA replication using the outer strands as templates, or to enzymic breakage of the bridging strands. Only those genotypes that include a wild-type molecule are shown.

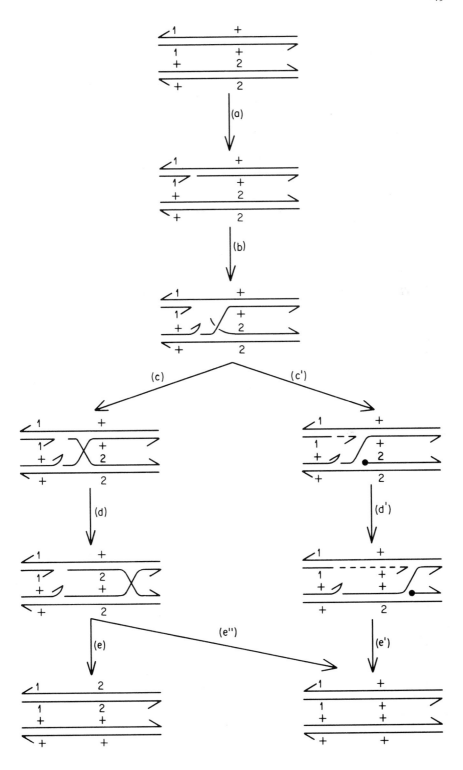

on single-strand aggression. They suggested that single-strand breaks, in the presence of the host *recA* gene product (Chapter VI, Section 10), are aggressive intermediates in the formation of recombinants. They called the process aggression because only one of the two recombining molecules must contain a break in order to stimulate recombination. The catalysis by the *recA* product would explain the greatly reduced frequency of recombination – indeed, the abolition of the primary mechanism – in a *recA⁻* host, despite the presence of simplex breaks. It is immaterial whether the break is formed at a specific site or a random position, nor does it matter what caused the break.

The proposed steps in the recombination process are outlined in Fig. 11 and are as follows:

(a) A single-strand break occurs in one of the two parental RF molecules.

(b) Activated by the *recA* gene product, the broken strand invades the homologous region of the other molecule and anneals with the complementary strand, leading to breakage of the displaced strand of the invaded molecule.

(c) The broken strand displaced from the invaded molecule anneals with the region of the other molecule vacated by the aggressing strand, giving a reciprocal strand exchange, that is, heteroduplex segments in both molecules.

(d) Branch migration extends the heteroduplex regions in both molecules.

(e) Mismatch correction removes mispaired bases at the sites of mutations in the heteroduplex regions, gaps are closed and sealed, and replication (which will require the gene *A* product) takes place. Monomers (such as those shown arising at Fig. 10(e)) would presumably result from replication that used as templates the strands not involved in the exchange.

The single-strand aggression would explain the occurrence of recombinant figure of eight molecules, since these would be generated by step (b). Benbow *et al.* (1974*b*, 1975) invoked branch migration in order to stabilize the heteroduplex regions, which they assumed would be quite short initially; as already discussed, branch migration had been demonstrated *in vitro* in ϕX174 DNA by Kim *et al.* (1972). The negative interference, that is, two (or more) recombination events in short intervals, would be attributed to the mismatch repair and to its occurrence in proximity to the sites of the initial breaks.

A model similar in a number of respects to that of Benbow *et al.* had been put forward earlier by Doniger *et al.* (1973), although they favoured nicks in both molecules in strands of the same polarity at nearly corresponding sites to start the recombination process (see Thompson *et al.*, 1975). This nicking is followed by the joining of the two parental RF molecules through reciprocal hybrid DNA formation to give a figure of eight molecule. The initial heteroduplexes are believed to extend as a result of branch migration. Doniger *et al.* (1973) suggested that the breakdown of the figure of eight occurs by nicks in homologous strands at the node. Assuming that the strands cut are the bridging ones, two circular monomers will arise or if, through isomerization, the outer strands can become the bridging ones (see below), then a circular dimer will be generated. Thompson *et al.* (1975) illustrated the genotypes that could arise from mismatch correction in these molecules, and discussed possibilities for the subsequent fate of the circular dimer.

If the bridging strands are severed, recombinant progeny may arise in the absence of mismatch correction. Consider, for example, the consequences of omitting the mismatch repair from steps (e) and (e') in Fig. 11.

One feature of these hypotheses of Doniger *et al.* (1973) and Benbow *et al.* (1975) does not seem to tally with the experimental data. If the mismatch correction occurs equally often in each direction, and independently in each heteroduplex segment, non-reciprocal and reciprocal recombination events that generate wild-type progeny will be expected to be equally frequent. Thus, if mutants 1 and 2 are crossed, the cross can be denoted as

$$\frac{1 \ +}{+ \ 2},$$

where the line separates the two parental molecules. After hybrid DNA formation at the site of mutant 2, the genotype becomes

$$\frac{1 \ +/2}{+ \ +/2}$$

and after mismatch correction this becomes

$$\text{(i)} \ \frac{1 \ +}{+ \ 2} \quad \text{or} \quad \text{(ii)} \ \frac{1 \ 2}{+ \ +} \quad \text{or} \quad \text{(iii)} \ \frac{1 \ +}{+ \ +} \quad \text{or} \quad \text{(iv)} \ \frac{1 \ 2}{+ \ 2}.$$

Alternative (i) is a restoration of the parental genotype and so will not be detected genetically, (ii) is reciprocal recombination, and (iii) and (iv) non-reciprocal, but (iv) lacks a wild-type product. The experimental data from bursts of single cells, discussed above, suggest that when a wild-type molecule is generated alternative (iii) predominates, that is, one parental and one recombinant genotype. The rarity of reciprocal recombination would be accounted for if step (c) – the formation of reciprocal heteroduplex segments – often fails to occur. Breakdown of the displaced strand by an exonuclease might take place, with the gap in the aggressing molecule filled by repair synthesis (Fig. 11(c')), as in the model proposed by Meselson and Radding (1975) – discussed in Chapter VII, Section 14. With hybrid DNA confined to one molecule, alternative (ii) is eliminated (see Fig. 11). This conclusion may need to be qualified if the outside strands and the bridging strands are interconvertible, as suggested by Sigal and Alberts (1972). But in a small circular genome, as Benbow *et al.* (1975) pointed out, the occurrence of this isomerization is questionable. Moreover, cutting the previously unbroken strands gives a circular dimer, and the genetic predictions depend on its fate.

6.   RECOMBINATION IN THE FILAMENTOUS PHAGES

(a) Introduction

Boon and Zinder (1970) studied recombination in phage f1. They took advantage of the fact that the parental replicative form (RF) DNA requires the product of phage gene *2* in order to replicate. In consequence, f1 genomes with an amber mutation in this gene are unable to multiply in a non-permissive host unless reversion occurs. It was found, however, that two such mutants can recombine, because mixed infections of *E. coli* cells generated progeny phage with a frequency much greater than the reversion rates of the individual mutants. It was evident that recombination could occur in the absence of replication, and that if the recombination event gave rise to a wild-type gene *2*, that is, without either amber mutant, then DNA replication could take place, leading to progeny particles. The frequency of recombination was very low.

(b) Breakage and rejoining

Boon and Zinder (1970) showed that a considerable fraction of the cells in which such a recombination event had taken place released a recombinant, wild type for gene *2*, whose genome consisted, at least in part, of parental DNA. The parental DNA, that is, the original virion or plus strand, was recognized because it had been modified by the host B restriction–modification system. The modification consists of the methylation of an adenine at (or near) each of two sensitive sites (SB$_1$ and SB$_2$) in the f1 DNA. This methylation takes place if the phage is grown in strain B of the host. By making mixed infections with the gene *2* amber mutants in another host strain (K), methylation of newly synthesized f1 DNA was prevented. The methylation of the parental DNA was recognized by its resistance to restriction, that is, endonucleolytic cleavage, in the B host. The demonstration that recombinant molecules contained parental atoms showed that recombination was occurring by breakage and rejoining.

(c) One wild-type and one parental genotype as recombination products

The particular advantage of the use of gene *2* mutants for the study of recombination in the filamentous phages is that, apart from the occurrence of revertants, the only genotypes present in the progeny phage will be products of the recombination process. Owing to the rarity of recombination, the occurrence of two independent recombination events in the same cell has a very low probability. Furthermore, the multiplicity of infection was such that the possibility of three or more phage particles infecting one cell could be ignored. It is presumed that whenever recombination between gene *2* mutants gives rise to a non-mutant gene *2*, thus allowing replication to proceed, other phage genotypes in the cell will be rescued by complementation, but that the cell contained only one DNA molecule of each parental genotype prior to recombination. Thus the phage yield of individual bacterial cells – the equivalent of a single burst in phages that cause cell lysis – should comprise all the emerging genotypes of individual recombination events, and no others.

Zinder and associates (Boon and Zinder, 1971; Hartman and Zinder, 1974*a*,*b*) made extensive studies of recombination in f1 by crossing amber mutants R21 and R86 of gene *2* and determining the genotypes of the progeny phage in individual cells. As unselected markers in these experiments they used mutations to resistance (*r*) at each of the two sites (SB$_1$ and SB$_2$) already referred to. These are the sites which, when not methylated, are recognized by the host B restriction–modification system. The SB$_1$ resistance mutant has a base change at SB$_1$, with the result that the site is not recognized by the host enzyme, and similarly with the SB$_2$ mutant. The crosses were made on a host of K specificity and the progeny tested on B to find if they were sensitive or resistant to the B enzyme. A temperature-sensitive mutant (*ts*) in gene *1* was also used as an unselected marker in the later experiments. The linkage map is as follows:

$$\underline{\quad \text{SB}_1 \quad \text{R21} \quad \text{R86} \quad \text{SB}_2 \quad \text{ts}\quad}$$

The map is really circular (see Fig. 12).

More than 90% of the phage yield from individual cells contained either (1) one wild-type genotype (i.e. wild-type for gene *2*) or (2) one such wild-type and one parental genotype. It was estimated that more than half the cells that released only the wild-type phage genotype derived it by reversion of one or other of the amber mutants. In the cells with a wild-type and a parental phage yield, particular combinations of genotype were favoured. This confirmed that the parental genotypes were products of the recombination event. The numbers of the various combinations of one wild-type (for gene *2*) and one parental genotype obtained by Boon and Zinder (1971) are shown in Table 5. The data refer to four different crosses and appear to be homogeneous. A remarkable feature is the high frequency of segregation for the unselected markers among the progeny phage from individual cells. In other words, the wild-type genotype is frequently associated with the parental genotype that differs in these characters. The data indicate that approximately 21%, 27% and 52% of the single-cell progenies showed segregation for 0, 1 and 2 unselected markers, respectively.

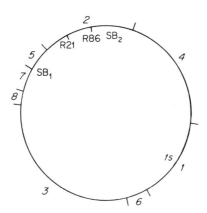

Fig. 12. Linkage map of phage f1, from Horiuchi *et al.* (1975). The numerals indicate the eight known genes, their lengths being proportional to the molecular weights of the corresponding proteins, apart from genes *6* and *7* where the size is unknown. The sequence of genes *5* and *7* is uncertain. The replication origin is believed to be near 12 o'clock on the map. The position of mutants referred to in the text is indicated inside the circle, those without lines being only approximate.

Table 5.   Results obtained by Boon and Zinder (1971) from crosses between mutants of phage f1 of *E. coli*. The lines show the genome, which is really circular. The numbers 21 and 86 indicate amber mutants in gene 2, the plus signs at corresponding positions indicating the wild-type nucleotide sequence. The letter *s* indicates sensitivity to the host restriction enzyme of B specificity. There are two sensitive sites, the left-hand one being no. 1 and the right-hand no. 2. The letter *r* at a sensitive site indicates resistance to the B restriction endonuclease as a result of a mutation at the site, though the phage will not be resistant unless both sites are mutant (*r*). In each cross one parent was sensitive and the other resistant. The genotypes of the two parents are shown above and below the line, respectively, and similarly with their contributions to the recombinant genotypes. The table shows the number of cells releasing particular combinations of a parental genotype and a wild-type one. The crosses were made on host strain K and the sensitivity to restriction tested by growing the progeny phage on strain B.

Cross	Progeny phage from individual cells		
	Wild-type genotype	Parental genotype	
I		$\dfrac{s\ 21 + s}{r + 86\ r}$	
	$\dfrac{s\ \ + s}{+}$	8	16
	$\dfrac{+ s}{r +}$	5	7
	$\dfrac{+}{r +\ \ r}$	13	5
II		$\dfrac{r\ 21 + r}{s + 86\ s}$	
	$\dfrac{r\ \ + r}{+}$	9	10
	$\dfrac{+ r}{s +}$	15	3
	$\dfrac{+}{s +\ \ s}$	9	1
III		$\dfrac{r\ 21 + r}{r + 86\ s}$	
	$\left.\begin{array}{c}\dfrac{r\ \ + r}{+}\\[4pt]\dfrac{+ r}{r +}\end{array}\right\}$	34	80
	$\dfrac{+}{r +\ \ s}$	28	19

Table 5. *Continued.*

	Progeny phage from individual cells		
	Wild-type genotype	Parental genotype	
Cross		$\dfrac{s\ 21 + s}{}$	$\dfrac{}{r + 86\ s}$
IV	$\dfrac{s\quad + s}{+}$	16	37
	$\dfrac{+ s}{r +}$		
	$\dfrac{+}{r +\quad s}$	61	23

Another important feature of the data was the failure to find the amber double mutant, that is, R21–R86. Boon and Zinder (1971) thought it was possible that this lack of observed reciprocal recombinants was because this genotype was not rescued, or failed to form a plaque. They also pointed out that reciprocal recombination between two circular molecules would generate a dimer, and this double-length circle might rarely undergo the resolution into unit-length circles necessary for packaging into phage coats. So the failure to detect reciprocal recombination might be from this cause. They thought it was likely that reciprocal recombination sometimes took place because L. B. Lyons and T. Boon had found that recombination in f1 was reduced about 100-fold in a $recA^-$ host. It was known from the work of Herman (1965) and Meselson (1967b) that the host-mediated recombination in an F' and in prophage λ, respectively, was sometimes reciprocal (see Chapter VI, Section 2).

In view of the absence of reciprocal recombinants among the progeny phage. Boon and Zinder (1971) thought that breakage and rejoining associated with reciprocal heteroduplex formation could not be the recombination mechanism, although, as already indicated, the possibility that such a process produced inviable or undetected reciprocal recombinants could not be excluded.

Two alternative mechanisms were considered to explain the occurrence of one parental and one recombinant genotype as recombination products: (1) asymmetric heteroduplex formation and (2) break-and-copy. For asymmetric heteroduplex formation a length of single strand from one parent would be transferred to the other, where it would replace the corresponding part of the strand of the same polarity. If the donor molecule were to be destroyed, replication of the heteroduplex recipient molecule would produce one parental and one recombinant molecule – the genotypes observed. To explain the high frequency of segregation for one or both unselected markers, however, it was necessary to suppose that the heteroduplex region was often quite long. Repair of the donor molecule and mismatch correction in the recipient heteroduplex offer variants of this hypothesis but still require a quite extensive heteroduplex region. For this reason, Boon and Zinder (1971) favoured break-and-copy as the recombination mechanism. On this hypothesis, which is discussed later, a broken strand is extended by DNA synthesis using a complementary strand from the other parent as template.

In the later experiments (Hartman and Zinder, 1974a,b) involving the additional unselected marker (ts) in gene 1, confirmation was obtained of the frequent occurrence of one parental and one recombinant genotype in the progeny recombinant phage released from individual cells. The results are shown in parentheses in Table 6. Again there was frequently a difference of genotype for the unselected markers between the parental and wild-type progeny from individual cells: approximately 21%, 11%, 5% and 63% of segregation for 0, 1, 2 and 3 unselected markers, respectively. Thus, nearly two-thirds of the cells giving one parent and one recombinant segregated for all three unselected character differences, an even higher proportion than for the two such differences in the earlier work.

Hartman and Zinder (1974a,b) investigated the genotype of the unselected markers in a random sample of progeny phage wild-type for gene 2. This means they were sampling wild types from cells that produced this genotype only, as well as cells that produced a wild-type and a parental genotype. The results are given in parentheses in the right-hand column of Table 6, and are similar to those from the single cell analyses. Separate experiments to measure the frequency of revertants, using single-mutant infections, indicated that less than 5% of the wild-type progeny from the crosses using the random sampling technique, were likely to be revertants.

(d) Extensive heteroduplex formation

Hartman and Zinder (1974a,b) also made crosses on *E. coli* strain B instead of K. The figures not in parentheses in Table 6 show the results. In crosses A–C, one parent was protected from B restriction by methylation at both SB sites, and the other parent was sensitive to restriction. When the cross was made on the B host, few cells liberated the sensitive parent along with the wild type. The relevant figures are marked with a superscript s in the table. Evidently the DNA of the sensitive parental genotype is cleaved by the B restriction endonuclease, though recombination is not entirely prevented since wild-type progeny are released along with the other (resistant) parental genotype.

An interesting feature of the data from these three crosses is the constancy of the relative frequencies of the different genotypes for the unselected markers, whether a cross was made on a B host or a K host. This supported the evidence of Horiuchi and Zinder (1972) for the B restriction enzyme and of Murray, Batten and Murray (1973) for the allelic K enzyme that the recognition sites are not the sites of cleavage, which seems to occur at random positions in the molecule. This would explain the surprising result that the degree of linkage of the f1 mutants is unaffected by the occurrence of B restriction.

Table 6.   The table shows the results obtained by Hartman and Zinder (1974a,b) from crosses between mutants of phage f1 of *E. coli*. The lines show the genome, which is really circular. The numbers 21 and 86 indicate amber mutants in gene *2*, and *ts* is a temperature-sensitive mutant in gene *1*, the plus signs at corresponding positions indicating the wild-type nucleotide sequence. The letter *s* indicates sensitivity to the host restriction enzyme of B specificity. There are two sensitive sites, the left-hand one being no. 1 and the right-hand no. 2. The letters *me* or *r* at a sensitive site indicate methylation of nucleotides or mutation, respectively, either of which gives resistance to the B restriction endonuclease (provided both sites are resistant). In each cross one parent was sensitive and the other resistant. The genotypes of the two parents are shown above and below the line respectively, and similarly with their contributions to the recombinant genotypes. The column headed 'Progeny phage from individual cells' shows the number of cells releasing particular combinations of a parental genotype and a wild-type one. The column headed 'Random sampling' shows the numbers of wild-type phage observed with various genotypes for the unselected markers, irrespective of their association with parental genotypes. The figures in parentheses are the numbers of progeny obtained on host K and the figures not in parentheses are the numbers on host B, that is, in the presence of the B restriction endonuclease and methylase. The superscript *s* against the number of progeny on host B indicates that the progeny phage of parental genotype correspond to a sensitive genotype in the original parent of the cross. The superscripts *me* and *r* against the number of progeny on host B indicate that the progeny phage of wild-type genotype have what was originally a sensitive genotype at an SB site, and that this site is either methylated (*me*) or mutated (*r*) in the phage of parental genotype released from the same cell.

	Progeny phage from individual cells		Random sampling
Cross	Wild-type genotype	Parental genotype	
A	*s* 21 + *r* + / *me* + 86 *me ts*		
	s + r + / + ; + r + / me +	2^s (84)	69 (180)	134 (209)
	+ r / me + ts	0^s (9)	1 (10)	3 (33)
	+ / me + me ts	3^s (95)	6 (20)	24 (159)
B	*r* 21 + *s* + / *me* + 86 *me ts*		
	r + s + / +	4^s (20)	44me (24)	60 (257)
	+ s + / me +	1^s (9)	0me (2)	4 (59)
	+ s / me + ts ; + / me + me ts	8^s (44)	9 (8)	24 (198)

Table 6. *Continued.*

Cross	Progeny phage from individual cells		Random sampling
	Wild-type genotype	Parental genotype	

Cross C

Parental genotype header: $\dfrac{me\ 21\ +\ me\ +}{s\ +86\ s\ ts}$

Cross	Wild-type genotype	Wild-type	Parental	Random sampling
C	$\begin{cases}\dfrac{me\quad +\ me\ +}{+}\\[2pt]\dfrac{+\ me\ +}{s\ +}\end{cases}$	25 (10)	0^s (20)	24 (27)
	$\begin{cases}\dfrac{+\ me}{s\ +\qquad ts}\\[2pt]\dfrac{+}{s\ +\quad s\ ts}\end{cases}$	32me (12)	0^s (7)	36 (33)

Cross D

Parental genotype header: $\dfrac{r\ \ 21\ +\ me\ +}{s\ +86\ s\ ts}$

Cross	Wild-type genotype	Wild-type	Parental	Random sampling
D	$\dfrac{r\quad +\ me\ +}{+}$	25	0^s	169
	$\dfrac{+\ me\ +}{s\ +}$	0^r	0^s	$1^{r,s}$
	$\begin{cases}\dfrac{+\ me}{s\ +\qquad ts}\\[2pt]\dfrac{+}{s\ +\quad s\ ts}\end{cases}$	0^r	0^s	$5^{r,s}$

Cross E

Parental genotype header: $\dfrac{r\ \ 21+\ r\ +}{s\ +86\ s\ ts}$

Cross	Wild-type genotype	Wild-type	Parental	Random sampling
E	$\dfrac{r\quad +\ r\ +}{+}$	—	$\_^s$	112 (48)
	$\dfrac{+\ r\ +}{s\ +}$	$\_^r$	$\_^s$	$0^{r,s}$ (10)
	$\dfrac{+\ r}{s\ +\qquad ts}$	$\_^r$	$\_^s$	$1^{r,s}$ (21)
	$\dfrac{+}{s\ +\quad s\ ts}$	$\_^r$	$\_^s$	$2^{r,s}$ (40)

Cross F

Parental genotype header: $\dfrac{r\ \ 21+\ r\ +}{r\ +86\ s\ ts}$

Cross	Wild-type genotype	Wild-type	Parental	Random sampling
F	$\begin{cases}\dfrac{r\quad +\ r\ +}{+}\\[2pt]\dfrac{+\ r\ +}{r\ +}\end{cases}$	10 (12)	1^s (22)	53 (20)
	$\dfrac{+\ r}{r\ +\qquad ts}$	6 (0)	0^s (4)	32 (3)
	$\dfrac{+}{r\ +\quad s\ ts}$	0^r (9)	0^s (8)	$4^{r,s}$ (7)

Crosses D–F (Table 6), reported by Hartman and Zinder (1974*b*), like crosses A–C, are between strains one of which is sensitive and the other resistant to B restriction. But unlike A–C, the resistant parent in crosses D–F derives its resistance, at least at one SB site, from mutation. Comparison of the results of crosses A–C with those of crosses D–F revealed an important difference between methylation and mutation as the cause of resistance. Mutation (*r*) as a source of resistance in one parent was associated with sensitivity to B restriction in any wild-type genotypes lacking *r* released from individual cells along with parental genotypes carrying *r*, while methylation (*me*) as a source of resistance was associated with resistance in the corresponding wild types. The relevant data in Table 6 are indicated by superscripts *r* and *me*, respectively. Without exception, those marked *r* are of lower frequency than without restriction, that is, when the cross is made on host K (numbers in parentheses), while those marked *me* are not of reduced frequency compared with when grown on K.

Hartman and Zinder (1974*b*) suggested that the cause of this difference between mutation and methylation in the fate of the progeny was in the difference of behavior of the host methylase to a mutation heteroduplex and a methylation heteroduplex. *In vitro* studies by Vovis *et al.* (1973) suggested that both kinds of heteroduplex were resistant to the B restriction endonuclease. The resistance of the methylation heteroduplex was discovered by Meselson and Yuan (1968) with the allelic K enzyme. They pointed out the adaptive value of this resistance in protecting newly replicated DNA of *E. coli* from attack by its own enzyme by allowing time for it to be methylated. Vovis, Horiuchi and Zinder (1974) showed that *in vitro* a heteroduplex (*me/s*) of one methylated (*me*) and one unmethylated (*s*) strand undergoes rapid methylation of the *s* strand in the presence of the B enzyme giving a methylated homoduplex (*me/me*), while a heteroduplex (*r/s*) consisting of a strand mutated at an SB site (*r*) and an unmutated strand (*s*) is not methylated in either strand (Vovis and Zinder, 1975). The consequence of this difference between the two kinds of heteroduplex is that, although both are resistant to restriction, the mutation heteroduplex (*r/s*), on replication, will give rise to one resistant (*r/s*) and one sensitive (*s/s*) molecule, vulnerable to restriction. On the other hand, the methylation heteroduplex (*me/s*) becomes *me/me*, which on replication gives two resistant molecules *me/s* and *s/me*, which likewise become methylated homoduplexes before the next replication. These conclusions from *in vitro* studies precisely match the *in vivo* data in Table 6, that is, the results marked *me* and *r* in relation to their counterparts on a K host shown in parentheses alongside.

An important implication of this explanation of the data in Table 6 is that heteroduplexes occurred at the SB sites in the recombination process that gave rise to these genotypes. Evidently, heteroduplex formation is extensive during the recombination process in phage f1.

(e) An asymmetric heteroduplex model

The inference that there are long heteroduplex segments in f1 recombination narrows the search for a satisfactory model to account for the data, and Hartman and Zinder (1974*b*) reconsidered the two alternatives, break-and-copy and asymmetric heteroduplex formation, that had been discussed by Boon and Zinder (1971) – see Section (c) above.

Break-and-copy had been favoured because it avoided the need to postulate long segments of hybrid DNA, but the new evidence conflicts with the break-and-copy hypothesis. Figure 13 shows the origin by this mechanism from cross C of Table 6 of a pair of recombination products comprising the parental genotype that is resistant to B restriction and a recombinant genotype wild-type for gene *2* and with the *ts* mutant in gene *1*. The steps in break-and-copy are shown in the diagram and described in the caption. They include the formation of a replication fork, allowing a strand from one parental molecule to be extended using the other parental molecule as template. Although progeny molecules of the appropriate genotypes result from the break-and-copy mechanism, the wild-type molecule is sensitive to B restriction. This conflicts with the results in Table 6. Furthermore, the predictions of this hypothesis are the same whether the resistance in the one parent arises by mutation or methylation, contrary to the experimental results.

The asymmetric heteroduplex model is illustrated in Fig. 14. By asymmetric heteroduplex is implied a non-reciprocal strand exchange, so one molecule can be called the recipient and the other the donor of this single-strand segment. The extensive heteroduplex region postulated would no doubt arise by branch migration from an initial short hybrid DNA segment. Fig. 14 shows the steps in the formation of the same recombinant molecules as in Fig. 13, that is, a particular pair of wild-type and parental progeny from cross C of Table 6. After the formation (step 1(a)) of an extensive heteroduplex region, methylation occurs at the SB sites in the strand derived from the sensitive parent (step 1(b)). Replication of this heteroduplex molecule (step 1(c)) then gives DNA molecules of parental and wild-type genotypes, and as a result of the methylation in the heteroduplex, the wild-type molecule is resistant to B restriction. This agrees with the data marked *me* for cross C in Table 6.

Zinder (1974) suggested that the release of wild-type recombinants unaccompanied by a parental genotype might be the consequence of mismatch correction in the heteroduplex. This possibility is shown in step 1(d) of Fig. 14.

If the resistance to restriction arose from mutation (*r*) at one or both of the SB sites, as in crosses D–F of Table 6, instead of from methylation (*me*), it would be necessary in Fig. 14 to substitute *r* for *me* and to omit step (b) because the sensitive strand is not methylated in a heteroduplex (*r/s*) for a mutation at an SB site (see Section (d) above). In consequence, the wild-type DNA molecules, following replication of the heteroduplex (step (c)) would be sensitive to the B restriction endonuclease. This agrees with the results in Table 6 (data marked with superscript *r*).

Thus the asymmetric heteroduplex model provides a satisfactory explanation for the resistance to B restriction of the wild-type genotypes in Table 6 marked *me*, and of the sensitivity of those marked *r*. Hartman and Zinder (1974*b*) attribute the scarcity of the release, under restriction conditions, of progeny phage with the genotype of the sensitive parent (data marked *s* in Table 6) to an inability

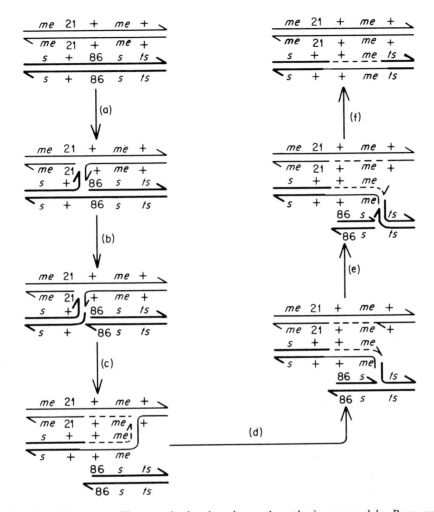

Fig. 13. Diagram to illustrate the break-and-copy hypothesis proposed by Boon and Zinder (1971) for recombination in phage f1, and further discussed by Hartman and Zinder (1974b). The diagram shows the origin of a pair of genotypes, one parental and the other wild-type for gene 2, from cross C of Table 6. Phage f1 has a circular DNA molecule but it is shown as a linear structure: the left- and right-hand ends are really joined. The numbers 21 and 86 indicate the sites of amber mutants in gene 2, and ts is a temperature-sensitive mutant in gene 1, the plus signs at corresponding positions indicating the wild-type nucleotide sequence. The letter s indicates sensitivity (when duplex) to the host restriction enzyme of B specificity. There are two sensitive sites, the left-hand one in the diagram being no. 1 and the right-hand no. 2. The letters me indicate methylation of nucleotides at a sensitive site and hence, if both sites are methylated, resistance to the restriction endonuclease. Thick and thin lines indicate the genetic origin of nucleotide chains. Broken lines indicate newly synthesized strands. Half arrowheads at internal positions indicate nicks or gaps. (a) Annealing occurs between broken strands of opposite polarity from the two molecules. (b) The other strand of the lower molecule is also broken, and the upper junction becomes a replication fork. (c) Replication takes place, two new strands being synthesized using the strands of the upper molecule as templates. (d) The strands which broke originally (one from each molecule) break again and anneal. (e) The second strand of the lower molecule is also broken. (f) Gaps are filled and nicks sealed.

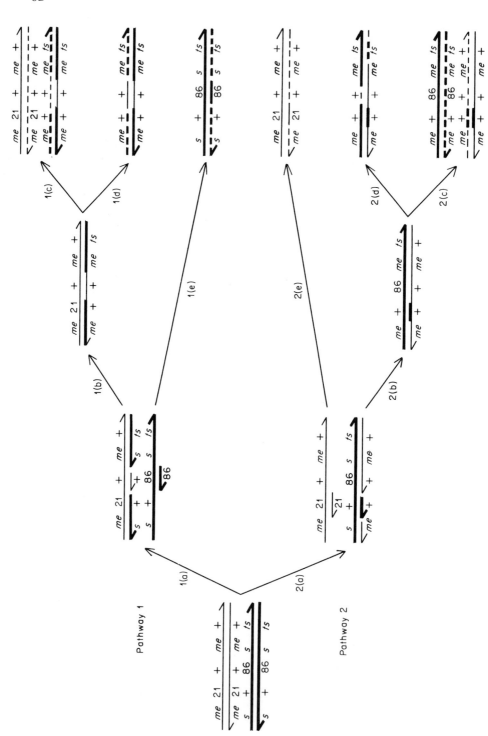

of the sensitve parent to act as recipient. They believe, however, that the duplex cleavage of the circular molecule at a random position by the B restriction endonuclease does not prevent the molecule from functioning as donor in hybrid DNA formation, as in pathway 1 of Fig. 14.

Hartman and Zinder (1974b) drew attention to a riddle posed by the derivation of the pair of genotypes, one parental and one wild-type, by replication of the heteroduplex (step 1(c) in Fig. 14), because for this step the product of gene 2 is required. This means that transcription must occur in the heteroduplex and furthermore the wild-type strand must be the minus strand, since it is from the minus strand as template that the messenger RNA is synthesized. But the experiments of Boon and Zinder (1970), described in Section (b) above, showed that the genome of the wild-type recombinants often consisted, at least in part, of parental DNA, that is, the virion or plus strand. To account for this discrepancy, Hartman and Zinder (1974b) suggested that a recombinant plus strand might be able to substitute for the gene 2 product for one round of replication, because Fidanián and Ray (1972) obtained evidence that the gene 2 product in phage M13 acts by nicking the plus strand (cf. Fig. 9(b)). A gap or nick in the plus strand arising through recombination might therefore produce the same result as the gene 2 product, though only for one round of replication. The replication, however, would produce a wild-type minus strand and so allow the gene 2 product to be synthesized.

Support for the asymmetric heteroduplex model was obtained from additional experiments. Enea, Vovis and Zinder (1975) prepared heteroduplex DNA of phage f1 in vitro. The RFI DNA was denatured with alkali and mixed with viral plus strands that differed from the RFI molecule in two genetic markers. As in previous experiments, these were a mutant giving resistance to B restriction, and either the ts mutant in gene 1 or an amber mutant in gene 2. The heteroduplex molecules were incubated with E. coli under conditions that allowed them to enter the host cells. As with transformation this transfection requires the cells to be competent, that is, at a particular phase of the cell cycle. Genetic analysis of the progeny phage, both in restricting and non-restricting hosts, gave results for the

---

Fig. 14. Diagram to illustrate the asymmetric heteroduplex model for recombination in phage f1 favoured by Hartman and Zinder (1974b). The letters and symbols for the genes have the same meaning as in Fig. 13 and the diagram illustrates the same cross. Pathway 1 shows the sensitive parent (that is, sensitive to the host B restriction endonuclease) acting as donor in the formation of a heteroduplex molecule, and pathway 2 shows the converse possibility, with the sensitive parent as recipient and the resistant as donor. (a) Strand from one molecule (donor) forms heteroduplex with the other molecule (recipient) extending over greater part of genome. (Molecules are really circular, so left- and right-hand ends are joined.) (b) Recipient molecule containing heteroduplex has sensitive sites methylated, gaps filled and nicks sealed. (c) Replication gives duplex molecules of wild-type and parental genotypes, respectively, both being resistant to restriction. (d) As an alternative to (c), mismatch repair gives a wild-type molecule (that is, wild type for gene 2); only one of several alternative genotypes is shown. (e) The donor molecule is repaired.

restriction site comparable to those obtained from phage crosses *in vivo* described above, and so supported the hypothesis of a segregating heteroduplex in recombination.

Further support for the asymmetric heteroduplex model was obtained by Enea and Zinder (1976) from study of recombination between a gene *2* amber mutant, no. 124, and (a) a deletion mutant, *d*, and (b) another gene *2* amber mutant, no. 60. In the latter cross, 124 was closely linked to a mutant, *r*, giving resistance to B restriction, the other parent being sensitive, *s*. In both crosses one parent could not act as a recipient of a strand from the other, in (a) because of the inviability of the deletion mutant when homozygous, and in (b) because of its sensitivity to B restriction. A single cell analysis in (a) revealed that the cross

$$\frac{+ \quad 124}{d \quad +}$$

was giving rise largely to + 124 progeny along with the wild-type recombinants. This is the result expected with segregation from a heteroduplex at the site of 124 in the + 124 parent. In (b) a large reduction in the yield of wild-type recombinants was observed on a B host compared with K. Again this is the result expected with segregation from a heteroduplex, the cross

$$\frac{+ \quad 124 \quad r}{60 \quad + \quad s}$$

leading to a heteroduplex at 124 and *r* in the + 124 *r* parent, from which the wild-type segregant (+ + *s*) would be sensitive to B restriction. Previous work with amber mutants more distant from the restriction recognition site (see Section 6(d) above) had not shown such a large reduction in yield of recombinants on a restricting host, presumably because the heteroduplex did not always extend as far as *r/s*. This relationship between the frequency of recombinants and the distance between mutant site and restriction recognition site is not related to the site of action of the restriction endonuclease since, as already pointed out (Section 6(d)), the enzyme moves from the recognition site to a random position in the molecule before cutting it.

7. CONCLUSIONS ABOUT RECOMBINATION IN SINGLE-STRANDED DNA PHAGES

The study of recombination in the filamentous phages has followed a different direction from that with the icosahedral phages. Where there has been overlap, as with the investigation of recombination products released from single cells, the results have been similar, with a preponderance of one parental and one wild-type genotype. It is not surprising, therefore, that the single-strand aggression model for the icosahedral phages and the asymmetric heteroduplex model for the filamentous phages have much in common. With the single-strand aggression hypothesis it was necessary to postulate that the hybrid DNA was often confined to one molecule, as in the Meselson–Radding model, in order to explain the rarity of reciprocal recombinants. A similar argument led Boon and Zinder (1971) to abandon symmetric, that is, reciprocal, heteroduplex formation as the basis of a model for filamentous phage recombination.

The chief difference between the models favoured for the icosahedral and filamentous phages, respectively, is over the fate of the donor molecule. Hartman and Zinder (1974b) favour donor loss, with the parental genotype that emerges along with a wild-type one being derived from the recipient molecule (step 1(c) in Fig. 14). On the single-strand aggression model, however, the donor molecule is repaired and contributes the parental genotype released in conjunction with the wild-type one (Fig. 11(c')–(e')). Donor repair has been included in Fig. 14 as step (e).

Hartman and Zinder (1974b) did not favour donor repair, no doubt because in conjunction with replication of the heteroduplex it would give rise to a wild-type genotype accompanied by both parental genotypes (Fig. 14, steps 1(c) and 1(e)) and this has not been observed. Donor repair, however, accompanied by mismatch correction in the heteroduplex molecule (steps 1(d) and 1(e)) would give one parental and one wild-type genotype, as so often observed, and would bring the hypothesis for the filamentous phages into line with that proposed for the icosahedral phages. On the donor repair hypothesis, however, it seems necessary to suppose that, under restricting conditions, the sensitive parent can act as recipient (pathway 2 in Fig. 14) as well as donor (pathway 1), but that no parental genotype will accompany the wild-type in pathway 1, since the sensitive donor, when repaired, will again be sensitive to the restriction endonuclease. Thus, in cross C of Table 6 under restricting conditions, Hartman and Zinder (1974b) attribute the lack of cells releasing the sensitive parental genotype along with a wild type (data marked s) to the inability of the DNA of this parental genotype to function as a recipient. On the donor repair hypothesis this lack would be attributed to the repaired donor DNA remaining sensitive and being cleaved by the B restriction endonuclease. Moreover, the progeny phage of resistant parental genotype, attributed by Hartman and Zinder (1974b) to replication of the heteroduplex (step 1(c) in Fig. 14) could arise through donor repair (step 2(e)).

The failure of host cells under restricting conditions to release appreciable numbers of sensitive parental genotypes (data marked s in Table 6) is independent of whether the resistant parent owes its resistance to methylation (crosses A–C) or at least in part to mutation (crosses D–F). This observation is in keeping with the idea that the released parental genotypes are not the product of a heteroduplex, though the same result is obtained if these genotypes are such a product and if, as Hartman and Zinder believe, the sensitive parent cannot function as a recipient. Thus, the question of mismatch correction with donor repair versus heteroduplex replication as the source of the pairs of parental and wild-type genotypes has not yet been resolved.

A further question concerns the figure of eight molecules that have been observed on electron micrographs of DNA molecules from the icosahedral phages. If recombination in the single-stranded phages is predominantly by asymmetric heteroduplex formation, following single-strand aggression, then only a single strand will bridge the two circles at the node of the figure of eight. This is contrary to most reports which have suggested a pair of bridging strands. Benbow *et al.* (1975) and Meselson and Radding (1975) have suggested, however, that the Sigal–Alberts isomerization can take place even when there is only one bridging strand (see Chapter VII, Section 14). This would interchange the positions of the outside and bridging strands and so would give rise to the physical appearance of a double-strand exchange although the heteroduplex would still be confined to one molecule. Whether, in fact, the Sigal–Alberts isomerization can take place in circular DNA molecules as small as those of these phages is another question. Nevertheless, clarification of the node structure in figure of eight molecules that are intermediates in the recombination pathway might help to reconcile the physical and genetic data.

Little is known at present about the secondary recombination mechanism in the single-stranded DNA phages, except that it is independent of the host $recA^+$ gene product and that, unlike the primary mechanism which takes place between molecules of the parental RF (replicative form), the secondary mechanism is believed to be confined to the progeny RF.

IV.   Recombination in phage T4

1. INTRODUCTION

Demerec and Fano (1945) collected seven phages that were active on a particular strain, 'B', of *Escherichia coli*. They called them Type 1, Type 2, and so on, abbreviated T1, T2, etc. Phage T2 had been isolated about 1924 by J. Bronfenbrenner of the Rockefeller Institute for Medical Research, New York, and called PC. The source does not seem to have been recorded, but was presumably faeces or sewage. Phages T4 and T6, which, unlike the odd-numbered T phages, are closely related antigenically and morphologically to T2, were separated by Demerec and Fano from a phage mixture supplied by T. L. Rakieten of The Long Island College of Medicine at Brooklyn. He had isolated them from sewage. Thus, all three T phages with even numbers appear to have been obtained from sewage in the New York area.

The even-numbered T phages have an elaborate head-and-tail structure (Fig. 15, Plate 1), first seen in detail by Brenner *et al.* (1959) following the use of phosphotungstate embedding for electron microscopy. The head has the shape of a bipyramidal hexagonal prism and measures 95 nm x 65 nm. It contains the DNA, which enters the host cell through a tail 95 nm long. The tail ends in a plate with six spikes 20 nm long and six fibres 150 nm long, which can bend at their mid point. The spikes and fibres serve for attachment to the host cell. The tail consists of a core of internal and external diameters 8 and 20 nm, respectively, surrounded by a sheath of external diameter 35 nm. After the phage has become attached to the outside of the cell, the sheath contracts and the core punctures the cell membrane. This allows the phage DNA to enter the cell.

Immediately after infection of an *E. coli* cell by a T2, T4 or T6 phage particle, synthesis of host nucleic acids and proteins is stopped and most or all of the host DNA is rapidly broken down. This breakdown is discussed in Section 8. By contrast, after the synthesis of various phage proteins, the viral DNA is replicated many times and the components of the phage head and tail synthesized and assembled. Ultimately, in less than 1 h at 37 °C, the cell undergoes lysis and about 200 mature phage particles are released.

2. DISCOVERY OF RECOMBINATION

Recombination in the even-numbered T phages was discovered by Hershey (1947) and by Delbrück and Bailey (1947) as a result of making mixed infections

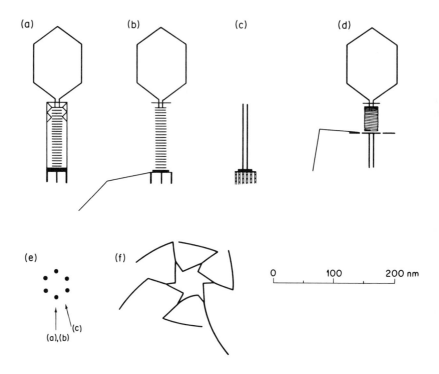

Fig. 15.   The structure of the T-even phage particle. (a) Head, tail jacket, and base plate with three of the six spikes. (b) Head, collar, tail, base plate and one of the six fibres. (c) Core and base plate with the six spikes. (d) Contracted sheath with spikes on base plate extended radially. (e) End view of the six spikes showing angle of view in (a)–(c). (f) End view of base plate and fibres. (Reproduced by permission of Professor E. Kellenberger.)

of *E. coli* with two strains that differed from one another in two characters. Hershey had found viral mutants of spontaneous origin with altered host specificity (*h*, host-range mutants) and others with distinctive types of plaque (*r*, rapid-lysis mutants). Delbrück and Bailey made a mixed infection with wild-type T2 and an *r* mutant of T4 and recovered among the progeny wild-type T4 and the *r* mutant in T2. T2 and T4 were distinguished by their host range, a particular strain of *E. coli* being attacked by T2 but not by T4. Hershey made mixed infec- tions with *h* and *r* mutants of T2 and recovered the double mutant and wild type in the progeny.

Using pairs of *r* mutants of T2, Hershey and Rotman (1948) found not only that recombinants (wild type and the double mutant) were formed, but also that there was genetic linkage between some of the mutants, certain pairs showing much lower recombination frequencies than others. Subsequently (Hershey and Rotman, 1949) indications were obtained that the recombination frequencies of linked mutants in T2 might fit a linear map, and this was established by Hershey and Chase (1952). Comparable experiments to those of Hershey and Rotman

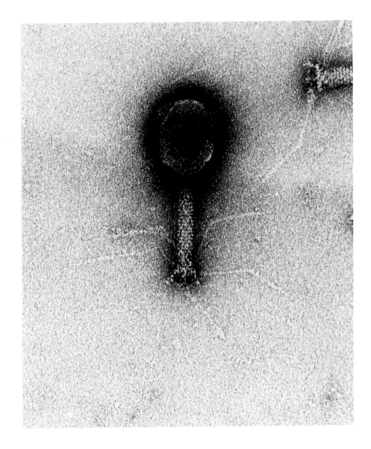

Plate 1. Electron micrograph of phage T4. Magnification 240 000×. Courtesy of B. ten Heggeler and E. Kellenberger.

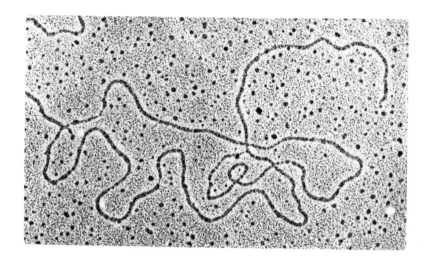

Plate 2. Electron micrograph of DNA from T4 of *pol⁻ lig⁻* genotype. Two branch points can be seen (and also three places where duplexes lie across one another). Single-stranded regions are visible near each branch point and at a point about halfway between them in the lower duplex. Magnification 63 800×. (Reproduced with permission from T. R. Broker, *J. molec. Biol.*, **81**, 8 (1973). Copyright by Academic Press Inc. (London) Ltd.)

were carried out by Doermann and Hill (1953) with phage T4, using mutants affecting plaque morphology. As with T2, the recombination frequencies fitted linear linkage maps. This linearity in T2 and T4 suggested that these viruses might have a genetic organization similar to that of higher organisms.

Hershey and Rotman (1948) found evidence that repeated exchanges (simultaneous or successive) took place during the multiplication of the virus in the host cell. This was shown by the finding of triparental recombination with the linked markers r4, r7 and r13. The technique was to make a mixed infection with the three double mutants, that is r4 r7, r4 r13 and r7 r13. Wild-type T2 were produced from such infections, implying that all three strains had interacted. Wild type was also obtained, though with lower frequency, in a similar experiment involving the more closely linked mutants r2, r3 and r6.

Study by Hershey and Rotman (1949) of the phage progeny from single bacterial cells infected simultaneously with two T2 mutants, one affecting host range and the other causing rapid lysis, indicated little or no correlation between the yields of the two kinds of recombinants (double mutant and wild type). Hershey and Rotman (1949) suggested there was genetic interaction between two sets (one derived from each parent) of independently multiplying chromosome-like structures, but pointed out that the interactions did not seem to be reciprocal exchanges.

A hypothesis to account for the genetic data obtained from mixed infections with T2 mutants was proposed by Visconti and Delbrück (1953). They suggested that in the host cell, before mature phage particles were produced, there was pairwise and repeated mating, at random with respect to partner and in time, of genetically complete structures. Symonds (1953) showed mathematically the consequences of random-in-time mating and Visconti and Delbrück (1953) treated the multiplication and interaction as a problem of population genetics with, as time elapses, a continual approach to genetic equilibrium. They found that the data fitted the theory if the average number of matings was five, before the mature particle was produced.

3.   HETEROZYGOSIS

Hershey and Chase (1952) reported that when *E. coli* was infected simultaneously with wild-type T2 and a rapid lysis (r) mutant, about 2% of the progeny particles gave rise to mottled plaques. Samples of phage from the mottled plaques were found to consist of about equal numbers of wild type and r mutant, a further 2% giving mottled plaques again. Hershey and Chase (1952) showed that the mottled plaques resulted from single particles, which must therefore contain both alleles (wild-type and mutant) and were therefore appropriately termed heterozygotes.

Study of heterozygotes when two mutants were present showed that each mutant considered individually gave about 2% of heterozygous progeny. If the mutants were unlinked, heterozygosis for one mutant was almost independent of that for the other. In general, it was only with closely linked mutants that a particle heterozygous for one mutant was also heterozygous for a second. It appeared as if regions of heterozygosity were rather short and hereditarily unstable.

With two unlinked markers the particles that were heterozygous for one of them were found to yield one parental and one recombinant genotype: for example, with a mixed infection by the double mutant $h\,r$ and wild type, the single heterozygotes for $r/+$ were found to be of two kinds yielding either (1) $h\,r$ (one parent) and h (one recombinant) progeny, or (2) wild-type (the other parent) and r (the other recombinant) progeny. Clearly there was some connection between heterozygosis and recombination.

A further observation by Hershey and Chase (1952) was important. The burst size – the number of progeny phage particles released from the host cell when it undergoes lysis – can vary considerably, but the fraction of the progeny showing heterozygosis showed no increase with increasing burst size. This implied that the heterozygotes did not reproduce themselves.

Levinthal (1954) considered two different ways of achieving a partial duplication or partial diploidy: to have a small piece attached to the side of a normal genome (model I in Fig. 16(a)), or to have an overlap of two pieces derived one from each parent (model II). If the phage were diploid, Fig. 16(b) would apply. He pointed out that models I and II can be distinguished in a three-factor cross, and put this to the test by making the cross

$$\frac{h\ +\ r7}{+\ r2\ +}\ ,$$

that is, he made a mixed infection using the $h\,r7$ double mutant of T2 as one parent and the $r2$ mutant as the other parent. These mutants are linked and map in the order shown. He selected particles showing heterozygosis for $r2$, the central marker, and looked at their genotype for the flanking markers h and $r7$. He found that the $r2$ heterozygotes were generally recombinant for h and $r7$, as predicted by model II but not by model I. Knowing from Hershey and Chase's experiments that heterozygotes lead to recombinants, Levinthal (1954) calculated the number of recombinants contributed in this way and found that it agreed with the observed number. He concluded that heterozygotes are intermediates in the production of recombinants. He pointed out that one possibility for the nature of the heterozygotes was provided by the structure of DNA that had just been proposed by Watson and Crick (1953), the overlap of pieces derived from each parent being at the level of the individual nucleotide chains of the DNA (model II in Fig. 16(b) with the lines representing the two strands of the duplex).

An experiment with T4 comparable to that of Levinthal's with T2 was carried out by Doermann and Boehner (1963). Rapid lysis mutants had been found to map in three regions of the genome, called rI, rII and $rIII$. A strain carrying six mutants in the rII region (four in gene A and two in the adjacent gene B) and a less closely linked plaque morphology mutant was mixed with wild type and the contents of mottled plaques analysed. When the heterozygous region was short, that is, lay within the marked rII region (and in fact commonly affected only one

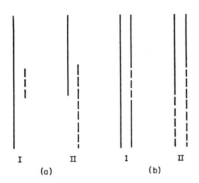

I   II   I   II
(a)   (b)

Fig. 16. Levinthal's models to explain the occurrence of heterozygosity for a genetic marker in a fraction of the progeny of phage T2 crosses. The contributions from the two parents are distinguished by full and broken lines, and corresponding parts of the genome are shown at corresponding positions. If the organisms were basically haploid, the heterozygosity would result from partial diploidy (a), but if it were diploid (b) there would be no need to postulate the occurrence of incomplete genomes. In either case, opposite ends of the genome might derive from the same (I) or different (II) parents. As an alternative possibility, the two lines in each diagram in (b) might represent the two polynucleotides in a single DNA molecule; in this case (b) would refer to a haploid organism. (Reproduced by permission of Cold Spring Harbor Laboratory from C. Levinthal. *Cold Spring Harb. Symp. quant. Biol.* 1953, **18**, 13 (1954).)

of the six mutants), markers on either side were usually in the parental configuration (33 out of 45). With longer heterozygous regions, extending beyond one end of the marked *r*II region, markers flanking the heterozygous region were usually recombinant with one another (53 out of 63). There was also another difference between the 'short' and 'long' heterozygotes: in the short ones the two alleles of heterozygous markers appeared in approximately equal numbers among the progeny, but with the long heterozygotes markers near the end of the heterozygous region showed unequal allele ratios in the progeny.

These puzzling results exemplify the uncertainty, which persisted for a decade following the publication of Levinthal's paper (Levinthal, 1954), about the nature of the partial heterozygotes in the T2 and T4 phages. Some of the results seemed to be in direct contradiction of others. How the problem was resolved is discussed in Section 6.

4. FALSE NEGATIVE INTERFERENCE

In three-factor crosses involving the linked rapid lysis mutants *r2*, *r3* and *r6* of phage T2, which map in numerical order, Hershey and Rotman (1948) obtained results that seemed to indicate that recombination in the *r2–r3* interval tended to be accompanied by a second exchange in the *r3–r6* interval. They discussed the data more fully later (Hershey and Rotman, 1949).

Visconti and Delbrück (1953) pointed out that if one were to interpret recombination values as in genetic studies with higher organisms, disregarding the complications resulting from repeated mating, one would call the positive correlation between the occurrence of recombination in one region and that in an adjacent region a 'negative interference'. As already mentioned (Section 2), Visconti and Delbrück proposed that mating occurred on a random-in-time basis during multiplication in the host cell and before the formation of mature virus particles. They assumed that mature particles were withdrawn from the pool of replicating and recombining structures (genomes) at a linear rate starting after a certain average number of rounds of mating had occurred. They showed that their hypothesis would explain the apparent negative interference found by Hershey and Rotman and also evident in their own experimental results with mixed infections of h and r mutants of T2. The basis of this explanation was that some phage particles represented lines of descent containing more matings than others – a necessary consequence of the assumption of randomness, both in time and as to partner. When allowance was made for differential mating frequency there appeared to be no interference between exchanges.

Similar apparent negative interference was discovered in T4 by Doermann and Hill (1953). As already referred to (Section 2), they obtained linear linkage maps using plaque-type mutants. The recombination frequencies were all derived from two-factor crosses. They found that with linked mutants the longer intervals on the map were lower than the values expected from the shorter intervals on the assumption that recombination in adjacent intervals was independent. Just as with Hershey and Rotman's three-point data for T2, it appeared as if recombination in one interval increased the likelihood of it occurring also in an adjacent interval. Like Visconti and Delbrück (1953), Doermann and Hill (1953) pointed out that the most plausible explanation of this result was that the individual genomes multiplying in the cell vary in the number of opportunities they have for recombination. For example, if only half take part in recombination, the proportion of double recombinants within that half might be that predicted from the frequencies of single recombinants within them, but when the whole population is used (as is necessarily done in practice) to estimate the frequencies of single recombinants, the predicted number of doubles would be only half the observed number. On applying Visconti and Delbrück's theory to calculate the recombination value per mating, the small positive correlation in the occurrence of neighbouring recombination events disappeared.

Thus this so-called negative interference would appear to be a consequence of the way phage multiplies: some matings will be selfings between genomes derived from the same parent, and successive opportunities for crossing will give double events that are false in the sense that they occurred at different times and are not related to one another. This phenomenon may be called *false negative interference* since it does not imply that one recombination event interferes in any way with the occurrence of another such event in its vicinity. It is sometimes called *low negative interference* to distinguish it from the phenomenon discussed in the next section, which is commonly called *high negative interference*. This terminology, however, obscures the fact that the two phenomena are quite unrelated to one another.

5. TRUE NEGATIVE INTERFERENCE

While studying crosses between mutants of the *r*II region of T4, with the object of testing whether the recombination frequencies fitted a linear map, Chase and Doermann (1958) encountered an unexpected result. A total of 10 *r*II mutants were used. The frequency of wild-type recombinants was measured from various three- and four-point crosses, double mutants being crossed with a third mutant or with another double mutant. A positive correlation was found between recombination events, and the correlation was greater the shorter the interval between the sites. A preliminary result of the same kind was obtained with *h* mutants of T2 by Streisinger and Franklin (1957). The false negative interference shown by Visconti and Delbrück (1953) to arise from the mating kinetics (Section 4) was unable to explain the magnitude of the negative interference observed in these experiments, nor could it account for the relationship with the recombination frequency. Chase and Doermann (1958) referred to the new phenomenon as *high negative interference*. From the four-factor crosses it was evident that not only double but also higher-order exchanges occurred over a short length of the genome with unexpectedly high frequency. The interference was very well marked: in the three-factor crosses in which the outermost mutant sites were less than 2 map units apart, the fraction of double recombinants exceeded the random expectation 15–30 times, and for four-factor crosses the interference was still greater.

Edgar and Steinberg (1958) concluded that the excess of double recombinants was produced in a single mating event. They reached this conclusion by systematically varying the relative frequency of the two parents in three-factor crosses of *r*II mutants of T4. By reasoning comparable to that used in chemical kinetics they inferrred that the three wild-type alleles selected for were derived from only two parental genomes. Moreover, they showed (Steinberg and Edgar, 1961) that in three-factor crosses of *r*II mutants *a*, *b* and *c* involving three parental genotypes (*ab* x *bc* x *ac*), such that the selected double recombinant derived its three wild-type alleles one from each parent, there was little or no negative interference. It was evident that the phenomenon was a true multiple exchange within a short region of the genome. The phenomenon is demonstrable only in crosses with closely linked markers.

Explanations of the clustering of multiple exchanges are discussed in Sections 6 and 10.

6. GENOME STRUCTURE AND THE CAUSES OF HETEROZYGOSIS

Streisinger and Bruce (1960) pointed out that linkage of character differences is not as readily demonstrated in phage crosses as it is in higher organisms because, as Visconti and Delbrück (1953) showed (see Sections 2 and 4), the production of recombinant phage in a mixedly infected bacterial cell can be understood only if it is treated as a problem in population genetics. Streisinger and Bruce (1960) realized, however, that when considering the frequencies of different genotypes for a pair of alleles at each of three loci, the occurrence of linkage can be tested without making any assumptions about the kinetics of mating or about the recombination mechanism. For example, in a cross between

$$\frac{a\ b \qquad\qquad c}{+\ + \qquad\qquad +}, \qquad \text{and}$$

where a and b are linked loci, if c is also linked as indicated, $a + +$ progeny will be more frequent than $a + c$ progeny which require two recombination events for their formation. On the other hand, if c is not linked to the other two loci, these two genotypes will be expected to be equally frequent. To help maintain differences in these frequencies, if present, opportunities for successive matings can be minimized by premature lysis, and the crosses can be made with unequal numbers (multiplicities) of the two parents, each in turn being in excess. Using this sensitive test, Streisinger and Bruce (1960) discovered that all the known genetic markers of phages T2 and T4 were linked. Previously, several linkage groups had been recognized.

Streisinger, Edgar and Denhardt (1964) showed from three- and four-factor crosses that markers near opposite ends of the linear linkage map of T4 were themselves linked and that the map was in fact circular. Circularity had been demonstrated by Foss and Stahl (1963) from a cross involving four well spaced mutants, and was confirmed by Edgar and Lielausis (1964) from two-factor crosses with a large number of temperature-sensitive mutants which, unlike wild type, were unable to grow at high temperatures (37–42 °C).

Despite the circular map, Streisinger *et al.* (1964) did not favour a circular genome for T4, since the DNA of T2 was already known from [³H]thymine autoradiography by Cairns (1961) to be a single rod-shaped molecule of about 52 μm in length. Instead, they proposed that the genome was linear but terminally repetitious:

$$a\,b\,c\,d\,e\,f\,g\,h\,i\,j\,k\,l\,m\,n\,o\,p\,q\,r\,s\,t\,u\,v\,w\,x\,y\,z\,a\,b\,c,$$

and that after several rounds of replication, following infection, progeny genomes arose which had circularly permuted genetic sequences such as

$$g\,h\,i\,j\,k\,l\,m\,n\,o\,p\,q\,r\,s\,t\,u\,v\,w\,x\,y\,z\,a\,b\,c\,d\,e\,f\,g\,h\,i$$

or

$$m\,n\,o\,p\,q\,r\,s\,t\,u\,v\,w\,x\,y\,z\,a\,b\,c\,d\,e\,f\,g\,h\,i\,j\,k\,l\,m\,n\,o.$$

Such an organization would lead to a circular genetic map. This bizarre hypothesis had been arrived at following discussions with F. W. Stahl, M. Meselson and M. Fox. The model implied that in a population of phage particles the ends of the genome would be randomly distributed along the linkage map and a particle carrying different alleles at a site that was repeated at each end of the genome would be heterozygous.

Nomura and Benzer (1961) had suggested that there were two structurally distinct classes of heterozygotes in the T-even phages. They were led to this conclusion from study of the frequency of progeny that produced mottled plaques following mixed infection with wild-type T4 and individual deletion mutants of the *r*II genes. All the deletion mutants were found to show approximately one-third the frequency of heterozygosity of point mutants. The deletions were of various lengths, covering the sites of from 20 to over 300 point mutants, but all behaved alike. Nomura and Benzer (1961) suggested that deletion mutants were unable to form an extended mismatch. In other words, the type of heterozygote that Levinthal (1954) had postulated with overlapping strands (see Section 3) did not arise because the two nucleotide chains, one from each parent, would not match. The second class of heterozygote – the only ones observed with deletion mutants – was evidently insensitive to mismatch.

Streisinger's hypothesis of terminal repetition of nucleotide sequence offered an explanation for this second class of heterozygote. He and associates (Séchaud *et al.*, 1965) pointed out how the two kinds of heterozygotes were expected to differ:

(1) Terminally repetitious heterozygotes (Fig. 17(i)(c)) for a mutant will arise if the mutant site is included at both ends of the genome and if the last recombination event, occurring somewhere within the genome, had been such that one end of the genome carries the mutant and the other end the wild-type allele. This type of heterozygosity is both formed and lost by recombination, so its frequency, as a function of time after infection, will depend on the recombination rate but not the replication rate. Because of the high frequency of recombination in T4, the frequency of terminally repetitious heterozygotes will be expected to rise during the early stages of infection and rapidly to reach the equilibrium value.

(2) Internal or heteroduplex heterozygotes, that is, one strand of the DNA carrying the mutant nucleotide sequence and the other strand the wild-type sequence, can arise either by the insertion of a segment of a strand from one parent into a molecule from the other parent in place of the corresponding segment, flanking markers remaining in a parental configuration (an *insertion heteroduplex*, Fig. 17(ii)(c)), or at the junction of duplex segments from each parent, flanking markers showing recombination with one another (a *crossover heteroduplex*, Fig. 17(ii)(d)). Unlike terminally repetitious heterozygotes, internal heterozygosity would be lost when the DNA replicates. So an internally heterozygous phage particle will arise if a recombination event giving heterozygosity has occurred at the mutant site and if the genome has been incorporated in a mature particle without further replication. The frequency of internal heterozygotes, as a function of time after infection, will therefore depend on both the recombination and the replication rates.

Séchaud *et al.* (1965) tested these predictions by inhibiting DNA synthesis with the thymidine analogue 5-fluorodeoxyuridine (FdUrd) and measuring the frequency of heterozygotes for point mutants and for deletion mutants of the *r*II region of the T4 genome at various times after infection. The time variation was achieved by blocking phage maturation for various lengths of time with chloramphenicol. It was found that point-mutant heterozygotes increased in frequency with time after infection whereas deletion-mutant heterozygotes did not. These results were in agreement with the hypothesis that Nomura and Benzer's

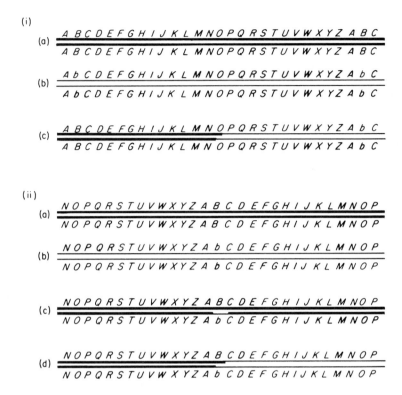

Fig. 17. The various kinds of heterozygotes found in phage T4. The lines represent individual polynucleotides of DNA, their thickness distinguishing their parentage. The letters of the alphabet indicate the entire genome, the parents, (a) and (b), of each kind of heterozygote differing only at site *B/b*. In (i) is shown the derivation of a terminally repetitious heterozygote (c) as a result of recombination occurring between the parental genomes at sites other than *B/b*. In (ii) internal, that is, heteroduplex heterozygotes ((c), (d)) arise as a result of recombination at *B/b* and may be of insertion (c) or crossover (d) type.

two classes of heterozygotes were the internal and the terminally repetitious, with deletion mutants showing only the latter. Another character difference was found by Séchaud *et al.* (1965) to behave like the deletion mutants. This concerned the host ranges of phages T2 and T4, the T2 genetic region having been transferred to T4 stocks by backcrossing. The frequency of heterozygotes for this character difference did not increase with FdUrd treatment. The presumption was that, like the deletions, a structural difference in the genomes was involved such that internal heterozygotes were not formed.

Additional evidence for two kinds of heterozygote in T4 was obtained by Shalitin and Stahl (1965). They used two deletion mutants at opposite ends of the *r*II *B* gene and two point mutants similarly placed, and made crosses between mutants at either end in the presence and in the absence of FdUrd. Wild-type recombinants were tested for heterozygosity at either mutant site. The FdUrd caused a fourfold increase in heterozygote frequency when the point mutants were crossed but had no effect with the deletion mutants. This result from two-factor crosses agreed with the previous results from one-factor crosses.

If it is true that T4 genomes are linear molecules that are terminally repetitious, and that a population of phage genomes comprises a circularly permuted series of these molecules, what determines the length of the genome? Streisinger, Emrich and Stahl (1967) suggested that the length is determined by some factor extrinsic to the genome itself, such as the amount of DNA that can be contained in a phage head. A prediction of this hypothesis is that deletion of part of the nucleotide sequence will be compensated for by an increase in the length of the terminally repeated segment. For example, if a normal genome is represented as

$$a\,b\,c\,d\,e\,f\,g\,h\,i\,j\,k\,l\,m\,n\,o\,p\,q\,r\,s\,t\,u\,v\,w\,x\,y\,z\,a\,b\,c,$$

a genome containing a deletion of m and n would be

$$a\,b\,c\,d\,e\,f\,g\,h\,i\,j\,k\,l\,o\,p\,q\,r\,s\,t\,u\,v\,w\,x\,y\,z\,a\,b\,c\,d\,e,$$

with a repetition of five letters instead of three.

Streisinger *et al.* (1967) tested this prediction by measuring the frequency of terminally repetitious heterozygotes in point mutants of the *r*II region of T4 and in deletion mutants of this region of various lengths. To measure terminally repetitious heterozygosis the frequency of heterozygotes for the T2 versus T4 host range was used. This was a character difference which, as they had already shown (Séchaud *et al.*, 1965; see above), produced no internal heterozygotes. Streisinger *et al.* (1967) found that the frequency of terminally repetitious heterozygotes was greater among the progeny of crosses involving a deletion (in both parents) than in those involving a point mutant (likewise homozygous). Furthermore, the frequency was greatest with the longest deletion and decreased as the length decreased. These results were in agreement with the hypothesis that each T4 particle contains a headful of DNA, irrespective of nucleotide sequence. Moreover, Kvelland (1969) showed that a deletion covering at least the whole *r*II region (which occupies about 17 units of the T4 genetic map) increased the length of the terminal repetition by a corresponding amount. She studied heterozygosity in seven amber mutants spread over about 30 units of the genetic map in a part of the genome remote from the *r*II region, and found that the average length of the heterozygous region increased from a normal value of 5 units on the map to 24 units in the deletion strain. Conversely, Weil and Terzaghi (1970) showed that duplications of the *r*II region were frequently associated with deletions elsewhere in the genome. This is predicted by the Streisinger model since duplications which exceed the normal length of the terminal repetition will be lethal unless compensated for by deletion of non-essential parts of the genome.

Confirmation of the truth of Streisinger's theory has also been obtained from physical studies. Thomas and Rubenstein (1964) passed a dilute solution of unbroken molecules of T2 DNA through a capillary at gradients that cause a single break in some of the molecules. The molecules were then fractionated in a serum–albumin column and the smallest fragments collected. These would all be less than half size, some coming from the left and some from the right end of the molecule. Thus all would lack a central region. But on re-annealing these fragments with denatured, that is, single-stranded, T2 DNA immobilized in agar, all the nucleotide sequences of T2 DNA were found to be present among the fragments. This showed that different molecules of T2 DNA begin at different points in the nucleotide sequence. Thomas and MacHattie (1964) denatured T2 DNA with alkali and renatured it at neutral pH. This allowed annealing to occur of strands from different molecules. Electron micrographs were prepared of the resulting duplex molecules and found to be circular. This showed that different molecules had circular permutations of the same nucleotide sequence. Furthermore, such annealing will result in two single strands projecting from the duplex circle at different positions, for example,

g h i j k l m n o p q r s t u v w x y z a b c d e f g h i

annealed with

m n o p q r s t u v w x y z a b c d e f g h i j k l m n o

will have one of the *g h i* segments of the former strand and one of the *m n o* segments of the latter projecting. The pairs of simplex branches were observed by MacHattie *et al.* (1967): the branch had a bush-like appearance, the strand being much contorted. MacHattie *et al.* (1967) found that all the possible distances between the two branches, measured along the duplex molecule, occurred with equal frequency, as expected if there is no preferred permutation of sequence. They also treated unbroken linear duplex DNA molecules of T2 with exonuclease III of *E. coli*. This enzyme removes nucleotides stepwise from 3′ ends in duplex DNA, exposing 5′-ended single strands at both ends of the molecule. After a short time the reaction was stopped and annealing of complementary strands allowed to occur. When examined on electron micrographs circular molecules had formed. This showed that the nucleotide sequences at the two ends of each original molecule were identical, while those at the ends of one molecule differed from those at the ends of others, since they did not anneal. All the results of these physical studies were in precise agreement with the Streisinger model for the structure of the genome of T2 and T4.

Evidence from a number of sources suggested that the 'short' and 'long' heterozygotes which Doermann and Boehner (1963) had found to show differences in behaviour (see Section 3) corresponded to internal and terminally repetitious heterozygotes, respectively. Thus, Kvelland (1969) found that the average length of the heterozygous region was greatly reduced in crosses carried out in the presence of FdUrd. It was evident that the heterozygotes that increase in number with FdUrd treatment (see above) are short; as already indicated, these are the internal heterozygotes. Mosig *et al.* (1971) made crosses between abnormal T4 particles with incomplete genomes. Such particles could be isolated because of their low buoyant density in caesium chloride. Owing to its genetic deficiency a small particle can produce progeny only if the bacterial cell is also infected with another particle that can provide the missing genes. In the aggregate the parental strains differed by 19 factors that mapped at widely scattered sites along the linkage map. Individual small particles, however, would lack some of the markers. The positions of the genome ends in the pair of particles infecting particular cells could thus be deduced from the presence or absence of parental alleles in the progeny obtained from the single burst. Mosig *et al.* (1971) found that recombination was five to 10 times more frequent when a marked region was near the end of the genome than when it was in the interior of the molecule. Frequent recombination at genome ends had also been suggested by A. H. Doermann and associates on the basis of their investigations. It would often lead to the elimination of one of the two alleles in a heterozygote of the terminally repetitious type. This would explain the unequal allele ratios in the progeny, shown by markers near the end of a long heterozygous region (see Section 3). Thus the long heterozygotes appeared to be the terminally repetitious ones.

It will be recalled that Doermann and Boehner (1963) found that short heterozygous regions were usually associated with a parental genotype for flanking markers (Section 3). With the knowledge that the short heterozygotes were the internal, that is, heteroduplex ones, it was evident that insertion heteroduplexes (the ones with parental flanking marker genotypes – Fig. 17(ii)(c)) were of frequent occurrence. This conclusion provides a partial explanation for true (high) negative interference, because insertion of a segment of a single strand of DNA from one parent into a duplex molecule of the other parentage, in place of the corresponding segment, implies two recombination events, that is, changes of parentage, in proximity – at the two ends of the inserted segment. Confirmation that insertion heteroduplexes contribute to negative interference is given in Section 10.

The occurrence of insertion heteroduplexes cannot be the whole explanation of negative interference because, as already pointed out (Section 5), Chase and Doermann (1958) had found that the phenomenon was not restricted to two events: from their four-factor crosses it was evident that three exchanges, one in each of the marked intervals, occurred with a higher frequency than was expected by chance. In order to explain the high efficiency of recovery of markers from incomplete genomes, Mosig *et al.* (1971), in the investigation described above, found it necessary to postulate that more than one round of recombination occurred in the region of the genome ends. This reiterated recombination provided an additional explanation of negative interference and one that was not restricted as to the number of changes of parentage in proximity to one another.

A further possible contributory factor to negative interference is discussed in Section 10.

7. JOINT MOLECULES

Evidence that recombination in phage T4 took place by breakage and rejoining, rather than during replication by copying partly from one parent and partly from the other, was obtained by Kozinski (1961). Host cells carrying 5-bromodeoxyuridine in their DNA as a density label were infected with T4 of normal density but with ^{32}P as a radioactive label. Phage DNA was extracted from the cells at various times after infection, and in other experiments progeny DNA was fragmented by sonication. The DNA from each treatment was then centrifuged in a caesium chloride density gradient. It was evident that only a small fraction of the parental phage DNA was incorporated into any one progeny DNA molecule. This fragmentary transfer from parent to progeny implied breakage and reunion: a copy-choice mechanism would not lead to fragmentation of parental DNA. Kozinski and Kozinski (1963) separated the strands of the progeny DNA by heat treatment and density-gradient centrifugation. They found that most of the radioactive material remained associated with the progeny strands, indicating that it was joined by covalent linkages. From the density it appeared that the parental contribution to progeny strands averaged 10–15%.

A more direct approach than hitherto to the question of the mechanism of recombination in the T-even phages was achieved by Tomizawa and Anraku (1964b) as a result of the discovery that recombination can occur in the absence of gross metabolism. J. Tomizawa had found that chloramphenicol (an inhibitor of protein synthesis) and 5-fluorodeoxyuridine (FdUrd, the inhibitor of DNA synthesis already referred to) did not reduce the frequency of recombination, and similar conclusions had been reached independently by Hershey, Burgi and Streisinger (1958) for chloramphenicol and by M. C. Frey (see Symonds, 1962) for FdUrd. Furthermore, Tomizawa and Anraku (1964a) found that potassium cyanide and other metabolic inhibitors increased the recombination frequency between rII mutants of phage T4, although synthesis of phage DNA, measured by the extent of incorporation of phosphate labelled with ^{32}P, was strongly suppressed. This discovery enabled them (Tomizawa and Anraku, 1964b) to study the immediate products of recombination, hitherto impossible owing to further replication before phage particles matured. E. coli was mixedly infected with two strains of T4, one carrying a a radioactive label (^{32}P) and the other a density label (bromouracil), in the presence of cyanide to inhibit DNA synthesis. After 60 min the DNA was extracted and centrifuged in a caesium chloride density gradient. The ^{32}P was found in fractions with densities greater than that of the input ^{32}P-labelled DNA, so Tomizawa and Anraku (1964b) concluded that the two parental molecules must associate in some way. They called this a 'joint molecule'.

The structure which they favoured for this joint molecule was the same as that originally proposed by Levinthal (1954) for an internal heterozygote, namely, intact duplex molecules one from each parent held together by hydrogen bonds between complementary base pairs in a homologous overlap region of single strands one derived from each parent, that is, heteroduplex (Fig. 18(a)). They suggested that the unpaired polynucleotides shown in the diagram might form a second paired overlapping region through hydrogen bonding of their complementary bases.

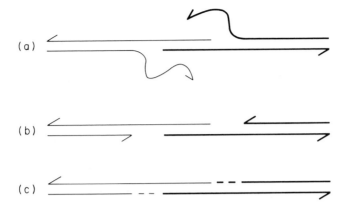

Fig. 18. Structures proposed by Tomizawa and Anraku (1964*b*) for (a), (b) joint molecules and (c) recombinant molecules of phage T4. The lines represent individual polynucleotides of DNA, their thickness indicating their parentage. Broken lines indicate newly synthesized strands to fill gaps.

They favoured this structure for the following reasons:

(1) The density of the joint molecules ranged from close to that of the heavy parent to close to that of the light parent. This implied that parental DNA molecules could be joined without separation of the original strands (except in the overlap region), the joining apparently occurring in any part of the molecule.

(2) Heat treatments which separate duplex DNA into single strands were found to separate the ^{32}P-labelled DNA of light density from the heavy DNA containing bromine atoms. In other words, the joint molecules were maintained only by hydrogen bonds.

(3) The sedimentation properties of the joint molecules in caesium chloride or sucrose density gradients were similar to those of normal DNA molecules of T4. This suggested that the joint molecules had a linear structure, and this was confirmed by quantitative analysis of the products of breakage of the joint molecules, both as a result of shearing forces derived from stirring at high speed, and also following treatment in a sonic disintegrator.

Similar experiments to those of Tomizawa and Anraku (1964*b*) were carried out by Kozinski and Kozinski (1964), but using FdUrd instead of potassium cyanide to restrict DNA synthesis. They found that the recombinant molecules, revealed by their radioactive label and intermediate density, still possessed contributions from both parents after strand separation by heating or treatment with alkali. On the other hand, breakage of the molecules by shearing separated the parental contributions. They concluded that stretches of parental nucleotide chains had been integrated into the recombinant molecules, and the continuity of the chains repaired.

Having established that joint molecules were formed from two parental molecules, the joining being by hydrogen bonds, Anraku and Tomizawa (1965a) went on to investigate how this structure became a recombinant molecule, such as Kozinski and Kozinski (1964) had demonstrated, with covalent bonds between the two parental contributions. As before, the host cells were infected with a mixture of ^{32}P-labelled and bromouracil-labelled T4 particles. DNA synthesis was largely inhibited with FdUrd. At various times after infection DNA was extracted and subjected to density-gradient centrifugation in caesium chloride. At early times (15–20 min) after infection the radioactive molecules that were heavier than the input radioactive DNA evidently had the polynucleotide containing ^{32}P joined by hydrogen bonds to that containing bromine atoms, because treatment with formaldehyde at 93 °C separated them. But at later times (30–45 min after infection) some of the joining was evidently by covalent bonds since, in a proportion of the molecules, no separation occurred with this treatment. This resistance to denaturing by heat treatment was also seen in the DNA of T4 particles that had formed in the presence of the FdUrd: the joining in these progeny particles was evidently almost exclusively by covalent bonds. As with the hydrogen-bonded joint molecules (see above), the covalently joined recombinant molecules can apparently keep their parental duplex structure, that is, without the separation of their strands. This was shown from the density-gradient data, which indicated that some of the recombinant molecules had conserved their heavy DNA.

These results with FdUrd inhibition of DNA synthesis led Anraku and Tomizawa (1965a) to suggest that recombination in T4 takes place in two successive steps: (1) joining of duplex parental DNA molecules by hydrogen bonds to form a joint molecule; (2) their joining by covalent bonds to form a duplex recombinant molecule. For this second step, they favoured the synthesis of polynucleotides to fill single-strand gaps (Fig. 18(b),(c)). The fairly late appearance of covalently bonded recombinant molecules was attributed to the inhibitory action of FdUrd on the postulated DNA synthesis needed to fill the gaps.

The conclusion that recombination in T4 involves successively these two kinds of joining – hydrogen bonding and then covalent linking – was reinforced by making use of mutants defective in specific T4 activities. With the establishment of the circular linkage map (Section 6), the genes known at that time – 47 in all – were arbitrarily numbered sequentially clockwise round the map by Edgar, Denhardt and Epstein (1964). Mutants of seven of these genes (nos *1*, *32* and *41–45*) produced no phage DNA and so could be used in recombination studies as an alternative to or a reinforcement of the block to DNA replication provided by FdUrd treatment. Conditional lethal mutants, that is, either amber mutants or temperature-sensitive mutants, of these genes were particularly valuable as they allow normal or mutant gene expression according as the host strain, or the temperature, is permissive or restrictive. Anraku and Tomizawa (1965b) used a temperature-sensitive mutant of gene *42*. The cytosine residues in T4 DNA carry a hydroxymethyl radical at position 5. Gene *42* codes for the enzyme responsible for inserting this radical. In a permissive host the gene-*42* mutant produces a temperature-sensitive deoxycytidylate hydroxymethylase. Anraku and Tomizawa used FdUrd to restrict DNA synthesis, and further inhibited it by incubation at the non-permissive temperature. Using the same technique as previously, that is, mixed infections of ^{32}P-labelled and bromouracil-labelled phage particles, they found that covalently joined recombinant molecules were not formed, but hydrogen-bonded joint molecules were produced in a larger amount than that observed in their earlier experiments (Anraku and Tomizawa, 1965a) when DNA synthesis was inhibited by FdUrd alone. This result confirmed that hydrogen-bonded molecules were formed earlier than covalently linked ones, the formation of the latter being suppressed when DNA synthesis is completely inhibited.

Weiss and Richardson (1967) found that T4 infection of *E. coli* induces an enzyme system that catalyses the repair of single-strand breaks (nicks) in a DNA duplex by the formation of phosphodiester bonds. The enzyme, polynucleotide ligase, was purified. The reaction was found to require adenosine triphosphate and magnesium ions. Fareed and Richardson (1967) showed that gene *30* of T4 is the structural gene for the enzyme: amber mutants in this gene failed to induce active ligase and temperature-sensitive mutants induced an enzyme with increased heat-lability.

Continuing the studies on the effects of inactivating specific phage genes, Anraku and Lehman (1969) found that a mutant defective in gene *43* (*pol*), which codes for T4-induced DNA polymerase, allowed both hydrogen-bonded and covalently linked recombinant molecules to form. This was without the use of additional blocks to DNA synthesis such as FdUrd. On the other hand a double mutant defective in gene *43* and in gene *30* (*lig*) gave rise only to hydrogen-bonded molecules. It was not possible to test the ligase mutant alone as it did not inhibit DNA synthesis. It was evident, however, that ligase is required for the formation of recombinant molecules. They found, moreover, that if the joint molecules from the *30–43* (*lig–pol*) double mutant were incubated with host DNA ligase and T4 DNA polymerase in the presence of the four deoxynucleoside triphosphates, covalently linked recombinant molecules were formed, but either enzyme alone was insufficient to bring this about.

From these experiments it was evident that both a polymerase and a ligase were required to produce covalently linked recombinant molecules from hydrogen-bonded joint molecules. It was also apparent that *in vivo* a DNA polymerase of the host can replace the phage-induced enzyme but that, unlike *in vitro*, the host ligase cannot substitute for the T4 ligase. In agreement with these observations, the level of ligase activity in host cells infected with the *30–43* double mutant of T4 was found to be very low – perhaps too low to meet the demands of T4. It is known that after T4 infection high levels of endonuclease activity are induced, leading to the destruction of the host genome, that of the virus being protected by the hydroxymethylation of its cytosine residues (see Section 8). The endonuclease makes single-strand breaks in duplex DNA and the resulting nicks would provide a substrate for DNA ligase, so it may be that the large number of nicks in the host DNA saturates the supply of the host ligase.

The need for polymerase as well as ligase to produce covalently linked recombinant molecules from hydrogen-bonded joint ones implied that the joint molecules contained gaps rather than nicks. Anraku, Anraku and Lehman (1969) estimated the size of these gaps by measuring the uptake of ^{3}H-labelled cytidine triphosphate into joint molecules by the T4-induced DNA polymerase. They used a gene-*43* (*pol*) amber mutant and a *30–43* (*lig–pol*) double mutant, and concluded that the average gap size was of 300–400 nucleotides.

Examination under the electron microscope of the joint molecules formed by the *30–43* double mutant (defective in T4 polymerase and ligase) showed branched structures with some unexpected features. These are discussed in the next section.

8. BRANCH MIGRATION

It was pointed out in Section 6 that when molecules of T2 DNA are denatured and the single strands then allowed to anneal, this will usually take place between strands from different molecules and, due to their circularly permuted sequence, will result in circular duplex molecules with two single-stranded branches corresponding to the terminal repetitions of nucleotide sequence in each original molecule. Lee, Davis and Davidson (1970) examined on electron micrographs DNA molecules of phage 15 of *E. coli* that had been treated in this way. This phage is like T2 and T4 in having in the mature particle a linear duplex DNA molecule that is terminally repetitious, and in different particles the nucleotide sequence is circularly permuted. Lee *et al.* (1970) found that when the molecules had been renatured in formamide the branches became extended on the electron micrographs instead of forming a bush. They discovered that some of the branches were double structures, with two single strands arising from the same point on the duplex. Each branch, when single, is expected to be of the same length, corresponding to the terminally repeated segment. When there were two branches arising at the same point, the sum of their lengths was equal to that of the single branches. Lee *et al.* (1970) realized that the double branches evidently arose as a result of movement of the position of the branch along the duplex (see Fig. 19). They called this newly discovered phenomenon *branch migration*. In this process, at the point where the single strand, or the pair of single strands, leaves the duplex, a change of pairing partner evidently occurs, one strand with a free end taking the place of the other one in the duplex. Thus, no breakage of covalent links is involved, merely severance of hydrogen bonds and their reacquisition in new pairings.

The process to which Lee *et al.* (1970) gave the name branch migration relates to movement of a simplex branch. Broker and Lehman (1971) extended the principle to a duplex branch. As already pointed out (Section 7), Anraku, Anraku and Lehman (1969) had found that a double mutant of phage T4 defective in genes *30* (*lig*) and *43* (*pol*), that is, in T4-induced polynucleotide ligase and in T4 DNA polymerase, gave rise only to hydrogen-bonded joint molecules, the further step in the recombination pathway to covalently joined recombinant molecules being blocked by these defects. Under the electron microscope branched duplex

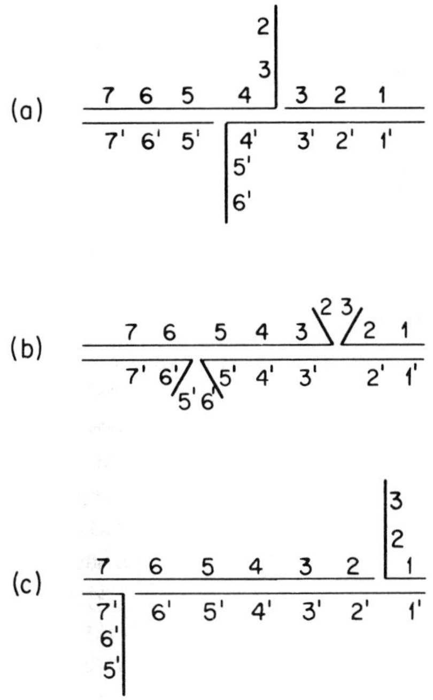

Fig. 19. Branch migration, as first described by Lee, Davis and Davidson (1970). The drawings represent part of the circular DNA molecule obtained by annealing single strands of DNA from a phage that has a linear molecule with a terminal repetition of nucleotide sequence, and in which different particles have circularly permuted sequences. The left- and right-hand ends (numbered 1 and 7) of the upper strand are supposed to be joined round the circle. This strand has a terminal repetition of segments 2 and 3. The lower strand, the ends of which (numbered 1' and 7') are supposed to be similarly joined, has a terminal repetition of segments 5' and 6'. Numbers with a prime indicate nucleotide sequences complementary to those with the same number without a prime. (a) Single-strand branches in proximity. (b) Pairs of shorter branches at an intermediate separation. (c) Single-strand branches at widely separated positions. The change from (a) to (b) to (c) or from (c) to (b) to (a) results from branch migration. (Reproduced with permission from C. S. Lee *et al.*, *J. molec. Biol.* **48**, 11 (1970). Copyright by Academic Press Inc. (London) Ltd.)

molecules were seen. Broker and Lehman (1971) showed that the branched DNA was an intermediate in recombination. Their evidence was threefold:

(1) A dilute infection, averaging one T4 particle per bacterial cell, failed to reveal branched molecules, implying that their occurrence was the result of an interaction between molecules, such as was more likely to occur when a large number were present.

(2) Additional genetic defects in T4 that led to fewer hydrogen-bonded joint molecules also gave rise to fewer branched molecules.

(3) Joint molecules were selected by their hybrid density in a caesium sulphate density gradient, one parent having been density-labelled with ^{13}C and ^{15}N. The molecules of hybrid density showed a threefold higher frequency of branching compared with those of either parental density, indicating that the branches were generated in the course of DNA exchange.

The structure of the branched molecules was of considerable interest. They were highly variable as to length, number and arrangement of branches. Simple forks (three duplex arms) were frequent and junctions with four arms also occurred. Two forks in proximity, giving an H-shaped configuration, also occurred regularly. The length of the cross-bar of the H was variable and was estimated to be of 200–2000 nucleotide pairs. Multibranched molecules were rather frequent, some apparently representing intermediates of triparental recombination, because the geometry of the cluster of branches required the same genetic region to be represented three times.

From these observations Broker and Lehman (1971) were led to propose a 'branch migration' model for DNA recombination. The novel feature of this hypothesis was that after the pairing of complementary single strands from different molecules had given rise to a branched structure (Fig. 20(b)), long hybrid regions may be generated as a result of branch migration (Fig. 20(c),(d)). In this process base pairs on the two parental limbs adjacent to the branch point ($D \cdot D$, $E \cdot E$, $d \cdot d$ and $e \cdot e$ in Fig. 20(b)) become unbonded and two sets of complementary hybrid base pairs are formed ($D \cdot d$ and $E \cdot e$; $d \cdot D$ and $e \cdot E$ in Fig. 20(d)), resulting in displacement of the branch. They pointed out that since there was no net dissociation of base pairs, such as would require energy, the partner exchange might continue indefinitely until a nick was reached in any of the four strands entering the exchange site. These four strands make up the two parental duplexes. If these differed structurally, for example, through a deletion in one of them, heteroduplex DNA could not form and this would also be expected to prevent further branch migration. The example of branch migration shown in Fig. 20 starts as a four-limb junction and becomes an H-shaped junction with a projecting cross-bar to the H, a configuration that was often seen on the electron micrographs.

Having established that the branched molecules seen in *lig⁻ pol⁻* T4 were intermediates in recombination, Broker (1973) went on to make a detailed electron-microscope examination of the frequency and structure of these molecules in about 20 phage genotypes carrying various combinations of mutants known to affect DNA metabolism, but all *pol⁻* to prevent replication.

Of particular interest was the effect of mutations in genes *46* and *47*. Mutants defective in either of these genes stop DNA synthesis prematurely. Furthermore, they are defective in the degradation of the host DNA.

The initial step in this breakdown requires T4 endonuclease II. Sadowski *et al.* (1971) isolated mutants of T4 defective in this degradation and found they mapped at the *denA* locus (*den* = DNA endonuclease) and were deficient in the induction of endonuclease II. They concluded that this enzyme, which makes

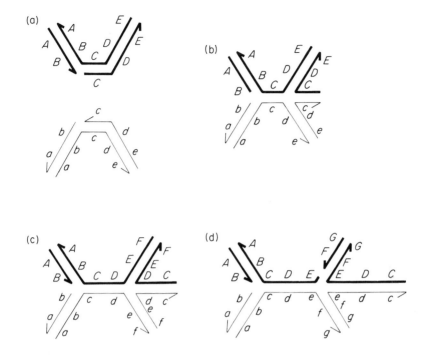

Fig. 20. Steps in Broker and Lehman's branch migration model for recombination in T4. The lines represent individual polynucleotides of DNA, a pair of parallel lines implying the duplex structure. Parentage of strands is indicated by the thickness of the lines, and by the capital letters and lower case respectively, corresponding letters implying corresponding nucleotide sequence. (a) Two parental molecules with single-strand breaks at corresponding positions in strands of opposite polarity. (b) Complementary strands derived from each parent have paired in the region (C/c) adjoining the break-points. (c) Branch migration has extended the heteroduplex regions to include segment D/d. (d) Branch migration has continued and has resulted in segment E/e being included in the heteroduplex regions. The cutting of one strand between segments E and F has prevented branch migration from continuing any further. In later stages of the recombination process it is presumed that further nicks lead to loss of the branches. The positions of these cuts will determine the genotype of the end-product. (Reproduced with permission from T. R. Broker and I. R. Lehman, *J. molec. Biol.* **60**, 143 (1971). Copyright by Academic Press Inc. (London) Ltd.)

single-strand breaks in duplex DNA, is involved in the breakdown of host DNA. T4 DNA is not attacked by it, probably because of he hydroxymethylation of its cytosine residues (see Sadowski and Hurwitz, 1969).

The second step in host DNA breakdown was shown by Kutter and Wiberg (1968) to involve the products of genes *46* and *47*, since in the presence of mutants of either of these genes there was no further degradation beyond the fragmentation caused by endonuclease II. Normally these fragments are rapidly broken down to an acid-soluble form.

Berger, Warren and Fry (1969) found that amber mutants in genes *46* or *47*, when grown in a semipermissive host, greatly decreased the recombination frequency in phage T4 compared with that in permissive cells, and similar results were obtained by Bernstein (1969) with temperature-sensitive mutants. It was evident from these results, and from the arrest of T4 DNA synthesis, that the products of these genes, unlike that of *denA*, affect T4 DNA as well as that of its host. Prashad and Hosoda (1972) studied the nucleolytic activity regulated by these genes. They used the *30–43* (*lig–pol*) double mutant, that is, defective in DNA ligase and DNA polymerase, and found that it accumulated single- and double-strand breaks, mostly gaps. The nature of the damage was determined by use of these enzymes, since nicks (phosphodiester bond interruptions on duplex DNA) can be repaired by ligase alone, while gaps (interruptions with missing nucleotides) require polymerase as well. The extent of repair was estimated from the sedimentation behaviour in alkaline sucrose. An additional mutation in gene *46* or *47* resulted in breaks that were mostly nicks. Furthermore, with the *30–43* double mutant there was considerable breakdown of the DNA to an acid-soluble form, but in the triple mutants (*30–43–46* or *30–43–47*) this did not happen. Replacing the *43* mutant by another mutant (in gene *32*, *44* or *45*) that blocked DNA synthesis, gave similar results. Prashad and Hosoda (1972) concluded that genes *46* and *47* regulate, either directly or indirectly, an exonuclease activity that can attack T4 DNA at nicks to create gaps.

Broker (1973) found that *pol⁻ lig⁻* T4 that also carried defects in genes *46* or *47*, or both, failed to form branched molecules. This was in keeping with the much reduced recombination frequency (see above) shown by *46⁻* or *47⁻* T4. Furthermore, Broker failed to detect any single-strand gaps or single-strand termini on the electron micrographs of *46⁻* or *47⁻* DNA, such as are present when these genes are active (Plate 2). On the other hand, Broker and Lehman (1971) had found that mutations in these genes did not affect the time course or the extent of nicking of the DNA as judged from centrifugation in an alkaline sucrose gradient. The inference from these results was clear: in agreement with Prashad and Hosoda's studies, an exonuclease that is dependent for its expression on the activity of genes *46* and *47* produces single-strand gaps from nicks, and these were prerequisites for recombination.

A T4 gene of quite exceptional interest which, like *46* and *47*, was the subject of investigation by Broker, is no. *32*. This is discussed in the next section.

From the electron micrographs of T4 DNA in the various genotypes, Broker (1973) was able to suggest a pathway for recombination under the conditions prevailing, that is, when DNA synthesis is prevented by a defect in the gene for T4 DNA polymerase. This pathway is discussed in Section 11.

9.   GENE *32*

As mentioned in Section 7, one of the T4 genes that is essential for DNA synthesis in no. *32*. The effect of a defect in this gene on the formation of recombinant molecules was investigated by Tomizawa, Anraku and Iwama (1966). Using the same technique as in their earlier work, that is, density-gradient centrifugation of DNA extracted from cells infected with a mixture of ^{32}P-labelled and bromouracil-labelled T4, they found that an amber mutant (A453) in gene *32* in a non-permissive host failed to give rise to any joint molecules, that is, containing ^{32}P and heavier than normal. They concluded that the product of gene *32* is necessary for their formation. Although, as already mentioned, this product is also needed for DNA replication, their earlier results (Section 7) had shown that the formation of a joint molecule does not require DNA synthesis. They inferred that the gene-*32* protein is separately involved in both processes. Kozinski and Felgenhauer (1967), also working with the A453 amber mutant in a non-permissive host, made the same discovery: there were no recombinant molecules; if present, they would have been revealed by their hybrid density.

Complementary to these results, Berger, Warren and Fry (1969) found that the same amber mutant (A453) decreased the recombination frequency between two *r*II mutants under semipermissive conditions, that is, when the gene-*32* activity of the mutant would be limited. Also, Broker and Lehman (1971) found very few branched molecules on electron micrographs of the DNA of a gene-*32* mutant – amber mutant E315. They believed that such molecules were intermediates in the recombination pathway (see Section 8).

Alberts (1970) reported on progress that he and associates had made in characterizing the product of gene *32*. They had devised a technique for isolating individual proteins that interact with DNA, such as polymerases and nucleases (Alberts *et al.*, 1969). The method was based on the fact that DNA-binding proteins are adsorbed when passed over a column consisting of DNA adsorbed to cellulose. They applied this DNA-cellulose chromatography to T4 because its DNA-binding proteins can be selectively labelled with radioactive isotopes after infection of *E. coli* with the phage. Host proteins are not labelled because their synthesis is turned off shortly after T4 infection. Furthermore, under non-permissive conditions a phage carrying an amber mutation synthesizes only a fragment of the product of that gene. If the product is a DNA-binding protein the fragment is likely to have lost the ability to bind to DNA. The product of a particular gene can then be identified by comparing the electrophoretic patterns on polyacrylamide gel of the wild type and the mutant DNA-binding proteins, isolated by the cellulose technique.

Alberts *et al.* (1969) found that after infection with the A453 amber mutant of gene *32*, a major DNA-binding protein was lacking. A peculiarity of this protein was that if duplex DNA was used in the cellulose column the protein was readily eluted by a solution of low ionic strength, but if simplex DNA was used it was easily removed when the protein was in low concentration but not when in high concentration. From this variation with concentration they inferred that individual molecules of gene-*32* protein bind in cooperative units to simplex DNA, lining up in close juxtaposition along the strand (Fig. 21). Alberts and Frey (1970) confirmed this by studying the interaction of ^{3}H-labelled gene-*32* protein with T4 DNA, using velocity sedimentation in sucrose gradients. They found that one protein monomer can be bound per 10 nucleotides.

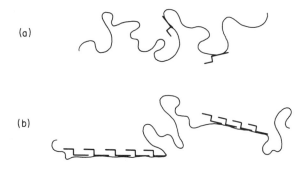

Fig. 21. Alberts' model for the cooperative binding of gene-*32* protein to single-stranded DNA. (a) Weak binding at low concentration. (b) Strong binding at high concentration. (Reproduced by permission of the Federation of American Societies for Experimental Biology from B. M. Alberts, *Fed. Proc., Fed. Am. Soc. exp. Biol.* **29**, 1157 (1970).)

Alberts (1970) argued that because gene-*32* protein binds more strongly to simplex than to duplex DNA, it should loosen the double helix. Evidence for action of this kind was obtained by Alberts and Frey (1970), who found that the gene-*32* protein brought about strand separation at 25 °C in a synthetic DNA containing only adenine and thymine bases with these alternating in each strand (poly(dAT)). This temperature was 40 °C below that at which separation would occur in the absence of the protein. Because of this effect the product of gene *32* is sometimes called a DNA-unwinding protein. Alberts now suggests that such proteins be called helix-destabilizing proteins.

Alberts (1970) argued also that gene-*32* protein, again because of its binding behaviour, should unravel the short, imperfectly paired, duplex hairpins believed to occur in single nucleotide chains through the chain folding back upon itself. The protein would thus not only favour strand separation in duplex DNA but also have the reverse effect and actually facilitate the annealing of complementary strands to form a duplex structure, by removing the regions of weak secondary structure that act as barriers to renaturing. Alberts and Frey (1970) confirmed this expectation by finding that single strands of T4 DNA will renature rapidly at 37 °C only when covered with gene-*32* protein. Thus, one of the major difficulties of all heteroduplex recombination models was removed, the effect of the gene-*32* protein on single strands of DNA indicating how the bases could be made accessible to another strand with a complementary sequence.

Snustad (1968) had obtained evidence that the gene-*32* product acts structurally rather than as a simple catalyst. His work is discussed below. Alberts produced some additional support for this conclusion:

(1) He found (Alberts, 1970) that a temperature-sensitive mutant (L171) of gene *32* gave rise to a gene-*32* protein that bound more weakly than normal to the DNA.

(2) Alberts and Frey (1970) found that large quantities of gene-*32* protein were produced. There were about 10 000 molecules per infected cell, though few if any of them were used in the construction of the mature phage particle. The molecules were stable and their synthesis continued until lysis.

Alberts surmised that likely functions for the gene-*32* protein would be binding to single strands in recombination and replication, thus accounting for the failure of these processes in its absence.

Support for these ideas about the way the gene-*32* protein acted has come from many sources. Snustad (1968) and Sinha and Snustad (1971) made gene dosage experiments with amber mutants in numerous T4 genes, mixing wild type and mutant in various proportions and infecting a restrictive host with the mixture. With mutants in genes coding for known structural components of the phage particle, the burst size – the number of progeny particles released from individual cells – was found to decrease rapidly as the proportion of wild type in the initial mixture decreased. In other words, the mutant showed considerable co-dominance. By contrast, there was little effect on burst size with mutants in genes coding for enzymes, presumably because the decrease in the amount of gene product was compensated for by an extended period of activity. The mutant was thus completely recessive. In this comparison, gene *32* behaved as if coding for a structural component.

Huberman, Kornberg and Alberts (1971) tested the effect of gene-*32* protein on the activity of the T4 DNA polymerase. This enzyme requires a simplex DNA template. They found that the rate of synthesis *in vitro* was increased five- to 10-fold by the protein. The stimulation was greatest at low temperature, high ionic strength, and when there was sufficient gene-*32* protein to bind most or all of the template DNA present. From these results they concluded that the gene-*32* protein acts by removing inhibitory secondary structure from the template.

Delius, Mantell and Alberts (1972) used electron microscopy to examine the structure of the complex which gene-*32* protein forms with single strands of DNA. With DNA in excess, the distribution of the protein along the DNA could be seen on electron micrographs and was found to be non-random, the protein tending to bind in long clusters leaving other regions of the DNA free of it. This confirmed the cooperative binding (Fig. 21(b)) inferred from the elution behaviour in a DNA–cellulose column. With protein in excess, the single strands were coated throughout with it, forming a flexible rod about 6 nm in diameter. As predicted, it was found that the gene-*32* protein induced denaturation of duplex DNA, pre-ferentially invading regions rich in adenine and thymine. The strand separation could be seen on electron micrographs by fixation in glutaraldehyde. The simplex strands appeared thicker than the duplex on account of the gene-*32* protein bound to them.

Carroll, Neet and Goldthwait (1972) studied the self-association of gene-*32* protein, using sedimentation equilibrium centrifugation and polyacrylamide gel electrophoresis. They found that stable dimers were formed under some condi-tions, for example, high ionic strength and pH 10, and under other conditions the dimers associated with one another to give larger aggregates containing at least 10 molecules. It was evident that the bonds between dimers, necessary for aggregation, have different properties from those that hold monomers in the dimer. On the basis of these results they proposed a model of how the gene-*32* protein might function in replication and recombination. According to their hypothesis the two parental strands in replication would be held partially separated from one another by a series of gene-*32* protein dimers. This would allow the strands to base pair with incoming nucleotide precursors. Recombination would similarly be facilitated, as already pointed out, because the exposed bases on the single strands would provide opportunity for annealing with a complementary strand from another molecule.

Wackernagel and Radding (1974) showed that gene-*32* protein would bring about annealing of complementary single strands of DNA as effectively at 37 °C as that resulting from heat treatment at 75 °C in the absence of the protein. As substrate they used molecules of phage λ DNA that had been broken approximately in half by shearing, and then partially digested by λ exonuclease at the broken ends. Some of the left-hand 'halves' would then have a 3'-terminal single strand at their right-hand end that was complementary to a similar strand in some of the right-hand 'halves'. The DNA of two different mutants of λ was treated in this way and then mixed. A measure of the annealing of complementary strands that took place was provided by the infectivity of the resulting preparation, since only joint molecules carrying the wild-type alleles of two mutants would lead to viable phage. It was evident that gene-*32* protein promotes the formation of joint molecules *in vitro*, perhaps simulating part of the T4 recombination pathway.

Thus, all these studies, using a diversity of techniques, gave support to Alberts' ideas about how gene-*32* protein acts. Knowledge of the activities of this remarkable molecule has been extended considerably by the work of Mosig and her associates. They found that various mutants of gene *32* differed to a surprising degree in phenotype. They studied three temperature-sensitive mutants (P7, L171 and G26) and the two amber mutants used in earlier investigations (A453 and E315). From the recombination frequencies in pairwise crosses the mutants mapped in the following sequence: P7, L171, A453 and G26 (so closely linked that their order could not be established), E315 (Mosig, Berquist and Bock, 1977). In amber mutants under restrictive conditions the polypeptide terminates at the position corresponding to the mutant site in the gene. The molecular weights of the gene-*32* proteins produced by A453 and E315 were estimated to be about 30% and 40%, respectively, of that of wild type. It follows that the direction of translation is that in which the mutants are quoted above, so P7 is evidently the nearest of the mutant sites to the promoter of the gene, and E315 the furthest from it. The technique which Mosig and her associates used in studying these gene-*32* mutants was to infect host cells under completely restrictive conditions. None of the gene-*32* mutants produced any progeny under these conditions. The infection was made with a mixture of two phage strains, one carrying a density label (^{13}C and ^{15}N) and the other a radioactive label (^{32}P), in the presence of ^3H-labelled thymidine, and then to monitor the DNA at various times after infection by extraction and density-gradient centrifugation. It was possible in this way to investigate not only DNA replication, but also the formation of joint and of recombinant molecules, and the degradation of the DNA that occurs with some of the mutants.

Breschkin and Mosig (1977a) found that all five gene-32 mutants allowed some DNA replication to occur. The most severely defective was the promoter-proximal temperature-sensitive mutant P7 and the least defective were the amber mutants. The P7 mutant seems to be defective in the binding of the gene-32 protein to the DNA, for Curtis and Alberts (1976) showed that the binding was temperature sensitive *in vitro*. Breschkin and Mosig also studied the production of joint molecules in the gene-32 mutants and found that no such molecules were produced by mutant P7, nor were covalently linked recombinant molecules formed. They concluded that the failure of the gene-32 protein of P7 to bind to DNA prevented recombination as well as replication. The slight replication with P7 was shown to be the result of host enzyme activity (Breschkin and Mosig, 1977b). Mutant G26 and the amber mutants were also lacking in joint and recombinant molecules but this could have been a secondary effect, as the DNA of these mutants is broken down (see below). When the effects of some of the host genes, such as that coding for DNA polymerase I, were removed by the use of mutant host strains, Breschkin and Mosig (1977b) found that a parallel appeared between the effects of individual gene-32 mutants on the initiation of DNA replication in T4 and their effects on the initial stages of T4 recombination. It seemed as if both processes were affected by the same part of the gene-32 protein molecule.

Mosig and Breschkin (1975) investigated the L171 mutant. They found that it produced some unsealed recombination intermediates, that is, joint molecules. No covalently linked recombinant molecules were formed, however. The inability to produce these molecules was partially suppressed by mutations in the rII A or B genes, provided that the host DNA ligase was functional. They inferred that the gene-32 protein is involved in the closure of nicks in duplex DNA, interacting in some way with the phage ligase. The gene-32 protein of mutant L171 was evidently defective in this interaction. The rII mutations allowed the host ligase to function instead. None of the other four gene-32 mutants investigated showed effects comparable to L171.

It was also evident that gene-32 protein can bind specifically to T4 DNA polymerase, because Huberman, Kornberg and Alberts (1971) found that, in the absence of DNA, it formed a weak complex with T4 DNA polymerase but not with DNA polymerase I of the host, and it stimulated the activity *in vitro* of the phage enzyme but not that of the host. Breschkin and Mosig (1977b) found that the gene-32 mutant G26 of T4 was defective in the initiation of DNA replication, but this defect was not evident unless the host lacked DNA polymerase I. They suggested that the gene-32 protein of mutant G26 was defective in an interaction with T4 DNA polymerase, and that the host DNA polymerase I can substitute for the complex of phage polymerase and gene-32 protein.

Mosig and Bock (1976) studied DNA degradation in the gene-32 mutants.

They measured it by observing the proportion of ^{32}P that became acid-soluble. They found that the DNA was degraded when the part of the gene-*32* protein towards its carboxy end was abnormal, that is, either missing (the amber mutants) or modified (mutant G26). The DNA was not broken down if this part of the polypeptide was normal (wild type, and mutants P7 and L171). As already indicated, these two temperature-sensitive mutants map near the promoter end of the gene, with the implication that their gene-*32* protein is normal towards the carboxy end. Mosig and Bock (1976) found that mutations in T4 gene *46*, or in the *E. coli recB* gene, reduced the degradation of the DNA that took place with the carboxy-end gene-*32* mutants, and if both the *46* and *recB* mutants were present DNA breakdown was abolished. The pair of phage genes, *46* and *47*, and the pair of host genes, *recB* and *recC*, control the production of a phage nuclease (Section 8) and a host nuclease (Chapter VI, Section 5), respectively. Mosig and Bock concluded that the carboxy part of the gene-*32* protein prevents the nucleolytic activity of both these enzymes *in vivo*. They surmised that this effect of the gene-*32* protein would prevent destruction of intermediates in the T4 recombination pathway. It was of particular interest that the P7 mutant, with failure of its gene-*32* protein to bind to DNA, was nevertheless able to prevent the nucleolytic attack on its DNA, presumably because, with a normal carboxy end, its gene-*32* polypeptide was able to bind to the *46/47*-controlled nuclease or the *recBC* nuclease – and this binding evidently inactivates the enzymes.

From these studies of gene-*32* mutants Mosig and associates concluded that the gene-*32* protein binds to several different enzymes as well as to DNA, and that different parts of the polypeptide are concerned with different functions. In sequence from the amino end, successive regions of the polypeptide seemed to be concerned with binding (1) to DNA, (2) to T4 DNA ligase, (3) to T4 DNA polymerase, and (4) to a nuclease controlled by T4 genes *46* and *47*, and to the host *recBC* nuclease. Mutations in the DNA-binding region affected the initial stages of both replication and recombination. It was clear that the importance of the gene-*32* protein in recombination was not just to facilitate unwinding of duplex DNA and exposing the bases of the single strands, but that it also interacted with the enzymes that were believed to function in T4 recombination. Mosig *et al.* (1977) suggested that gene-*32* protein coordinated and controlled these enzymes in recombination, and they concluded that this was achieved through the formation of a complex between gene-*32* protein and the recombination enzymes. It had already been suggested that DNA replication in T4 requires a large multi-enzyme complex involving the gene-*32* protein and the products of genes *41*, *43–45* and *62* (Morris, Sinha and Alberts, 1975). Since gene-*32* protein interacts with many components of the recombination complex, Mosig *et al.* (1977) inferred that gene-*32* mutations may differentially affect various recombination steps.

10. MISMATCH REPAIR

It was suggested by Nomura and Benzer (1961), following their discovery of two classes of heterozygotes in T4, that deletion mutants were unable to form an extended mismatch (see Section 6). In other words, internal (that is, heteroduplex) heterozygotes did not appear to arise. Drake (1966) investigated mutants intermediate in extent between such deletions and base-substitution mutants, which form internal heterozygotes readily, in order to establish whether there was a sharp discontinuity of behaviour depending on whether or not the two strands in a heteroduplex differed in length. He studied frameshift mutants of the rII region. These were estimated to have between one and about 20 nucleotides added or removed. Drake (1966) classified the frameshift mutants as 'small', 'medium' and 'large'. Those in the 'medium' class failed to recombine with two single-site mutants that nevertheless recombine with one another. The 'small' mutants occupied only a single site, and the 'large' mutants more than two sites. Drake crossed each mutant with an rII amber mutant and examined mottled plaques for the presence of the frameshift mutant. Its recovery would indicate heterozygosis. He found there was a gradual change with increasing length of the mutant, recovery dropping from frequent (as found with substitution mutants) to rare. Despite this inverse correlation between recovery frequency and genetic extent, some of the 'small' frameshifts were recovered only rarely. Nevertheless, it was evident that some frameshift mutants could form internal heterozygotes, presumably containing short looped-out single strands, but that it was only those of limited length that did so frequently. Drake pointed out that the failure of larger additions or deletions to produce internal heterozygotes might be primary or secondary, that is, the heteroduplex might fail to form in the first place or, having formed, the mismatch might be repaired.

This question was examined by Benz and Berger (1973). They infected cells with a mixture of wild-type T4 and rII deletion mutants and tested the progeny for selective loss of the wild-type allele. In order to favour DNA repair, crosses were made in the presence of fluorodeoxyuridine to slow DNA replication, and of chloramphenicol to delay phage maturation. In control experiments with mixed infections of individual rII base-substitution mutants and wild type, they found that the proportion of mutant progeny was slightly less than in the infecting mixture, apparently as a result of mixed growth of rII and wild-type phage. With each of five deletions, however, the frequency of recovery of the mutant in the progeny was significantly higher, relative to wild type, than with the substitution mutants. Benz and Berger (1973) suggested that this loss of the wild-type allele was caused by excision of the looped-out strand in the heteroduplex. With a deletion mutant the looped-out strand would necessarily be the wild-type one (Fig. 22(i)(c)).

This hypothesis predicts that an addition mutant will show the converse effect, because now the mutant strand will be the one that is looped out (Fig. 22(ii)(c)). This prediction was tested by making comparable experiments with four frameshift mutants of the lysozyme gene that were known to be additions of from one to five nucleotides. As predicted, all four showed fewer mutant progeny, relative to wild type, than in the parental mixture.

Harm (1963) isolated mutants of T4 that showed increased sensitivity to ultraviolet light in the dark. The mutants were found to belong to two genes, v and

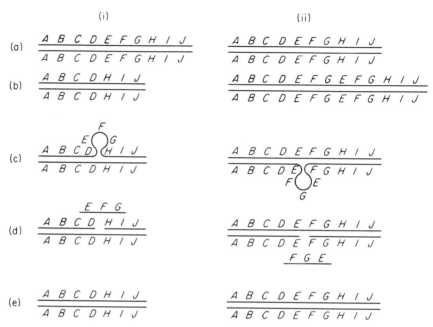

Fig. 22. Diagram to illustrate Benz and Berger's conclusions from study of crosses between wild-type T4 and mutants with (i) a deletion and (ii) an addition of nucleotides (Benz and Berger, 1973). The lines represent individual polynucleotides of DNA, pairs of parallel lines implying complementary strands forming a duplex. The letters *A–J* indicate part of the normal genome. (a) The wild-type parent of the cross. (b) The other parent, with segment *E F G* deleted in (i) and duplicated in (ii). (c) A heteroduplex molecule has formed through the annealing of strands, one from each parent, the unmatched segment forming a single-strand loop. (d) The loop is excised, giving a deletion-mutant genome in (i) and a wild-type genome in (ii). (e) The nick or gap is repaired.

x. Yasuda and Sekiguchi (1970) and Friedberg and King (1971) purified an enzyme, T4 endonuclease V, controlled by the *v* gene. They showed that it induces single-strand breaks in ultraviolet-irradiated DNA. Sekiguchi *et al.* (1975) found that the breaks were on the 5′ side of pyrimidine dimers resulting from the ultraviolet treatment. They isolated two exonucleases each of which was induced by T4 and acted in the 5′ to 3′ direction at the break, thereby excising the pyrimidine dimer. Sato and Sekiguchi (1967) studied a temperature-sensitive *v* mutant and found that the mutant enzyme was abnormally thermolabile. This demonstrated that *v* is the structural gene for endonuclease V. The enzyme is evidently responsible for the first step in *excision repair* of DNA containing pyrimidine dimers (cf. Chapter VI, Section 9).

Benz and Berger (1973) tested the possibility that the *v* gene was also involved in mismatch loop repair. The frequency of recovery of *r*II deletion mutants, relative to wild type, following mixed infections, was examined in the presence of a *v* mutant in both parents. No loss of wild-type progeny, relative to mutant, was observed. It was evident that gene v functions in mismatch repair of deletion mutants. They found, however, that in the presence of a homozygous v mutant the frequency of internal heterozygotes for the deletion mutants remained low, despite the lack of mismatch repair. They concluded that the failure to obtain phage progeny with internal heterozygotes for deletions could not be explained my mismatch repair alone, and they suggested that the heteroduplex DNA containing a simplex loop may not be packaged in the phage head, or may not be ejected from the head on infection of a host cell.

Berger and Benz (1975) observed the frequency of mottled plaques, indicating heterozygosity, in the progeny of crosses between rII base-substitution mutants and wild type. They found no difference in the frequency of rII heterozygotes whether or not the cross was homozygous for a mutant of gene v. They concluded that the v gene product (T4 endonuclease V) is not involved in the repair of mismatched bases (as distinct from loop repair in an addition- or deletion-mutant heteroduplex). Repair of mismatched bases, if it occurs in T4, must be random repair of either strand, since there is no selective allele loss with substitution mutants (see above).

Benz and Berger (1973) also examined the effect of a mutation in gene x and found that, like the v mutant, it prevented loss of wild type in the progeny of mixed infections of rII deletion mutants and wild type. The inference from this result was that the product of the x gene plays an essential part in mismatch loop repair. Boyle and Symonds (1969) isolated a radiation-sensitive mutant, y, of T4 that reduced recombination by about a half when homozygous. In this and other respects it was similar to x, though mapping in a different gene. They concluded that x and y functioned in the same DNA-repair pathway. Maynard Smith and Symonds (1973) measured the ability of ultraviolet-irradiated phage to complement phage carrying a specific amber mutation. They did this for mutations in various T4 genes, and compared the results obtained in the presence of the y mutant with those in its absence, using various doses of ultraviolet light. The mutants that were unaffected by the y mutant were presumed to be in genes that functioned in the y-mediated repair pathway. In this way Maynard Smith and Symonds established that the pathway included genes needed for DNA replication. They concluded that y repair was analogous to *post-replication repair* in *E. coli* (see Chapter VI, Section 9).

Following treatment with a mutagen, Hamlett and Berger (1975) tested minute plaques for sensitivity to ultraviolet light and obtained several sensitive mutants. Some showed decreased recombination and mapped in a new gene, w; others showed increased recombination and mapped in gene 58. Double mutants with x or y were no more sensitive to ultraviolet light than the w or 58 mutants alone, and it was evident that both classes of new mutants were defective in the same pathway for the repair of ultraviolet-induced damage as the x and y mutants. The nature of this w–x–y–58 repair pathway is not known in detail but is presumed to involve recombination and, from Benz and Berger's data (Benz and Berger, 1973), to have at least one step common to heteroduplex loop repair.

Luria (1947) discovered that preparations of phages T2, T4 and T6 which had been inactivated by ultraviolet light could be reactivated by making infections at high concentrations such that each bacterial cell was infected with at least two particles. This *multiplicity reactivation* was described in more detail by Luria and Dulbecco (1949) and was attributed to recombination, lethal mutations or damage being replaced by non-defective material derived from the other parent. Luria (1947) showed, however, that it could not be random recombination between the parental genomes, because active particles were never inactivated by co-infection with an inactivated one, even when large doses of radiation were used (20–50 hits per particle): random recombination between parental genomes would be expected frequently to transfer lethal mutations to the non-irradiated parent and so inactivate both. Sturtevant (in Luria and Dulbecco, 1949, p. 119, footnote) suggested that there was 'selective recombination of active units', an idea that is close to the current view that one of the sources of multiplicity reactivation is repair by recombination initiated at the sites of damage (see below). Studies by many authors, particularly of the kinetics of inactivation, have led to the conclusion that two factors contribute to multiplicity reactivation: (1) functional complementation, each parent providing genes that are defective in the other, and so allowing replication, recombination and DNA repair to take place; and (2) repair processes, including recombination, allowing the formation of an intact genome. Symonds, Heindl and White (1973) studied multiplicity reactivation in single and double combinations of ultraviolet-sensitive mutants and concluded that both excision repair and post-replication repair were involved. Symonds (1975) and Nonn and Bernstein (1977) investigated multiplicity reactivation in T4 inactivated by treatment with ultraviolet light and with nitrous acid, respectively. Mutants that reduce the recombination frequency, such as those defective in genes *32, 46, 47, x* or *y*, showed lowered frequencies of multiplicity reactivation. Conversely, mutants of several other DNA-metabolism genes had no effect on either process. It was concluded that a repair pathway involving recombination contributes to multiplicity reactivation. It was pointed out that if the two complementary strands of the duplex were both damaged at the same or nearby positions, recombination between the parental genomes would be required, whereas if the damage at a particular site was confined to one strand, post-replication repair by recombination between daughter duplexes from one parental genome would suffice.

Mismatch repair has provided confirmation of one of the sources of true (or high) negative interference between recombination events in short intervals of the genome. Negative interference was described in Section 5 and explanations of it were given in Section 6 in terms of the occurrence of insertion heteroduplexes (Fig. 17(ii)(c)) and of recombination two or more times at genome ends.

Vigier (1966) and Berger and Warren (1969) made three-factor crosses involving *r*II mutants, double mutants being crossed with a third mutant that mapped between the sites of the other two. They recorded the frequency of wild-type recombinants, that is, + + + from

$$\frac{1 + 3}{+ 2 +},$$

where 1, 2 and 3 represent the three mutants. They found that when the central marker (2) was a deletion the negative interference was reduced compared with when it was a point mutant. On the other hand, Doermann and Parma (1967), in comparable experiments in which they recorded the frequency of the triple-mutant recombinants (1 2 3), found that the degree of negative interference was more or less the same whether the central marker was a point mutant or a deletion. Efficient repair of deletion heteroduplexes by excision of the simplex loop was proposed to account for this differential effect on negative interference, and such repair has now been confirmed by the work of Benz and Berger (1973) already described. When the central mutant is a deletion, the occurrence of an insertion heteroduplex, followed by excision of the single-strand loop, can generate the triple mutant (Fig. 23(i)(c)–(e)) but not wild type (Fig. 23(ii)(c)–(e)), thus explaining why negative interference is evident when the recombinants are the triple mutant but not when they are wild type. Thus, the evidence for heteroduplex loop

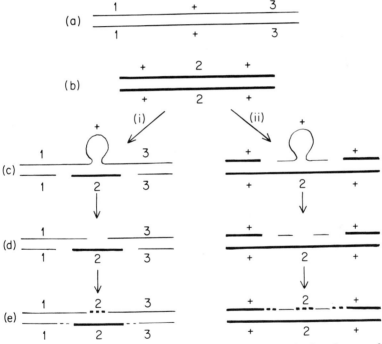

Fig. 23. Proposed explanation for the differential effect on negative interference of selection for wild type and for the triple mutant in a cross between a double mutant and a deletion mutant that maps at an intervening site. The mutants are denoted by 1, 2 and 3, respectively, and their wild-type alleles by plus signs. Lines indicate nucleotide chains, a pair of parallel lines implying the complementary strands of duplex DNA. The thickness of the lines shows their parentage. (a), (b) The parental genomes. (c) The two kinds of insertion-heteroduplex molecule ((i), (ii)) that can arise if a strand segment that includes the central site but not the other two is inserted into a molecule of the other parentage. Molecules (i) and (ii) are potentially in the pathways to the triple mutant and wild type, respectively. (d) The result of excision of the single-strand loop: the pathway to the triple mutant is maintained but that to wild type is lost. (e) The resulting molecules when repaired. Broken lines show newly synthesized strands that fill gaps.

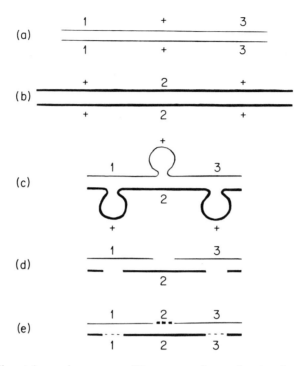

Fig. 24. Mismatch repair as a possible source of negative interference. The lines represent individual polynucleotides, their thickness indicating their parentage. Two parallel lines imply complementary strands forming a duplex. The parents (a) and (b) of a cross differ from one another at three sites, the mutants (1–3) all being deletions. If a heteroduplex (c) should arise that overruns all three sites, separate excision (d) of each single-strand loop, followed by new synthesis (broken lines) to fill the gaps, will give rise to the triple-mutant recombinant (e).

repair in T4 has indirectly provided confirmation that insertion heteroduplexes are a source of negative interference.

A further possible source of negative interference is mismatch repair itself, which could explain the occurrence of two or more exchanges in proximity. Thus, if a double mutant is crossed with a single mutant, the mutants being deletions that map alternately from one parent and the other (Fig. 24(a),(b)), and if the mutant sites are sufficiently near to one another for a heteroduplex region to overrun all of them (Fig. 24(c)), excision of each simplex loop (Fig. 24(d)), followed by filling of any gaps, will lead to the triple-mutant recombinant. This source of negative interference would also apply to base-substitution mutants if they undergo mismatch repair in heteroduplex DNA.

11. THE RECOMBINATION PATHWAY

It was pointed out at the end of Section 8 that the study by Broker (1973) of electron micrographs of the DNA of T4 defective in various genes enabled him to propose a pathway for recombination. This pathway relates to the situation when DNA synthesis is blocked. It is outlined in Fig. 25 for recombination between the end of one molecule and a homologous region at an internal position in another molecule; Broker presumed that similar interactions can occur between homologous segments in other situations. The proposed steps are as follows: (1) single-strand nicks are produced by an unidentified endonuclease; (2) the nicks are converted to gaps by the exonuclease dependent for its expression on genes *46*

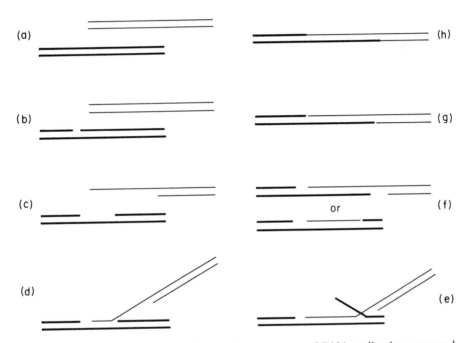

Fig. 25. The pathway for recombination, in the absence of DNA replication, proposed by Broker to account for the experimental results of various kinds obtained when DNA synthesis is prevented, particularly his own study of electron micrographs of the DNA of T4 defective in the gene (*43*) for T4 DNA polymerase and in other genes affecting DNA metabolism. Lines represent nucleotide chains of DNA, two parallel lines implying the duplex structure, and the thickness of the lines indicating their parentage. Homologous regions are placed above one another. For some of the steps in the pathway host enzymes may substitute for the phage ones. (a) The parental molecules, which differ in the position of their genome ends. (b) An endonuclease has produced a nick in one molecule. (c) The exonuclease controlled by the products of genes *46* and *47* has produced a gap at the site of the nick and has eroded one strand of the other molecule from its end. (d) The single strands exposed by the exonucleolytic action are complementary and anneal to give a branched molecule. This pairing is facilitated by the gene-*32* protein. (e) Branch migration extends the heteroduplex segment displacing a strand in the lower molecule. (f) Branches are eliminated by nuclease action. Depending on which strands are cut or eroded, the resulting linear molecule may have a crossover heteroduplex (upper alternative) or an insertion heteroduplex (lower). (g) Gaps are filled by the action of the gene-*43* product: T4 DNA polymerase. (h) Nicks are sealed by the gene-*30* product: T4 DNA ligase. (Reproduced with permission from T. R. Broker, *J. molec. Biol.* **81**, 10 (1973). Copyright by Academic Press Inc. (London) Ltd.)

and *47*, and in the same way one strand is eroded at the ends of the molecules; (3) the gene-*32* protein binds to the single-stranded regions and promotes their complementary pairing; (4) the extent of heteroduplex DNA is enlarged by branch migration; (5) the branched molecules are converted to linear ones by the action of one or more unidentified nucleases; (6) gaps are filled by the action of the product of gene *43*, that is, T4 DNA polymerase; and, finally, (7) the nicks are sealed by the gene-*30* product, T4 DNA ligase.

Much of the evidence for this pathway has been outlined in Sections 7 and 8. The existence of a virus-coded endonuclease responsible for making single-stranded breaks (nicks) in the duplex DNA of T4, as a first step in recombination, had been proposed by Kozinski and associates. Kozinski, Kozinski and Shannon (1963) found that in the presence of chloramphenicol the fragmentary transfer of DNA from parent to progeny which they had discovered (see the beginning of Section 7) did not take place. They concluded that the synthesis of an enzyme responsible for recombination was being inhibited by the drug. Kozinski, Kozinski and James (1967) gave chloramphenicol treatment at various times after infection with T4, extracted the DNA, and made density-gradient studies in caesium chloride. They obtained evidence that an enzyme synthesized shortly after infection, and therefore coded by T4, introduced single-strand breaks into the DNA. They concluded that this endonuclease was the enzyme inferred from their earlier work to be necessary if recombination was to take place. They suggested that the existence of this enzyme might explain why recombination occurred with such high frequency in T4 compared with phage λ. The extent to which nicking of the DNA triggers recombination is illustrated by the work of Krisch, Hamlett and Berger (1972). They found that deficiency for T4 DNA ligase increased the recombination frequency more than fourfold and they suggested, from studies on the time course of recombinant production, that with low levels of ligase DNA synthesis generates highly nicked molecules, and these promote recombination.

The main results from the electron microscopy of DNA extracted from cells infected with T4 mutants were given in Sections 8 and 9 and, in summary, indicated that mutations in genes *32*, *46* and *47* greatly inhibited the formation of branched molecules, while mutation in other genes necessary for DNA replication and repair such as *30* (*lig*) and *43* (*pol*) enhanced it. In addition, mutation in genes *46* and *47* prevented the formation of single-stranded regions that otherwise were seen in the *lig⁻ pol⁻* phage. It was partly on the basis of these observations that Broker (1973) proposed that genes *46* and *47* acted early in the pathway.

Hosoda (1976) tested this idea that genes *46* and *47* act early. She determined whether a mutation in gene *46* prevents the formation of joint molecules. This test was necessary because the virtual absence of branched molecules with a gene-*46* mutant might be the consequence of accelerated removal of the branches rather than reduced incidence. The technique she used was to make phage crosses in which one parent carried a radioactive label ([³H]thymidine) and the other a density label (¹³C, ¹⁵N), and to measure the density on a caesium sulphate gradient of the DNA extracted from the infected cells. She found that in *lig⁻ pol⁻* phage a considerable fraction of the ³H label was at a hybrid density, but with gene *46* also defective this fraction was much reduced. She concluded that the activity of gene *46* was needed for the formation of joint molecules, in agreement with her earlier conclusions (Section 8) and Broker's model. She pointed out, however, that her results did not exclude the possibility that the products of genes *46* and *47* also had other nucleolytic functions later in the pathway.

As indicated in Section 9, Mosig and collaborators obtained evidence that the gene-*32* protein binds to several of the recombination enzymes, and so they suggested that it may affect many of the steps in the recombination pathway in addition to step (3) of Broker's model. Indeed, if it is true, as Mosig *et al.* (1977) believe, that the gene-*32* protein and the recombination enzymes form a complex, then a mutation in any of the relevant genes might affect many steps.

As discussed at the end of Section 6, Mosig *et al.* (1971) and others had found it necessary to postulate that genome ends often underwent more than one round of recombination. Broker (1963) obtained support for such reiterated events from his electron micrographs of the DNA of T4 defective in genes affecting DNA metabolism. He found that molecules with multiple branches were much more frequent than expected from the numbers with no branch and with one branch. This result implied that an exchange was often followed by further exchanges in the same region: indeed, about one-third of all the branched molecules appeared to have undergone one or more additional exchanges. The reiteration of recombination could be represented diagrammatically by a recycle in Fig. 25 from (f) back to (c), exposed single strands re-engaging in pairing. Broker pointed out that there was such a variety of branched structures that many alternatives seemed likely in the details of the recombination pathway that he had outlined.

The generation of either an insertion heteroduplex or a crossover heteroduplex, following the invasion of an internal region of one molecule by a partially single-stranded terminal region of another molecule, has been discussed by Mosig *et al.* (1971) and is illustrated in Fig. 26. As was indicated in Section 3, short heterozygous regions (that is, heteroduplex heterozygotes – see Section 6) are more often of the insertion type (with flanking markers non-recombinant) than of the crossover type. It may be, however, that crossover heteroduplexes are more frequent under normal conditions than in the special situation, with DNA replication blocked, under which most of the experiments have been carried out (see below). Mosig *et al.* (1977) found that the temperature-sensitive mutants P7 and G26 in gene *32* (see Section 9), when grown under permissive conditions (30 °C), caused a twofold reduction in recombination between closely linked *r*II mutants, but had no effect on recombination over longer intervals. In other words, P7 and G26 appeared to be defective in the formation of insertion heteroduplexes (Fig. 26(c)(i)) but not of crossover heteroduplexes (Fig. 26(c)(ii)). As a possible explanation they suggested that the gene-*32* protein of these mutants, as a result of its interaction with the recombination enzymes, increased the ligase activity or decreased the endonuclease activity. If this change particularly affected strands that had already been broken, it would reduce the likelihood of alternative (i) in Fig. 26(c).

Although Broker's pathway (Fig. 25) describes recombination in the absence of

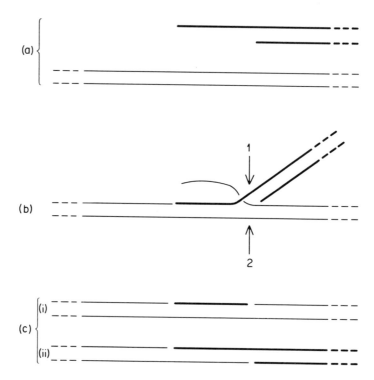

Fig. 26.   Mosig's hypothesis of the invasion of an internal region of one molecule by a single-stranded terminal but homologous region of another molecule. Lines represent polynucleotides, two parallel lines indicating the duplex structure. Homologous regions are placed above one another. The thickness of the lines distinguishes the contributions of the two parents. (a) The parental molecules, the upper ending in a single-stranded region. (b) The single-stranded region has invaded an internal but homologous region of the other molecule, forming a heteroduplex segment. The numbers 1 and 2 indicate alternative sites of endonuclease action to generate an unbranched molecule. It is assumed that in either case the unpaired strand, displaced by the invading strand, is removed or broken down. (c) The resulting linear molecules. The alternatives, (i) an insertion heteroduplex and (ii) a crossover heteroduplex, correspond to cutting at 1 and 2, respectively, in (b).

replication, he was of the opinion that, whatever the conditions, the structural requirements for T4 recombination would be similar. Replication and recombination in T4 seem to be interrelated in several ways. Miller, Kozinski and Litwin (1970) and Shah and Berger (1971) used sedimentation analysis in sucrose density gradients to study the intracellular DNA after infection with T4. Kozinski and associates had already found that chloramphenicol treatment early in infection prevented recombination from taking place (see above). Miller et al. (1970) now found that such treatment also prevented the formation of molecules longer than those of mature phage, such as are produced late in infection in the absence of the drug. Shah and Berger (1971) obtained the same result without chloramphenicol treatment when the cells were infected with a double mutant of genes 46 and 47.

Both groups suggested that the long molecules were concatemers arising by recombination and Shah and Berger concluded that the products of genes *46* and *47* were involved in their formation. Concatemer production by recombination at terminal repetitions had been suggested by Streisinger, Edgar and Denhardt (1964) when proposing their model of T4 genome structure, for the headful hypothesis (see Section 6) requires pieces of DNA large enough to fill the phage head to be cut from longer concatemers.

Watson (1972) and Broker (1973) independently suggested a specific hypothesis of what might happen. Since DNA replication seems always to be in the 5' to 3' direction, it will necessarily be in the direction away from the replication fork on one of the strands, and hence discontinuous on that strand. If replication is bidirectional from an internal origin, this discontinuous synthesis will be expected to leave the 3'-hydroxyl ends of the parental templates unreplicated (Fig. 27(a)). Because of the terminal repetition of nucleotide sequence, these two simplex regions, if long enough, will be complementary and so might anneal to give a linear dimer genome. Broker suggested that the function of the gene *46–47* exonuclease might be to remove redundant single strands that would arise, following this annealing, if the simplex regions at the genome ends were shorter than the region of repeated nucleotide sequence (Fig. 27(b)). Failure to remove these

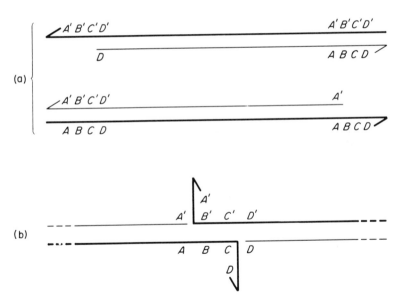

Fig. 27.   Recombination of single-stranded ends during DNA replication, as proposed by Broker (1973). Parental strands are shown with thick lines and newly synthesized ones with thin lines. Half-arrowheads are at 3' termini, *A B C D* indicates the extent of the terminally repeated nucleotide sequence, *A' B' C' D'* being the complementary sequence. (a) Replication has left single-stranded regions at the 3' ends of the parental strands. (b) Annealing of these simplex regions leaves 3'-terminal redundant strands if the single-stranded regions in (a) are shorter than the terminally repeated sequence *A B C D*.

branches would prevent covalent joining of the two molecules and could explain Shah and Berger's results. Watson (1972) suggested, with particular reference to another phage with terminal repetitions, T7, that the single-stranded branches might be removed by an exonuclease acting on the 5' ends in the duplex (that is, those labelled A' and D and without arrowheads in Fig. 27(b)), so allowing the 3'-ended unpaired strands (those similarly labelled, but with arrowheads, in Fig. 27(b)) to be assimilated into the duplex in their place. This is comparable to the known behaviour, at least *in vitro*, of the exonuclease of phage λ (see Chapter VI, Section 10). Watson also pointed out that a 5'-exonuclease might be needed in T4 to increase the length of the 3'-ended single strands sufficiently for complementary nucleotide sequences to be exposed; in other words, it would have acted on the ends without arrowheads labelled A' and D in Fig. 27(a).

Broker (1973) suggested that single-stranded ends resulting from incomplete replication (as in Fig. 27(a)) might invade homologous regions of other molecules, particularly where nicked, gapped or replicating, and this could account not only for the favouring of genome ends for recombination, but also for the reiteration of recombination events in these regions, if daughter molecules end at the same position in the genome as their parents. Concatemer production by annealing of single-stranded ends on newly replicated DNA, if it occurred between molecules of differing parentage, would give rise to crossover heteroduplexes. It is possible, therefore, as indicated above, that experiments carried out under conditions where replication is prevented, have given a false impression of the frequency of crossover heteroduplexes.

The occurrence of duplications of the rII region of the T4 genome was referred to in Section 6. These had originally been obtained by Weil, Terzaghi and Crasemann (1965) by crossing overlapping deletion mutants, defective respectively in the A and B genes, and selecting for wild-type progeny. Such progeny cannot arise by recombination because the deletions overlap, but do appear with extremely low frequency through structural changes. Symonds *et al.* (1972) inferred from genetic studies that a duplication called M4, which they had obtained in this way, consisted of a tandem repeat of the whole rII region with the A deletion mutant (no. 1589) in the left-hand and the B deletion mutant (no. 638) in the right-hand repeat:

A defective $\ \ B$ functional	A functional $\ \ B$ defective
P_1 $\qquad\qquad\qquad\qquad\qquad$ Q_1 P_2	Q_2

Denoting the terminal regions of the repeated segment by P and Q, and distinguishing the two copies by numbered suffixes, the occurrence of recombination between region P_1 in one molecule and P_2 in the same or another molecule, or similarly with Q_1 and Q_2, can give rise to the original rII mutant strains lacking the duplication.

Van den Ende and Symonds (1972) used this behaviour of their tandem repeat to select for mutants with recombination deficiency. Such mutants, if recessive, would not be detected in an ordinary T4 cross, but this drawback does not apply if the deficiency can be recognized in single infections. They treated M4 with a mutagen and looked for plaques that gave a lower than normal frequency of recovery of the rII mutants that are present but not manifest in M4: in other words they searched for a reduced frequency of recombination between region P_1 in one molecule and region P_2 in the same or another, and similarly with Q_1 and Q_2. One of the mutants isolated in this way was called 1206. It showed a fourfold reduction in segregation frequency and was found also to have increased sensitivity to ultraviolet light. Symonds, Heindl and White (1973), from their study of phage survival and multiplicity reactivation in ultraviolet-sensitive single and double mutants (see Section 10), concluded that mutant 1206 was defective in the same post-replication repair pathway as mutants defective in genes w, x and y.

Davis and Symonds (1974) used the M4 duplication to investigate which of the genes that affect DNA metabolism play an essential part in recombination. The use of the duplication has an advantage over a normal cross, because recombination frequency can be measured without the necessity for pairing between genomes of differing parentage. Such pairing will be influenced by the concentration of genomes in the cell, and so is likely to be affected in mutants with defective DNA metabolism. An amber mutant in a gene to be tested was introduced into the M4 strain, and wild-type revertants of the amber mutant were selected by infecting a suppressor-negative host. Some of these revertants will be expected to produce an enzyme with different properties from those of wild type, and if the enzyme acts in the recombination pathway, a change in the frequency of recovery of the rII mutants from M4 will result. Using this method, Davis and Symonds (1974) found that some of the revertants of amber mutants in genes *32*, *46* and *47* gave low rII recovery frequencies, as expected since these genes were already known to be involved in recombination (see Sections 8 and 9). They also found, however, that revertants of amber mutants in genes *44* and *59* behaved similarly: these two genes had not previously been implicated in recombination. Mutants of gene *44* show no DNA synthesis. With amber mutants of gene *59*, which maps between genes *32* and *33*, DNA synthesis is arrested in cells of a non-permissive host about 7 min after it had begun. The nature of the products of these genes and where they act in the recombination pathway are not known.

For all five genes (*32*, *44*, *46*, *47*, *59*), the amber revertants with low recombination frequency showed increased sensitivity to ultraviolet light and increased sensitivity in multiplicity reactivation, similar to that of mutants defective in genes *x* and *y*. As Symonds (1975) pointed out, there seems therefore to be at least a considerable overlap in the pathways of genetic recombination and of post-replication repair.

In summary, to the recombination pathway outlined in Fig. 25 must be added the part played by the gene-*32* protein in binding to the recombination enzymes; the probable involvement in recombination of genome ends left single-stranded in the course of DNA synthesis; the unknown contributions of the products of genes *44* and *59*; and the excision of heteroduplex loops by mismatch repair enzymes (see Section 10). This excision involves enzymes that also function in excision repair and in post-replication repair of DNA containing pyrimidine dimers, following exposure to ultraviolet light. As indicated above, there seems in addition to be sharing of other enzymes between the recombination pathway and the post-replication repair pathway. Moreover, multiplicity reactivation appears to result from a repair pathway involving recombination between parental genomes, as well as from post-replication repair by recombination between sister genomes (see Section 10). How far these repair pathways differ from one another and from the recombination pathway remains to be resolved.

V. Recombination in phage λ

1. INTRODUCTION

The strain of *Escherichia coli* called K-12 had been used for research for nearly 30 years without any suspicion that it carried a latent virus. Then, following irradiation of the bacterial cells with ultraviolet light, Esther Lederberg (1951) isolated a mutant that failed to ferment galactose and, unexpectedly, was a sensitive indicator of the presence of a virus that had evidently been latent in the host strain. The virus caused breakdown (lysis) of the mutant cells. It was given the name lambda (λ) standing, no doubt, for lysogen (meaning a phage carrier). A characteristic feature of lysogeny is that the cells are immune to infection by further particles of the specific virus which they carry in latent form.

The discovery of λ and the associated lysogenic character of K-12 were reported more fully by Lederberg and Lederberg (1953). They were surprised to find that lysogeny for λ showed linkage in inheritance to a host gene concerned with galactose fermentation; the same discovery was made by Wollman (1953). It had been anticipated that lysogeny would be inherited cytoplasmically.

The λ phage particle has been found from electron micrographs to consist of an icosahedral head about 54 nm in diameter and a slender tubular tail about 150 nm long ending in a fibre. The duplex DNA molecule which forms the genome has a molecular weight of about 30.8×10^6 daltons, corresponding to about 46 500 nucleotide pairs or perhaps 60 genes. Thirty three were known by 1971. Within the phage head the DNA is linear, though of course much folded. The two ends of the molecule are cohesive, having projecting single strands 12 nucleotides long and complementary to one another. When the DNA enters the host cell, which it does through the phage tail, the ends of the molecule anneal spontaneously by base pairing of the complementary simplex segments, and the joints are sealed by a DNA ligase of host origin to give a covalently closed circular molecule (Fig. 28(a),(b)).

In the presence of proteins coded by two closely linked λ genes, *O* and *P*, the circular molecule replicates to generate further circles. The replication is bidirectional and is initiated near gene *O*, ending on the opposite side of the circle where the two replication forks meet. The replication origin is about one-fifth of the circumference away from the closure-point of the linear molecule (see Fig. 29). When about 20 circular molecules have been formed, the method of DNA replication changes and concatemers are formed by rolling circle (Fig. 28(d)). A concatemer is a linear molecule of more than monomer length: in other words, two or

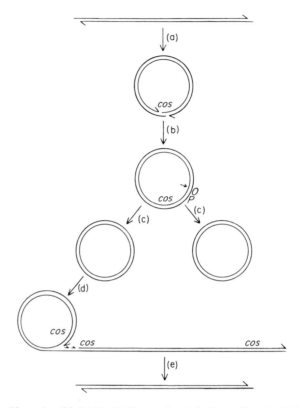

Fig. 28. The life-cycle of λ DNA. Half arrowheads indicate 3′ ends. Cohesive end sites are indicated by *cos*. *O* and *P* are λ genes concerned with replication. (a) Spontaneous annealing of simplex ends. (b) Sealing of nicks by host polynucleotide ligase. (c) Replication, using products of genes *O* and *P*. Arrow indicates replication origin. The process is bidirectional. (d) Replication by rolling circle to produce a concatemer. Replication is continuous round the circle, and discontinuous on the tail in the direction away from the fork. (e) Severance of the concatemer at the *cos* sites to produce the linear genome with complementary simplex ends.

more λ genomes joined end to end. In the process of packaging the DNA into the phage head the concatemer is cut into linear monomers (Fig. 28(e)) by a protein called *ter* (short for termini), the product of λ gene *A*. *ter* acts at a site called *cos* (cohesive end site) cutting the two strands 12 nucleotides apart and generating the single-stranded cohesive ends.

   Recombination in λ was discovered independently by Jacob and Wollman (1954) and Kaiser (1955). They worked with mutants that affected the morphology of the plaques and found that all the mutants showed linkage in inheritance and fitted a linear map. The frequency of recombination was quite low. Wollman and Jacob (1954) and Kaiser (1955) studied the progeny phage released from single bacterial cells and found no correlation between the frequencies of reciprocal recombinants from these single bursts. Owing to variability

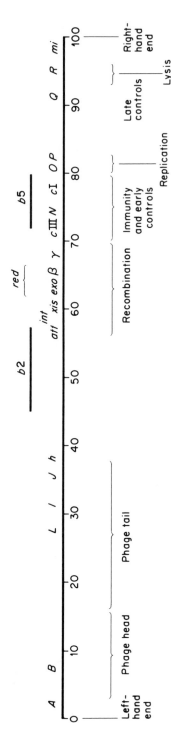

Fig. 29.   Genetic map of phage λ. The figures are percentages of the total length. The genes marked are those referred to in the text.

between cells in the time and number of recombination events during the multiplication of the phage genomes, the proportion of recombinants differs from cell to cell. If the individual event is reciprocal, giving for example ab and $++$ progeny from a cross between markers a and b, there should be a correlation between numbers of reciprocal recombinants within single bursts. Wollman and Jacob (1954) and Kaiser (1955) concluded that recombination in λ is non-reciprocal, any one event producing only a single recombinant. From crosses involving three mutants they discovered that double crossovers were more frequent than was expected from the frequency of single crossovers. In other words, the exchanges showed negative interference.

2. RECOMBINATION BY BREAKAGE AND JOINING

Meselson and Weigle (1961) carried out experiments to see whether there is parental DNA in recombinant λ. Crosses were made with one parent labelled with two heavy isotopes: ^{13}C and ^{15}N. The parents differed in two genetic characters: c (now called cI) giving a clear plaque versus cI^+ with a turbid plaque; and mi with a minute plaque versus mi^+ with a large plaque. These mutants had been obtained by Kaiser (1955) by irradiation of wild-type λ with ultraviolet light. Density gradient centrifugation in caesium chloride solution was used to find if there were heavy atoms among the progeny phage of parental and recombinant genotypes.

Fig. 30.   The density distributions of progeny phage of various genotypes obtained ((a), (b)) by Meselson and Weigle (1961) and ((c)–(f)) by Meselson (1964) from crosses between λ strains differing in density ((a), (b)) or both containing heavy atoms ((c)–(f)). All crosses were made on bacteria containing normal (light) atoms. The progeny were centrifuged to equilibrium in a caesium chloride gradient, and the numbers per millilitre of phage particles of various genotypes in each fraction were determined. These numbers are plotted on a logarithmic scale against the density, given as the drop number. The parental genotypes were as follows:

(a)	$\underline{+\ \ +}$	heavy
	$cI\ mi$	light
(b)	$\underline{cI\ mi}$	heavy
	$+\ \ +$	light
(c)–(f)	$\underline{h\ \ cI}$	heavy
	$+\ \ +$	heavy

where cI, h and mi are clear-plaque, host-range and minute-plaque mutants, respectively. Each curve refers to progeny of the genotype given alongside it. C, conserved; LS, light shoulder; NS, newly synthesized; SC, semi-conserved. CL and SCL indicate that a small amount of light DNA is attached to otherwise C or SC material. In (d)–(f) the results are given of further density-gradient centrifugation of material from (c), as follows: (d) fully labelled (drop numbers 10–13); (e) three-quarters labelled (drop numbers 14–17); (f) half labelled (drop numbers 19–21).

In the first cross (I), wild-type λ was labelled with the heavy atoms and crossed with the *cI mi* double mutant containing normal atoms, that is, ^{12}C and ^{14}N. The cross was made on bacteria also containing the normal isotopes. The results for the progeny phage of + + and + *mi* genotypes are shown in Fig. 30(a), where the number of phage particles per millilitre is plotted against the drop number from the centrifuge tube. The first drop is the heaviest and the last the lightest.

The wild-type progeny, that is, with the genotype of the heavy parent, show a trimodal curve, the left-hand peak (marked C) corresponding to conserved DNA, that is, DNA which has retained the heavy atoms in both strands. This conserved

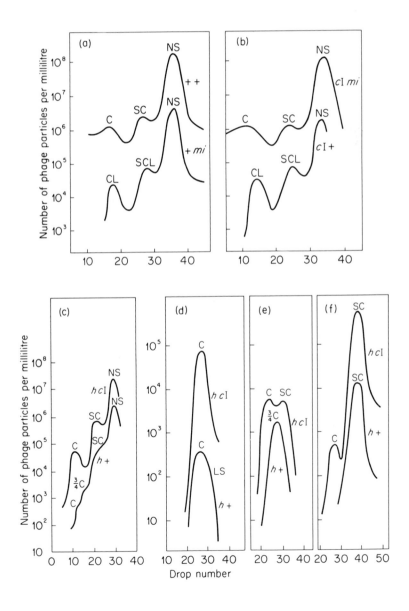

peak is found only with large numbers of phage particles infecting each cell – high multiplicities of infection – and so is believed to represent infecting phage genomes that have not replicated but become incorporated directly into progeny virions. The central peak, marked SC, represents semi-conserved DNA, and the right-hand peak, marked NS, DNA in which both strands are newly synthesized.

The curve for the + *mi* recombinant progeny is also trimodal, but the left-hand and middle peaks, marked CL and SCL respectively, are slightly displaced to the right compared with the phage of wild-type genotype. This is the result expected if a small amount of DNA from the light parent is attached to a large amount, either conserved (CL) or semi-conserved (SCL), from the heavy parent.

It is significant that the genetic map of λ indicates that *mi* is to the right of *cI* on the map and that the mid-point between their map positions is at a distance of about 15% of the total length of the map from the right-hand end (Fig. 29). In other words, to generate + *mi* from + + would require, on average, the insertion of only 15% of the light genome, replacing the right-hand end of the genome contributed by the heavy parent. Thus, the data are in agreement with the hypothesis that recombination occurs by breakage and rejoining (Fig. 31(c)). They contradict the copy-choice hypothesis which predicts that recombinants will contain no heavy atoms (Fig. 31(a)). Furthermore, the occurrence of the CL peak reveals that recombination can occur in the absence of replication of the genome. Meselson and Weigle (1961), however, pointed out that their results do not rule out break and copy (Fig. 31(b)) as the recombination mechanism, that is, breakage of the genome of one parent followed by its restoration by copying the missing region from the homologous part of the genome of the other parent.

In a second cross (II) Meselson and Weigle (1961) reversed the labelling, giving the *cI mi* double mutant the heavy atoms and crossing it on unlabelled (light)

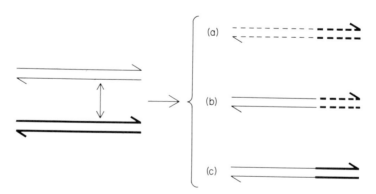

Fig. 31. Diagram to show different types of recombination. The thickness of the lines shows the source of the genetic information, broken lines indicating newly synthesized strands and unbroken lines parental material. The genetic exchange takes place at the point indicated by the double-headed arrow. (a) Copy choice. (b) Break and copy. (c) Break and join.

bacteria with wild-type λ of normal (light) density. The results for the progeny phage of cI mi and cI + genotypes are shown in Fig. 30(b). They confirm the findings from the first cross.

The results from both crosses for the progeny of the unlabelled (light) parental genotype (cI mi in cross I and + + in cross II) and for the predominantly light recombinant genotype (cI + in cross I and + mi in cross II) were unimodal curves at the position corresponding to light DNA in both strands, but with a shoulder on the heavy side. These curves are not shown in Fig. 30.

Wild-type λ has a buoyant density, as revealed in a caesium chloride density gradient, of $1.508 \, \text{g cm}^{-3}$. Kellenberger, Zichichi and Weigle (1960) found that existing stocks of λ contained a mutant, b2, which had a buoyant density of only $1.491 \, \text{g cm}^{-3}$, equivalent to a loss of about 17% of the DNA. The mutant was mapped in the linkage group and shown to be unable to form a stable lysogen. Subsequently, they found (Kellenberger et al., 1961) another deletion mutant, b5, mapping at a different position and having a buoyant density of $1.501 \, \text{g cm}^{-3}$ (a 7% DNA loss). This mutant, unlike wild type, forms plaques on lysogenic bacteria; in other words, the normal immunity of lysogens to the virus that they carry had broken down.

The explanation of this behaviour was discovered later: b5 is really a substitution of the immunity region of the λ genome by the corresponding region of the genome of phage 21 with, as observed, a net loss of DNA. Phage 21 is one of a number of temperate phages allied to λ which were isolated by Jacob and Wollman from strains of E. coli of human faecal origin (see Jacob and Wollman, 1961). The immunity region of the genome is that which codes for the repressor, its regulators and its sites of action (operators). The repressor, by its action of binding to the operators of any genome of that specific phage which invades the host cell, renders the cell immune to that phage. This is the mechanism by which a lysogenic cell maintains its immunity to the specific phage. The b5 deletion is thus a recombination product of hybridization between phages λ and 21 and is sometimes denoted as imm^{21}, the immunity region of λ, absent from b5, being called imm^{λ}.

Since the b2 and b5 deletions are at different positions in the genome, the b2 b5 double mutant has the buoyant density expected from the sum of the two DNA losses: only $1.483 \, \text{g cm}^{-3}$. Kellenberger et al. (1961) took advantage of the differing densities of the four genotypes (wild type, b2, b5, and double mutant) to see if there was a physical exchange of DNA in recombination.

They crossed the double mutant with wild type, one parent being radioactively labelled with ^{32}P, and studied the radioactivity of the progeny phage of each parental and each recombinant genotype; these could be separated on a caesium chloride density gradient. Each parent in turn was labelled, and both experiments indicated that many of the recombinants contained a large amount of material from the labelled parent. The total amount of ^{32}P in the progeny with the genotype of the unlabelled parent was only about 12% of the total amount in the two recombinant genotypes, showing that the results cannot be explained by breakdown of parental DNA and its re-utilization by the newly synthesized phage. Indeed, the amount of label in the progeny with the genotype of the unlabelled parent was that expected from recombination outside the $b2–b5$ interval: $b2$ is near the mid-point of the genome and $b5$ halfway from $b2$ to the right-hand end – see Fig. 29.

As with the results obtained by Meselson and Weigle (1961), the simplest explanation of the data of Kellenberger *et al.* (1961) is that recombination occurs by breakage and joining, although break and copy is not ruled out. These experiments by Meselson, Kellenberger, Zichichi and Weigle were the first demonstration in any organism that recombination involves breakage of the parent molecules and the physical incorporation of parts of them into the recombinant molecules.

Meselson (1964) carried out further experiments with heavy isotope labelling in order to distinguish between the break-and-join and break-and-copy hypotheses. Unlike the earlier experiments, both parents were labelled with ^{13}C and ^{15}N. The genetic markers used were a host-range mutant, h, and clear-plaque mutant in the cI gene. These mutants were chosen because they span the centre of the map (see Fig. 29), unlike cI and mi used before. The labelled double mutant, $h\,cI$, was crossed with labelled wild type on bacteria containing the normal (light) atoms ^{12}C and ^{14}N. The progeny phage were centrifuged in a caesium chloride density gradient and the number of particles of the $h\,cI$ and $h\,+$ genotypes plotted against the drop number.

The results are shown in Fig. 30(c). The progeny of the parental genotype, $h\,cI$, formed a trimodal curve, the peaks corresponding to conserved (C), semi-conserved (SC) and completely newly synthesized (NS) DNA. As before, the C peak is attributed to non-participating phage in which the DNA is carried passively from parent to progeny without replicating. The $h\,+$ recombinants showed the NS peak, but no others. There were inflexions, however, at the positions expected for fully labelled (C), three-quarters labelled ($\frac{3}{4}$C) and half labelled (SC) DNA, but lack of resolution prevented positive identification.

This dilemma was overcome by further density-gradient centrifugation, which revealed peaks at each of these positions (Fig. 30(d)–(f)). The occurrence of fully labelled recombinant molecules (Fig. 30(d)) implied that these recombinants were arising by breakage and joining of parts of unreplicated parental molecules of each kind, and ruled out break and copy. The $\frac{3}{4}$C molecules could have arisen by breakage and joining near the middle of the genome of a conserved molecule from one parent and a semi-conserved from the other.

The presence of a light shoulder (LS) to the $h\,+$ conserved peak (Fig. 30(d)) was an indication that a small amount – up to 5–10% – of DNA was being removed and resynthesized in the formation of the recombinant molecules.

3. HETEROZYGOTES

Heterozygotes in λ were first found by Wollman and Jacob (1954). For a morphological character of the plaque their effect may be observed directly, for example, a clear-plaque (c) mutant crossed with wild type (turbid plaques) will give a few mottled plaques having clear and turbid sectors. Experiments established that not all such plaques could be attributed to mixed infections by two or more phage particles.

Heterozygotes are rare (less than 10^{-4}) but Kellenberger, Zichichi and Epstein (1962) were able to study them by ultraviolet irradiation of the host before infection; this increases their frequency to about 2%. The heterozygous particle is revealed by the segregation of the parental character differences in the progeny, for example, the clear and turbid sectors in the mottled plaques originating from a single particle. Evidently the heterozygous condition is unstable and is not maintained as such in later generations.

As an explanation of the heterozygotes, Kellenberger *et al.* (1962) favoured a heteroduplex segment at the site of the mutation showing heterozygosity. On replication a heteroduplex will give rise to the respective homoduplexes. This would account for the instability. Moreover, there is little replication of phage DNA in ultraviolet-treated cells, so this might explain the increased frequency of heterozygotes. Investigations using several mutants simultaneously, for example, a double mutant crossed with a third mutant, showed that the heterozygous region was often between exchanged markers on either side, as expected if this region was the part of the DNA where segments from each parent were joined through the annealing of complementary nucleotide sequences to form a heteroduplex (Fig. 32(a)). About one-third of the heterozygotes, however, had a parental genotype for the markers on either side of the heterozygous region (Fig. 32(b)).

From the frequency of double heterozygotes, that is, two mutants heterozygous simultaneously, it was estimated that the heterozygous regions were variable in extent and had an average length of one-tenth of the genetic map.

Kellenberger *et al.* (1962) showed that in non-irradiated bacteria the frequency of heterozygotes was too low to account for the formation of all the recombinants observed. This conclusion was based on the assumption that the heterozygous segments were of the same length as in the irradiated cells and that recombination occurs at random in time during the phage DNA multiplication. A possible explanation for the higher frequency of heterozygosity, relative to recombination,

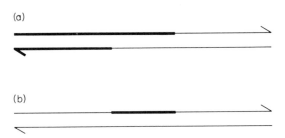

Fig. 32. The two primary classes of heteroduplex molecules. The thickness of the lines shows the parentage. (a) Associated with a crossover. (b) Associated with a parental genotype for markers on either side.

in the irradiated than in the non-irradiated host cells concerns mismatch correction. Such correction, which removes heterozygosity arising from a heteroduplex region, may be inhibited in irradiated bacteria (see Section 13).

In order to test whether or not the heterozygotes were formed in association with recombination, Meselson (1967a) crossed on bacteria of normal density a clear-plaque mutant (gene cI) of normal density with wild-type (turbid plaque) λ carrying the ^{13}C and ^{15}N density labels, and studied the distribution in a caesium chloride density gradient of the progeny of the three kinds: those with clear, turbid and mottled plaques, respectively. The significant observation was that the progeny with mottled plaques gave a trimodal curve with peaks of frequency at buoyant densities corresponding to (a) three-quarters label, (b) three-eights label, and (c) no label. Since the position of gene cI divides the genome into three-quarters and one-quarter (Fig. 29), these results are those expected from breakage and joining at or near the cI locus, peak (a) corresponding to a conserved and peak (b) to a semi-conserved contribution of three-quarters of the genome from the heavy parent with, in each case, the other quarter unlabelled. Meselson concluded that the heterozygous region was frequently located at or near the recombination joint.

4. CLUSTERING OF EXCHANGES

The discovery of negative interference between crossovers in λ was described in Section 1. This phenomenon was investigated by Amati and Meselson (1965) using crosses with up to five mutants present at various intervals on the linkage map.

If there is no interference between one exchange and another in an adjacent interval, the frequency of double recombinants (R_D), with one exchange in each interval, will equal the product of the frequencies, R_1 and R_2, of recombination in the individual intervals. The ratio

$$\frac{R_D}{R_1 \times R_2}$$

defines the *coefficient of coincidence* or the *interference index*, i (cf. Chapter VII, Section 1). A value for i of 1 indicates no interference; i less than 1 means positive interference; and i greater than 1 means negative interference.

Amati and Meselson (1965) confirmed the earlier findings of Jacob and Wollman (1954) and Kaiser (1955) for long intervals: a value for i of about 3, indicating negative interference. For short intervals ($R_1 + R_2$ less than about 6%) Amati and Meselson found higher values for i. Indeed, the value increased rapidly with successively shorter intervals, rising to about 10 at $R_1 + R_2 = 3\%$, to about 30 at $R_1 + R_2 = 1\%$ and to near 100 with even shorter intervals.

Amati and Meselson pointed out that these results could arise in more than one way. One possibility is that the exchanges are randomly distributed along the genetic map, but a few of the progeny have much more recombination than the average. Another possibility is physical clustering of the exchanges.

Amati and Meselson were able to distinguish these alternatives by selecting for recombinants in one interval of the linkage map and examining the frequency of recombination in an adjacent and in a non-adjacent interval. The genetic markers for this experiment were the h and cI mutants used by Meselson (1964) and referred to in Section 2, together with two suppressor-sensitive (*sus*) mutants, nos 3 and 80, in gene P (see Fig. 29). These mutants are now known to be amber mutants. Progeny in which recombination had occurred in the short interval between the P alleles could be selected by making a mixed infection with the two amber mutant phages on an amber suppressor host strain, and infecting a normal (non-suppressor) host strain with the progeny phage. On this strain only the am^+ recombinants could multiply. In these P gene recombinants, the frequency of recombination in the adjacent cI–P interval was raised over 30-fold, from 1.1% to 38%, whereas that in the non-adjacent h–cI interval was raised less than fivefold, from 6.0% to 27%. It was evident that there is real clustering of the exchanges.

The existence of physical clustering of exchanges does not mean that the other explanation of negative interference does not also apply. Indeed, the occurrence of an interference index of 3 with widely spaced mutants points to a few of the progeny having much more recombination than the majority.

Amati and Meselson found that with four or five mutants present simultaneously, recombination sometimes occurred in all three or four marked intervals, showing that the clusters of recombination events were not limited to two exchanges. The intensity of negative interference with three or four close exchanges was as high as with two.

5. PROPHAGE

The term *prophage* is used for a phage genome in the latent state in a lysogen. As indicated in Section 1, early studies showed linkage of the λ prophage to a galactose (*gal*) host gene. Campbell (1962) suggested that the prophage genome was integrated into that of the host as a result of a reciprocal exchange at a specific site in each, the λ genome having first taken on a circular form (Fig. 33). He was led to this hypothesis by the work of Calef and Licciardello (1960), who had made crosses between lysogenic strains of *E. coli* differing in genetic markers in the λ prophage as well as in the host genome.

The prophage mutants used were h (host range), c (clear plaque) and mi (minute plaque). The c mutant used was subsequently shown to be in gene $cIII$ (Fig. 29). The host mutants affected galactose fermentation (*gal*) and tryptophan synthesis (*try*). Calef and Licciardello (1960) had made three significant discoveries:

(1) Selection for recombination between *gal* alleles greatly increased the frequency of recombination between the prophage genetic markers, suggesting physical connection between host and phage genomes.

(2) The recombination frequencies for the prophage and bacterial mutants fitted a single linear map with the prophage genes lying between those of the

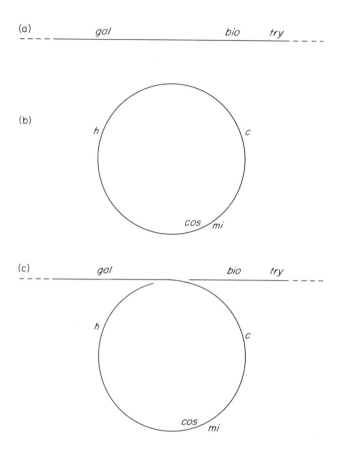

Fig. 33. Campbell's model for integration of the λ genome into that of the host (Campbell, 1962). (a) Part of the bacterial genome: *bio*, biothin, *gal*, galactose, and *try*, tryptophan operon. (b) The circular genome of λ after closure at the *cos* site: *c*, clear plaque, *h*, host range, and *mi*, minute plaque. (c) The λ genome integrated into that of the host following a crossover at the integration site in each.

bacterium. The map sequence was

$$gal - c - mi - h - try$$

(3) The *h* mutant mapped at a different position from that found in phage crosses, where the sequence is

$$h - c - mi$$

(see Fig. 29).

Campbell (1962) pointed out that the prophage gene sequence found by Calef and Licciardello (1960) is a circular permutation of the phage sequence. He inferred that the integration site on the phage genome was between the loci of the *h* and *c* mutants (Fig. 33).

Rothman (1965) studied the relation between λ prophage and host genes, using phage P1 as a means for transferring, by transduction, portions of the DNA of the lysogenic cells to another host strain. In this way the host gene *bio* (biotin synthesis) was found to be situated on the opposite side of the λ prophage from *gal* (Fig. 33). The frequency with which *gal* and *bio* were transferred simultaneously (co-transduction) was found to be much lower when both strains (donor and recipient) were lysogenic than when they were non-lysogenic, as expected if in lysogeny the λ prophage genome was inserted between the *gal* and *bio* genes. A number of prophage markers were mapped in relation to *gal* and found to be in a sequence that was a circular permutation of the sequence of the phage map, in agreement with Campbell's hypothesis.

Franklin, Dove and Yanofsky (1965) prepared a map of prophage mutants using a series of deletions. The deletion lysogens had been obtained by selecting for loss of a particular *E. coli* gene (the loss confers resistance to phage T1) located close to the prophage. It was found that the deletions simultaneously eliminated segments of the prophage genome. The deletions extended for various distances into the prophage, indicating that it must be linearly inserted into the bacterial genome. Furthermore, as in Rothman's studies, the prophage gene map was found to be a circular permutation of the phage map.

Campbell (1962) had also suggested that excision of the λ prophage from the host genome was brought about in a similar way to integration, as a result of a reciprocal exchange (Fig. 33 read in reverse).

6. INTEGRATION-DEFECTIVE MUTANTS

As discussed in Section 2, the *b2* mutant of λ has a deletion of part of the genome and is unable to form a stable lysogen. In consequence, when *E. coli* is infected with this mutant, the surviving cells continue to segregate sensitive ones as cell division proceeds. The *b2* mutant might lack an enzyme required for integration of the λ genome into that of the host, or it might lack the attachment site. By making mixed infections with *b2* and wild-type λ, Campbell (1965) found that the presence of the wild-type λ would still not allow *b2* to integrate normally. He concluded that *b2* had lost the attachment site.

Zissler (1967), Gingery and Echols (1967) and Gottesman and Yarmolinsky (1968) searched for point mutants of λ that had the *b2* phenotype, on the assumption that such mutants would be more likely to be defective in an enzymic step required for integration, rather than defective in the attachment site. To obtain point mutants, hydroxylamine was used as mutagen by Zissler (1967) and Gingery and Echols (1967) and nitrosoguanidine (i.e. *N*-methyl-*N'*-nitro-*N*-nitrosoguanidine) by Gottesman and Yarmolinsky (1968). Mutants defective in integration (*int⁻*) were isolated, and found also to be defective in excision of the prophage from the host genome. The mutants were not defective, however, in normal recombination between λ genomes.

7. MUTANTS DEFICIENT IN GENERAL RECOMBINATION

Van de Putte, Zwenk and Rörsch (1966), Takano (1966) and Brooks and Clark (1967) discovered that recombination in λ was approximately normal in recombination-deficient *E. coli*. It seemed likely, therefore, that one or more phage-directed enzymes were involved in λ recombination.

Mutants defective in general recombination in λ (as distinct from the special recombination process involved in integration of the phage genome into that of the host) were isolated by Echols and Gingery (1968) and Signer and Weil (1968), using nitrosoguanidine and hydroxylamine, respectively, as mutagen. The technique used by both groups to recognize recombination deficiency depended on the discovery by Fischer-Fantuzzi and Calef (1964) of a λ prophage deficient in a large part of the λ genome including the immunity region. In consequence, further infection by λ was not prevented, and recombination between the infecting phage and the prophage could be detected by the appearance of prophage markers among the progeny phage. If the prophage deletion included the site of a gene active in general recombination of λ, an infecting phage defective in the same gene, as a result of mutation, would be revealed because it would give no recombinants among the progeny. The recombination-deficient mutants were called *red*, this being an abbreviation for recombination deficient.

All the *red* mutants were found to map in the region of the λ linkage map known from the work of Radding, Szpirer and Thomas (1967) to contain the structural genes for λ exonuclease and β-protein. Subsequently, it was shown (Signer *et al.*, 1969; Radding, 1970) that the *red* mutants fell into three complementation groups, two (A and C) corresponding to the structural gene or genes for λ exonuclease and the third (B) the structural gene for β-protein. This was established from the discovery of mutants with altered immunologic properties of these proteins.

8. THE INT AND RED FUNCTIONS

Echols, Gingery and Moore (1968) and Weil and Signer (1968) found that *red* mutants of λ can integrate their genome into that of the host with high efficiency and, conversely, that *int* mutants show normal general recombination. They concluded that the Red and Int functions specified by the wild-type alleles of these genes, are different from one another. Using *red* mutants to eliminate general recombination they were able to demonstrate the existence of an integrative recombination process mediated at least in part by the product of the *int* gene and occurring at a specific site. Recombination was measured between genes *A* and *L*, *L* and *J*, *J* and *N*, and *N* and *P* (see Fig. 29) in *red*⁺ and *red*⁻ phage. Recombination in *red*⁻ phage was found to be of very low frequency in all the intervals except *J–N*. The presumptive integration site was thought to lie in this interval. It was evident that the Int product is active at a specific site on the phage genome, bringing about recombination at this site between two λ genomes, as well as between a λ genome and that of the host (integration). Weil and Signer (1968) pointed out that the site specificity of the Int system meant that there was no necessity for homology of nucleotide sequence at the phage and bacterial attachment sites, since an enzyme that can recognize one site might just as easily recognize two non-homologous sites.

As pointed out in Section 1, early investigations of the progeny of single cells suggested that recombination in λ was non-reciprocal. On the other hand, the Int system is known to cause reciprocal recombination. Weil (1969) therefore carried out single-burst experiments in which the Red and Int systems were examined simultaneously. Using a recombination-deficient host, the b2 mutant of λ was crossed with λ carrying a cIII mutant and amber mutants in genes A and B, and the phage progeny from single host cells examined. For the interval between the AB and b2 mutants on the λ linkage map, recombination is exclusively Red-promoted and very little correlation was found between the frequencies of reciprocal recombinants in the single bursts (correlation coefficient 0.16). The b2–cIII interval, on the other hand, contains the site of integration of the phage genome into that of the host. In consequence, recombination in this interval is pre-dominantly Int-promoted. Weil (1969) found considerable correlation in single bursts between reciprocal recombinants (correlation coefficient 0.64). He con-cluded that the Int-promoted recombination, even between one λ genome and another, is a reciprocal exchange, and that the Red-promoted event is a non-reciprocal one. Evidently the two processes may be quite different molecularly.

9. INTEGRATION AND EXCISION

Excision of prophage λ from the E. coli genome sometimes occurs abnormally such that bacterial genes located near to the prophage are included in the phage particle. This process allows host genes to be transferred from one cell to another (transduction). Guerrini (1969) used two transducing strains of λ to study the process of integration and excision. One strain, called λdg, contains the host galactose genes which are located just to the left of the prophage integration site on the E. coli linkage map. The other strain, called λdgb, is capable of transducing the galactose and biotin genes of the host. The biotin genes are on the right-hand side of the prophage. From comparisons of the efficiencies of integration and exci-sion of these strains and normal ones in various circumstances, Guerrini con-cluded (1) that the attachment region of the phage (attP) is straddled by two recognition elements, P and P', and similarly, the attachment region in the bacterial chromosome (attB) is straddled by two different recognition elements, B and B' (Fig. 34), and (2) that integration can occur only when one molecule is P–P' and the other B–B', or when both are B–P', or both P–B'. The transducing phages λdg and λdgb have the attachment genotypes B–P' and B–B' respectively. From his results Guerrini (1969) concluded that the Int product recognized the diagonally opposite PB' and BP' elements, which are present in the configurations that allow integration, and one or both of which are absent in the genotypes where integration does not occur.

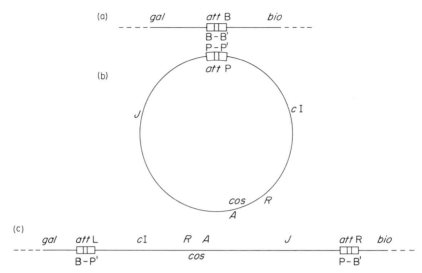

Fig. 34. The attachment regions of the *E. coli* and phage λ genomes. (a) The bacterial attachment region, B–B′ or *att*B. The positions of the galactose (*gal*) and biotin (*bio*) operons are shown. (b) The phage attachment region, P–P′ or *att*P. The positions of genes *A*, *c*I, *J* and *R* are shown, and of the *cos* site. (c) The left-hand attachment site, B–P′ or *att*L, and the right-hand attachment site, P–B′ or *att*R, of the prophage. The same bacterial and phage markers are shown.

An implication of these results is that excision is not just a simple reversal of integration, because the genotypes of the attachment regions when the phage genome is integrated in the host chromosome are B–P′ for the left-hand attachment site (*att*L) and P–B′ for the right-hand attachment site (*att*R), and these do not meet the requirements for recognition by the Int product, the diagonals for excision being PP′ and BB′. Confirmation that excision is different from integration came with the discovery by Kaiser and Masuda (1970) and Guarneros and Echols (1970) of a prophage excision gene.

A bacterial cell that is lysogenic for an abnormal strain of λ that carries the immunity genes of phage 434 instead of those of λ is not immune to further infection by λ, and this infection leads to excision of the prophage. Strains of λ able to transduce the host biotin genes carry deletions of part of the λ genome extending for various distances to the right on the λ genetic map from the attachment region (Fig. 35). Kaiser and Masuda (1970) found that the λ *bio* transductant with the shortest deletion (λ *bio* 16A) would bring about excision of the prophage with the 434 immunity, but *bio* transductants with longer deletions would not. The deletion in λ *bio* 16A extends into the *int* gene. Kaiser and Masuda concluded that there was a gene *xis* just to the right of *int* which was present in 16A but not in the other *bio* strains, and which was necessary for excision.

Guarneros and Echols (1970) isolated mutants of λ that were able to integrate efficiently into the host genome, but were unable to excise from it when released from the repression that prevents viral activity in the lysogenic state. Inability to excise was recognized by a failure of cell lysis after heating to 43 °C for 5 min. The heat treatment allows a brief derepression. The excision-defective (xis^-) mutants complemented *bio* 16A but did not complement the *bio* transductants with longer deletions, that is, extending further to the right of the attachment site (Fig. 35). Since all the *bio* transductants are int^-, Guarneros and Echols inferred that the *xis* mutations defined a gene to the right of *int*. They suggested that the Xis product might either recognize the prophage ends, just as the Int product appears to recognize the phage and bacterial attachment sites, or it might alter the Int protein so that excision was favoured.

Hradecna and Szybalski (1969) and Davis and Parkinson (1971) prepared electron micrographs of heteroduplex DNA derived from various λ deletion mutants and transductants. For this technique, two phage strains are mixed and the DNA extracted and denatured with sodium hydroxide. Renaturation (re-annealing of complementary strands) was carried out by neutralizing in for-mamide. Many of the molecules will be heteroduplexes with one strand derived from each parent. Under the electron microscope, regions of lack of homology in such heteroduplexes appear as two much-folded fine strands, which are replaced by a single stouter straighter strand in duplex regions where annealing has occurred. Deletions are readily recognized in heteroduplex molecules, as they appear as simplex loops. Their positions in the molecule were determined by measuring the duplex DNA lengths from the ends of the molecule. In heteroduplexes for two non-overlapping deletions, the length of duplex DNA between their sites can likewise be measured. Transducing strains of λ commonly have a deletion extending from the attachment region. From measurements on

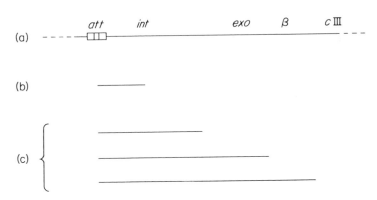

Fig. 35.   The extent of deletions in strains of λ able to transduce host biotin genes. These strains arose by excision of λ prophage at abnormal positions, with the result that inclusion of the biotin genes, which are situated to the right of *att*R (Fig. 34(c)) is associated with a corresponding loss to the right of *att*L. In consequence, all the deletions have their left-hand end at the phage attachment site. (a) Part of normal genetic map of λ. (b) The line shows the segment deleted in λ *bio* 16A. (c) The lines show the segments deleted in other λ biotin transductants.

heteroduplexes involving such strains, it was found that the attachment region for integration was at a fixed position 57.3% of the genome from the left-hand end. No measurable homology could be detected between *att*B and *att*P. From the lack of annealing at the attachment region in heteroduplexes of two deletion mutants, one in which the deletion extended to the left and the other to the right from the attachment region, Davis and Parkinson (1971) estimated that the region of homology, if any, between *att*B and *att*P could not be more than 20 base pairs long.

In an attempt to test whether there is, nevertheless, some homology between the bacterial and phage attachment sites, Shulman and Gottesman (1973) obtained attachment-site mutants. A transducing strain of λ was used that carries *att*L and *att*R separated by a few bacterial genes (Fig. 36(a)). Excision of these genes yields phage of lower buoyant density in a caesium chloride gradient. Attachment mutants were obtained by selecting phage that failed to yield such lower density molecules. As expected with *att* mutants, they depressed *int–xis*-promoted recombination only in *cis*. This indicates that the effect is structural and cannot diffuse to another molecule. It was found that the mutations could be transferred between all four normal *att* genotypes, that is P–P', B–B', P–B' and B–P', indicating that the mutations were located in a nucleotide sequence shared by all four, and not in the unique B, B', P and P' sequences. It was inferred that there was a core sequence, represented by the hyphen in P–P', etc., which was common to all four *att* genotypes, and that the mutations were situated in this region. This hypothesis was supported by the discovery that about 7% of recombinants were heterozygous for the *att* mutation. The heterozygosity was observed in *att*P recombinants resulting from exchange between wild-type *att*L and mutant *att*R and was revealed by isolating the recombinants under conditions which do not permit λ replication, and then testing the progeny of each for the presence of both mutant and wild-type phage. Heterozygotes are expected if there is a shared nucleotide sequence where staggered nicks can lead to hybrid DNA.

The nucleotide sequence of the central parts of the *att*P and *att*B regions were determined by Landy and Ross (1977). Four phage strains, each carrying a different attachment site, that is, *att*P, *att*B, *att*L or *att*R, were used. Digestion of the DNA of each strain with restriction endonucleases gave one unique fragment in each gel profile. These fragments evidently contained the crossover region and so were analysed for nucleotide sequence. The core sequence common to *att*P and *att*B was found to be 15 nucleotides long, in good agreement with the earlier prediction (see above). The sequence in the *l* strand (defined in Section 12) was

$$5'\ G\ C\ T\ T\ T\ T\ T\ T\ A\ T\ A\ C\ T\ A\ A\ 3'.$$

The sequences of the P, P', B and B' regions that span the core were determined for distances up to 200 nucleotides from the core. As expected, each region had a unique sequence. Davies, Schreier and Büchel (1977) also obtained the sequence for about 300 nucleotides of the central region of *att*P.

Gottesman and Gottesman (1975*a*) made further use of the *att*L–*att*R transducing strain of λ in order to study the effects of inhibitors of DNA replication, transcription, and translation on excisive recombination, using the sedimentation analysis as an assay. The inhibitors tested were nalidixic acid, rifampicin, and chloramphenicol respectively. The Gottesmans found no effect of these inhibitors and concluded that synthesis of DNA, RNA and protein was not

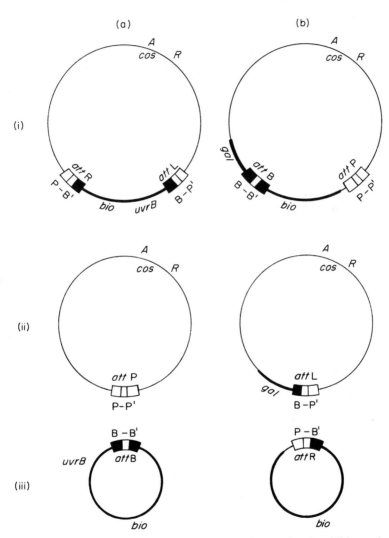

Fig. 36. Transducing strains of λ carrying two attachment sites in addition to bacterial genes. The bacterial DNA is shown with a thick line. In (a) the attachment sites are the prophage ones (*att*L and *att*R) and in (b) they are the bacterial one (*att*B) and the normal phage one (*att*P). The double-attachment-site strains are shown in (i), and the products of recombination between the two attachment sites in the same molecule in (ii) and (iii). The phage *A* and *R* genes and the *cos* site are given to show the genome orientation.

required for excision, contrary to some earlier claims that appeared to implicate transcription in the process.

The biochemical investigation of the integration mechanism was taken a stage further by Nash (1975a) who devised a technique for examining the recombinant DNA molecules rather than the recombinant progeny. He worked with a transducing strain of λ containing both attB and attP, with some bacterial genes near attB (Fig. 36(b)). Preliminary experiments showed that the integrative recombination between attB and attP was usually between these sites in the same molecule. The consequence of this was that the intervening segment of DNA, which amounted to about 13.4% of the initial genome, was detached as a circular molecule. At various times after there had been an opportunity for recombination between the att sites in the attB–attP molecules, the DNA was extracted and used to infect int⁻ spheroplasts of E. coli. As a result of their integration-deficient genotype there would thus be no opportunity for these cells to modify the recombination event that had occurred. In this way the formation of recombinant DNA was separated from the production of viable progeny.

The phage yield of the transfected int⁻ spheroplasts was studied by examining the plaque-forming activity in the presence of pyrophosphate, because mature phage particles with partially deleted genomes were found by Parkinson and Huskey (1971) to survive treatment with chelating agents such as pyrophosphate better than normal particles. Thus, the particles containing a recombinant DNA molecule survived this treatment, whereas those containing the larger parental molecule did not. After initial experiments had shown that the transfection method was reliable as an indicator of the presence of recombinant DNA, the effects of (1) inhibitors of transcription and translation and (2) the phage repressor were studied. Nash found that the inhibitors (rifampicin plus chloramphenicol) or the repressor led to early appearance of recombinant DNA. It was evident that a delay in recombination under normal conditions is caused by a process that depends on expression of the infecting phage molecule. In the absence of the inhibitors of gene expression or of the repressor, the presence of xis gene product had no effect, but in their presence the Xis function was found to prevent integrative recombination altogether. Clearly, the relationship between recombination and gene expression is complex but, as with the Gottesmans' study of excision (Gottesman and Gottesman, 1975a), there does not seem to be an absolute requirement for transcription.

Recombination by integration or excision has now been demonstrated in a cell-free system. Syvanen (1974) added λ DNA of one genotype to a cell-free extract from an induced E. coli lysogen of another λ genotype, and found that the phage particles formed in vitro included recombinants for parental character differences. He found that recombination took place between markers in gene J and the immunity region of the genome (Fig. 29), but not elsewhere. These markers span the attachment region and so he presumed that the recombination was by the Int system. He found that inhibition of the host polynucleotide ligase had no effect on the formation of recombinants, so this enzyme is evidently not involved in the covalent resealing presumed to occur after the exchange.

Nash (1975*b*) and Gottesman and Gottesman (1975*b*) have taken advantage of their methods for investigating, respectively, the integration process using λ *att*B–*att*P, and the excision process using λ *att*L–*att*R, and have adapted these for *in vitro* study of the recombination event. In Nash's experiments the parental and recombinant DNA were distinguished, as before, following transfection to mature phage in spheroplasts. Both Nash and the Gottesmans found that the reaction, in addition to needing enzymes from *E. coli* containing λ gene products, required ATP. Nash who, unlike the Gottesmans, used partially purified extracts, found that the integration reaction required magnesium ions and spermidine, whereas the Gottesmans found that these were not essential but that they stimulated the excision reaction three- to fivefold. The Gottesmans also found that RNA was not involved in the excision process, since the addition of pancreatic RNase had no effect, and that the addition of nalidixic acid, rifampicin or chloramphenicol to inhibit synthesis of DNA, RNA and protein, respectively, did not inhibit excisive recombination. Thus the behaviour *in vitro* was similar to that *in vivo* (see above). As expected, they found that expression of the *int* and *xis* genes was essential for the excision process, whereas Nash found that the integrative recombination was completely inhibited by extracts containing *xis* gene product.

A further advance in the *in vitro* study of λ integration, made by Mizuuchi and Nash (1976), was to assay the DNA itself rather than the progeny phage. This was achieved by the use of a restriction enzyme that cuts both strands of the DNA at specific sites, and so breaks it into pieces of characteristic size. Restriction enzyme no. I of *E. coli* (called *Eco*RI) was found to cleave the DNA of λ *att*B–*att*P into five fragments. On account of their differing sizes these gave bands at specific positions following electrophoresis on agarose gel. When integrative recombination occurs between the two attachment sites in λ *att*B–*att*P, the intervening segment of DNA is detached (Fig. 36(b)). The remainder, as expected, was found to give a different pattern of fragments following *Eco*RI treatment and electrophoresis: the fragments were those expected if the breaking and rejoining steps of integrative recombination had gone to completion *in vitro*. Electrophoresis of restriction fragments thus provides a sensitive test of the occurrence of the integration process.

Using this technique, Mizuuchi and Nash (1976) compared three different forms of λ *att*B–*att*P, namely, closed circular, linear, and hydrogen-bonded circular. The hydrogen-bonded circular form, unlike the closed circular, has the *cos* site unsealed, so there is a nick there in each strand (12 nucleotides apart). Unexpectedly, the linear DNA produced no detectable recombinants, and the hydrogen-bonded circles produced only small numbers; these evidently arose following synthesis *in vitro* of closed-circular DNA from the hydrogen-bonded circular form, because in the presence of nicotinamide mononucleotide (NMN), which inhibits the host ligase, no recombinants could be detected. Recombinants were formed, however, from closed-circular DNA in the presence of NMN, which evidently does not inhibit the ligation step in integration. Thus, it appears that closed-circular DNA is the only effective substrate for integrative recombination *in vitro*. This surprising conclusion led Mizuuchi and Nash (1976) to raise the possibility that integration requires supercoiled DNA, because supercoiling is not possible in hydrogen-bonded circles, as the single covalent bond opposite a nick can act as a swivel.

Strong support for the idea that supercoiled DNA is needed for integrative recombination was provided by the discovery of an enzyme, DNA gyrase, which functions in the integration reaction and which produces negatively supertwisted DNA from relaxed closed circles (Gellert *et al.*, 1976). Relaxed closed-circular λ DNA was prepared by sealing hydrogen-bonded circular DNA with DNA ligase. The supercoiling in this DNA caused by DNA gyrase was detected by the altered electrophoretic mobility of the molecules in agarose gel. All naturally occurring closed-circular DNA molecules have been found to be negatively supercoiled, that is, with extra twists (to the extent of between two and five twists per thousand nucleotide pairs) in the direction corresponding to uncoiling the double helix (see Chapter VI, Section 11). Ethidium bromide has been found to intercalate between the base pairs in the double helix and, if present at the time when the molecule is closed by polynucleotide ligase, to cause a reduction in the amount of supercoiling. From equilibrium centrifugation in a caesium chloride–ethidium bromide mixture, Gellert *et al.* (1976) inferred that the supercoiling induced by DNA gyrase was negative in direction and, with a high level of enzyme, showed about 1.5 times as many twists per thousand nucleotide pairs as were found *in vivo*. The action of the enzyme required ATP and magnesium ions and was stimulated by spermidine. Gellert *et al.* presumed that the reaction was driven by the hydrolysis of ATP, mechanical strain energy being stored in the DNA when it is supercoiled.

Using λ *att*B–*att*P, Mizuuchi, Gellert and Nash (1978) found that there was no DNA synthesis in integrative recombination *in vitro*. This implied that there was no filling of gaps and hence no trimming of strands by exonucleases. Evidently the cutting and rejoining were precise. Negatively supercoiled closed-circular DNA was the primary substrate and efficient recombination did not occur with closed-circular λ DNA lacking supercoils. These could be introduced, however, by the action of *E. coli* DNA gyrase. Mizuuchi *et al.* (1978) also found that there was no detectable difference in degree of supercoiling of the parental DNA molecules and of the products following recombination. Moreover, as previously shown, the host DNA ligase was not required. Thus, it appeared that the cutting of strands in the recombination process, unlike normal nicking, did not allow the supercoiling to be released, and the resealing of the broken strands was an integral part of the mechanism. As an explanation of why negatively supertwisted DNA was required for recombination, Mizuuchi *et al.* favoured its increased affinity for proteins, nucleic acids or small molecules that separate the strands of the duplex. They pointed out that the richness of the attachment site regions in A · T base pairs (see, for instance, the nucleotide sequence of the core, given above) would make them preferred regions for strand separation. They also suggested that the integrative recombination enzymes might resemble the DNA-untwisting enzymes (topoisomerases) described from *E. coli* by Wang (1971) and from *Mus musculus* by Champoux and Dulbecco (1972). These enzymes carry out a coupled breakage and rejoining of phosphodiester bonds in DNA, becoming covalently attached to the strand in the process, but do not bring about recombination, presumably because their action is not associated with close homologous pairing of two DNA molecules (but see Chapter VIII).

The conclusion that integrative recombination required supercoiled DNA was qualified by Pollock and Abremski (1979). They found that both integrative (*att*B x *att*P) and excisive (*att*L x *att*R) recombination could occur *in vitro* with non-supercoiled DNA at low concentrations of potassium chloride. DNA gyrase activity was absent, so supercoils were not being introduced. Intracellular ionic strengths *in vivo*, however, are such that there would be a preference for super-twisted DNA.

Another approach to understanding λ integration and excision was taken by Enquist and Weisberg (1976) who devised a new technique for finding mutants defective in either of these processes. The technique depended on the use of a host strain that carried a cryptic λ prophage inserted within gene *T* of the galactose operon. The cryptic prophage is bracketed by the attachment sites Δ–P′ and P–Δ′, where Δ and Δ′ represent the adjacent bacterial DNA sequences of the *gal* operon. The prophage lacks all the strong phage promoters, with the result that spontaneous excision is not possible. In the presence of normal λ, however, excision occurs with a frequency of 10^{-5}–10^{-4} per infection, which means many times in each plaque. The excision gives rise to a normal galactose operon in the host cells, and this can be detected by supplementing the growth medium with galactose and an indicator dye: the result is a plaque with a red centre. On the other hand, mutants defective in excision will fail to release the cryptic phage, galactose fermentation will not take place, and the plaques will be colourless. Since excision requires the *int* gene product as well as that of *xis*, mutants of either gene will be detected.

Several hundred colourless-plaque mutants were investigated (Enquist and Weisberg, 1977). Complementation studies with existing *int* and *xis* mutants indicated that all the mutations had occurred in these genes, and mapping the mutant sites confirmed this. It was evident that no other λ gene was required specifically for prophage excision. The mutants were mapped with the aid of deletion mutants ending at various places within the *int* gene, and confirmation was obtained that the map sequence is *att–int–xis*. The size of the *int* and *xis* genes was estimated by preparing heteroduplexes of pairs of non-overlapping deletions, and measuring the length on electron micrographs of the duplex region lying between the simplex loops that arise at the positions of the deletions one in each parent. Since the extent of the deletions on the genetic map was known, genetic distances could be converted into molecular distances. It was inferred that the *int* gene was about 1240 nucleotides long, in agreement with estimates of the size of the Int polypeptide, while *xis* appeared to be only 110 nucleotides long. The *att–int* and *int–xis* intervals were found to be quite short.

Using a DNA-binding assay, Kotewicz *et al.* (1977) purified the Int protein. The DNA binding was attributed to Int on the basis of three criteria: (1) the binding activity was not found after infection by phage carrying a stop mutation or a deletion of the *int* gene; (2) the activity was thermolabile in extracts of cells infected by λ carrying a temperature-sensitive mutation in the *int* gene; and (3) the binding was specific for DNA containing an appropriate attachment site. The binding specificity indicated that sequence recognition by the Int protein is crucial for the site specificity of integrative recombination. Moreover, the Int protein did not bind effectively to the right-hand prophage attachment site (*att*R); this may explain why Xis protein is required, in addition to Int, for excision.

Kikuchi and Nash (1978) also purified the Int protein, but using a different technique. The protein was assayed by its ability to bring about integrative recombination *in vitro*. The circular λ *att*B–*att*P molecule was prepared in a form containing a single target site for *Eco*RI restriction endonuclease. Following Int action, two circular DNA molecules will be produced (Fig. 36(b)), only one of which will contain the target site. Treatment with *Eco*RI will therefore break open all the parental DNA circles, but only half the recombinant ones. The presence of circles was revealed by their retention on nitrocellulose filters, which fail to trap the linear molecules resulting from *Eco*RI action.

Using the purified Int protein, they were able to confirm that it was a topoisomerase (Kikuchi and Nash, 1979). When the protein was incubated with superhelical DNA, relaxation of the supercoils took place. This was shown by centrifuging the DNA to equilibrium in a caesium chloride–ethidium bromide gradient, and by electrophoresis in agarose gel. The nicking–closing activity appeared to be non-specific, the *att* nucleotide sequence not being required. This activity of the *int* product was found to be inhibited by spermidine and by magnesium ions, unlike the behaviour of DNA gyrase (see above). Also, this Int activity was not affected by nalidixic acid, which inhibits DNA gyrase. The topoisomerase activity was evidently a property of Int because this function cosedimented in a glycerol gradient with the recombination activity, and the two properties were inactivated together by heat and by a specific antiserum. Furthermore, both activities were abolished as the temperature was raised to 33 °C in crude extracts of a strain carrying a temperature-sensitive *int* mutant. As to how the nicking–closing activity is used in recombination, Kikuchi and Nash (1979) assumed that within the 15-nucleotide core of *att* all four nucleotide chains are closely associated. They postulated that two Int molecules then act simultaneously to nick two strands of the same polarity at corresponding positions, each nicked strand swivelling round its unnicked complementary strand until it makes contact with the Int molecule covalently attached to the nicked strand of the other parent. Ligation would give a pair of crossing strands, as in the recombination model proposed by Holliday (1964*b*) – see Chapter VII, Section 7. Evidence for such an intermediate in *int*-promoted recombination is discussed below. To complete the recombination process, Kikuchi and Nash (1979) supposed that the other two strands underwent a similar nicking–swivelling–ligation event.

Miller *et al.* (1979) searched for *E. coli* mutants defective in λ integration. A phage strain defective in gene *N*, when inserted in the host genome, will replicate *in situ* and kill the cell, because phage repression cannot be established in the absence of the *N* gene product. Thus, mutant cells unable to integrate λ will survive. Another method of selecting for host mutants defective in λ site-specific recombination took advantage of the fact that in a particular λ *att*B–*att*P strain the host galactose operon lies between the two attachment sites (cf. *bio* in Fig. 36(b)) and is therefore excised in a circular DNA molecule when recombination occurs. In the presence of the λ *int* gene product, only mutants unable to delete the bacterial DNA will form colonies able to utilize galactose. Using these methods, a number of host mutants were obtained defective in λ integration. They mapped at several loci, including *hip* (host integration protein) and *himA* (host integration mediator). Miller *et al.* (1979) showed that cell-free extracts from these strains lacked factors needed for λ integrative recombination *in vitro*, but not required for general recombination.

An important discovery about *int*-promoted recombination in phage λ was made independently by Enquist, Nash and Weisberg (1979) and Echols and Green (1979). They found that in phage crosses this recombination is not restricted to the attachment region but can occur in neighbouring regions as well. Markers in the *int* and *xis* genes were used, as these are close to *att*. The phage Red and host Rec recombination systems were blocked by appropriate mutants. Confirmation that the recombination was determined by *int* was evident from the need for the host gene *hip*: Enquist *et al.* (1979) found that the recombination, both within *att* and in neighbouring regions, was abolished in a *hip*$^-$ host. To account for the occurrence of *int*-promoted recombination outside *att*, it was suggested that, following strand exchange between the two λ DNA molecules, branch migration took place. This would extend the heteroduplex region in each molecule for a variable distance beyond *att*. In keeping with this hypothesis, multiple exchanges were found to occur. These are expected if there is repair of mismatched bases in heteroduplex DNA. Branch migration would not be possible when integration of the λ genome into that of the host took place, because of the lack of homology between the phage and host attachment regions on either side of the central core: Enquist *et al.* (1979) found that the introduction of *att*B in place of *att*P in one of the parents of the phage cross did indeed eliminate the recombination outside *att*. This external recombination was also eliminated (together with that at *att*) in crosses involving a point mutation in *att*P. This showed that the recombination was not arising from secondary attachment sites elsewhere in the λ genome. The frequency of *int*-promoted recombination outside *att* was low. Echols and Green concluded that, following the exchange of one pair of strands, branch migration is usually restricted as a result of the concerted occurrence of the exchange of the other pair of strands. From the behaviour of flanking markers, Enquist *et al.* (1979) found that the strand exchange outside *att* may be resolved to give no overall exchange. In other words, as in the model for eukaryote recombination proposed by Holliday (1964*b*), either the crossing strands or the non-crossing strands of the recombination intermediate may be cut, leading to either a parental or a non-parental flanking marker genotype, respectively. Only the latter outcome would integrate the λ genome into that of the host when recombination occurs between *att*P and *att*B. If the two pairs of strand exchanges are normally concerted, however, there may then be control, such that cutting the same pair twice does not occur.

10. SINGLE-STRAND ASSIMILATION

As pointed out in Section 7, *red* mutants may be defective in the structural gene for λ exonuclease or in that for β-protein. This protein is associated with the exonuclease. Little (1967) found that the exonuclease cleaves 5′-mononucleotides stepwise from the 5′ phosphoryl ends of the strands in duplex DNA (Fig. 37(i),(ii)).

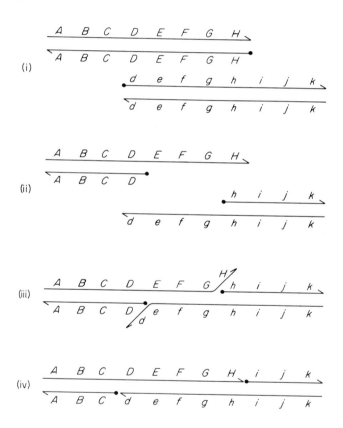

Fig. 37. Hypothetical steps in genetic recombination brought about by λ exonuclease, according to Cassuto and Radding (1971). The letters identify particular regions of the DNA, the contributions from the two parents being distinguished by upper and lower case. Half arrowheads indicate 3′ ends and the dots show the site of action of λ exonuclease. (i) Intact duplexes. (ii) The λ exonuclease has broken down single strands from 5′ ends. (iii) Annealing of complementary strands has occurred, leaving redundant single strands with 3′ ends. (iv) The λ exonuclease has continued to erode the 5′ ends, allowing the redundant single strands to be assimilated; its action has then stopped leaving only a nick.

Cassuto and Radding (1971) discovered a new substrate for the enzyme, namely, 5' ends at internal positions in the molecule where there is a redundant strand with a 3' end (Fig. 37(iii)). They called such a junction a *redundant joint*. The λ exonuclease used in the experiments that demonstrated this action was complexed with β-protein. It was found that the redundant single strand became hydrogen-bonded into the duplex structure of DNA as the λ exonuclease excised nucleotides ahead of the branch point. This action was demonstrated by means of experiments involving radioactive branched substrates. Cassuto and Radding called the entry of the redundant strand into the duplex *strand assimilation*. They found evidence that the action of the enzyme stopped when the whole of the redundant single strand had been assimilated, leaving only a nick in the duplex (Fig. 37(iv)).

This behaviour of λ exonuclease led Cassuto and Radding to a model for its action in recombination, outlined in Fig. 37. Covalent sealing of the nicks by a ligase is all that is needed to complete the process of recombination brought about by the enzyme, in conjunction with the spontaneous annealing of complementary strands (Fig. 37(iii)). Cassuto *et al.* (1971) elaborated this hypothesis.

Radding (1973) pointed out that λ exonuclease will slowly degrade an unpaired strand from its 5' end, so assimilation of a 3'-ended strand into a duplex (Fig. 37(iii)) could still occur even if the tip of the 5'-ended strand that was being eroded was unpaired.

The evidence that the exonucleolytic breakdown stopped precisely at the point reached by the redundant strand as it entered the duplex was of three kinds:

(1) The sedimentation rate of the product of the reaction corresponded to that of a hydrogen-bonded circle.

(2) The length of the redundant strand determined the amount of degradation of the 5'-ended strand by the exonuclease.

(3) Polynucleotide ligase could seal the joint after λ exonuclease had acted.

Cassuto *et al.* (1971) showed that the sequential action of λ exonuclease and polynucleotide ligase produced intact biologically active λ DNA. They concluded that λ exonuclease was capable of executing in a single concerted reaction most of the steps required to splice together DNA from two sources, such as occurs in recombination.

11. RED-MEDIATED RECOMBINATION AND ROLLING CIRCLES

Stahl and Stahl (1971a) pointed out that the experiment of Meselson (1964), described in Section 2, led him to conclude that recombination was by break and join rather than break and copy, but that this conclusion was open to criticism on two grounds:

(1) In the experiment, both parents were density labelled and allowed to multiply in host cells not so labelled. The recombinants that were fully dense, that is, inherited heavy atoms from both parents and hence by break and join not break and copy, may have been only a small minority of the recombinants.

(2) The central region of the phage genome chosen for study includes the interval between genes *J* and *c*I known to be acted on by both the Int and Red recombination systems as well as the host Rec system. Meselson's conclusion might not apply to all of these.

In order to meet these criticisms, Stahl and Stahl (1971*a*) repeated Meselson's experiment under conditions when DNA replication was minimized, and when particular recombination systems were eliminated. The reason for reducing the replication of the DNA was in order to judge the frequency of break-and-join recombination, since the evidence for it is lost when the DNA subsequently replicates. The limitation of DNA replication was achieved by using a host strain that carried a temperature-sensitive mutant of the *dnaB* gene – one of the host genes required for λ DNA replication. By making crosses at the non-permissive temperature (40 °C) synthesis was greatly reduced.

In the first experiment, amber mutants of genes *J* and *R* were crossed and the wild-type recombinants studied by density-gradient centrifugation. The *c*I marker was also present in the cross, so the *J–R* recombinants could be divided into two classes depending whether the recombination was in the *J–c*I or *c*I–*R* interval. It was found that the *c*I–*R* recombinants banded at a slightly lighter position than the total progeny, and the *J–c*I recombinants at a much lighter position (about midway between positions corresponding to wholly conserved and semi-conserved DNA). Stahl and Stahl (1971*a*) concluded that some DNA synthesis is associated with recombination, more in the one interval than the other. Repetition of the experiment using an integration-deficient mutant gave a similar result, so the contribution that the Int system is known to make to recombination in the *J–c*I interval was evidently not the cause of the differential behaviour of the two regions. Other markers were used in subsequent experiments and indicated various levels of DNA synthesis, the highest being in the *J–c*I interval, where at least 10% of the genome may be replicated in association with each recombination event, and the lowest between *P* and *R*, where there seemed to be little or no synthesis when recombination takes place.

Stahl and Stahl (1971*b*) used a different method for reducing DNA replication. The λ crosses were carried out in a strain of *E. coli* carrying λ prophage, which represses duplication of the infecting phage. Another lambdoid phage with a different immunity system was allowed to infect the cells at the same time, and this 'helper' phage provided the recombination system and the means for phage maturation. The results obtained were similar to those using the previous method for limiting replication. By the use in turn of *int⁻* and *red⁻* λ in a normal or a *recA⁻* host, Stahl and Stahl (1971*b*) found that the variable requirement for DNA synthesis in recombination, depending on the region studied, was shown primarily by the Red system.

An unexpected finding by Stahl and Stahl (1971*b*) was that elimination of the Red system greatly reduced the yield of fully conserved phage. This discovery was further investigated by Stahl *et al.* (1972*a*) by blocking DNA synthesis by the repressor method, density labelling both phage parents, and making crosses with various combinations of the recombination systems. They concluded that recombination is obligatory for the production of fully conserved phage under conditions when DNA replication is prevented. They pointed out that maturation of phage particles is preceded by a rolling-circle type of DNA synthesis that generates concatemers containing two or more λ genomes joined end to end. In the absence of such replication, recombination between simple monomeric circles would provide an alternative route to the dimeric or polymeric molecules required, if packaging of the DNA into the phage head is necessarily associated with cutting a monomer out of a polymer.

Using the temperature-sensitive *dnaB* mutant of *E. coli* to reduce λ DNA synthesis, Stahl *et al.* (1972*b*) made crosses at four different temperatures over the range 36–44 °C between phage mutants which had been density labelled, and studied the progeny in a caesium chloride density gradient. With the increase of temperature, DNA synthesis is progressively inhibited. They found that the blockage of DNA synthesis reduced Red-mediated recombination more strongly in those regions, for example, between *J* and *c*I, with the larger amounts of DNA synthesis associated with recombination than in those regions, for example, between *c*I and *R*, with a smaller amount. They concluded that DNA synthesis enhances the formation of recombinant particles more in some map intervals than others.

In order to restrict λ DNA synthesis to a greater extent, McMilin and Russo (1972) experimented with a double block to replication, namely, the presence of a mutant of the host *dnaB* gene, as used by Stahl and Stahl (1971*a*) and Stahl *et al.* (1972*b*), in conjunction with a mutant of gene *O* or gene *P* of λ. Using a density label, as before, in both parental phage strains, McMilin and Russo (1972) were able to demonstrate that recombinants were formed and that, at least for the middle region of the linkage map (*J*–*c*I), they were associated with a density shift corresponding to DNA synthesis equal to about 5% of the genome.

Stahl *et al.* (1972*a*) took advantage of the very low level of DNA synthesis permitted by McMilin and Russo's double block to carry out an experiment in which only one parent carried the density label. The assumption was that new synthesis was negligible, so that the density of the progeny was a measure of the contributions from each parent. Temperature-sensitive mutants in genes *A* and *R*, near opposite ends of the genome, were crossed, and a marker in gene *c*I was also present. The parents were

A	*c*I	+	heavy
+	+	*R*	light

Among the recombinants that were wild type for the temperature-sensitive *A* and *R* mutants, those that had the + + + genotype, that is, a crossover in the right-hand interval, were nearly all heavier than those that had the + *c*I + genotype, that is, a crossover in the left-hand interval (see Fig. 38). Thus, the density of the recombinants was an indication of where the crossover took place. It was of interest, also, that the *c*I mutant gave rise to mottled plaques for recombinants of

138

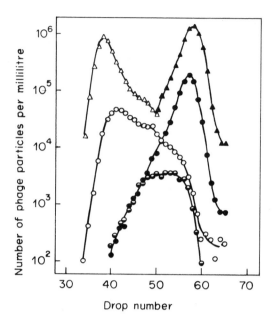

Fig. 38. The density distributions of progeny phage of various genotypes obtained by Stahl *et al.* (1972*b*) from a cross between λ strains differing in density. The heavy parent had a temperature-sensitive mutation in gene *A* and a mutation in the *c*I gene giving clear plaques. The other parent, which was of normal density, had a temperature sensitive mutation in gene *R*:

A	*c*I	+	heavy
+	+	*R*	light

The cross was made on cells and culture medium of normal density. Replication of the phage DNA was prevented by means of mutations in the *dnaB* gene of the host and the *O* gene of λ. The progeny were centrifuged to equilibrium in a caesium formate gradient, and the numbers per millilitre of phage particles of various genotypes in each fraction were determined. These numbers are plotted on a logarithmic scale against the density, given as the drop number of the fraction.

	A	*c*I	*R*	
▲		*c*I		Total phage irrespective of
△		+		genotype for *A* and *R*
●	+	*c*I	+	
◒	+	*c*I/+	+	Mottled plaques
○	+	+	+	

(Reproduced by permission of F. W. Stahl from F. W. Stahl *et al.*, *Proc. natn. Acad. Sci. U.S.A.* **69**, 3599 (1972).)

density corresponding to a crossover in the neighbourhood of cI, as expected if the mottled phenotype is a consequence of a heteroduplex at the site of the cI mutant (see Section 3). The peak in the density gradient for mottled plaques was broad (see Fig. 38), implying a large, possibly variable, length for the heteroduplex segment (Stahl *et al.*, 1972*b*).

Further experiments with the double block to DNA synthesis and with only one parent carrying a density label, were reported by Stahl *et al.* (1974). The effects of λ *red* and host *rec* deficiency were studied, and it was found that the recombination determined by the host Rec system occurred more or less equally often in different parts of the genome. This was in contrast to the recombination controlled by the λ Red system where, as already indicated, recombination in parts of the genome near the middle of the linkage map was depressed by the inhibition of DNA synthesis more than at the ends. This difference between the Rec and Red systems was no doubt related to the earlier findings that there is little DNA synthesis associated with recombination by the Rec system, whereas for the Red system there is much synthesis when recombination occurs in regions of the genome corresponding to the centre of the map, little or none near the ends.

Enquist and Skalka (1973) studied the rate of incorporation of [^{3}H]thymidine into λ DNA, measured by annealing it with nitrocellulose membranes containing an excess of λ DNA. They found that DNA replication in *red*$^-$ mutants took place at only one-third to one-half the rate in *red*$^+$ and concluded that the *red* proteins function in λ replication. This was a finding complementary to that of Stahl *et al.* (1974). It seems as if the replication and recombination pathways are partially interrelated, Red-mediated recombination being dependent, at least in the middle of the genome, on DNA replication, which itself requires the Red activity if it is to proceed normally. To accommodate this interdependence of replication and Red recombination, Skalka (1974) and Stahl and Stahl (1974) each proposed models relating the two processes.

Skalka (1974) suggested that the λ Red recombination system and the host polymerase and ligase repair system may each provide a pathway for generating the late replication mode for λ, that is, a rolling circle, from the early replication mode (bidirectional replication in a monomeric circle). Enquist and she had discovered (Enquist and Skalka, 1973) that when host cells are infected with a *gamma* mutant (*gam*$^-$) of λ, the transition from the early to the late replication does not occur and, if λ Red recombination is also inhibited, circular monomers accumulate. The *gam* gene had been discovered by Zissler, Signer and Schaefer (1971), who were led to it by the observation that certain *red*$^-$ deletions would not grow on a *recA*$^-$ host. This was unlike *red* point mutants, whether *exo*$^-$ or β$^-$, that is, defective in the structural gene for λ exonuclease or in that for β-protein, respectively. Zissler *et al.* (1971) found that the γ-protein (the product of the *gam* gene) was required for growth of *red*$^-$ λ in a *recA*$^-$ host. Unger and Clark (1972) and Sakaki *et al.* (1973) showed that the γ-protein inhibited the host *recBC* recombination pathway by interacting with the product of the *recB* and *recC* genes (*recBC* DNase, also called exonuclease V – see Chapter VI, Section 5).

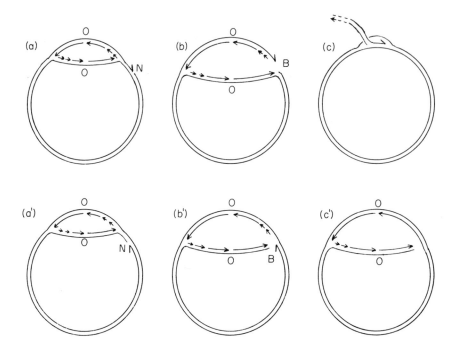

Fig. 39.   Pathways for generating a rolling circle from a simple circle according to Skalka (1974) by the λ Red system ((a)–(c)) and by the host repair system ((a')–(c')). Arrowheads are at 3' termini, half arrowheads for parental strands and whole arrowheads for daughter strands. B, broken molecule; N, nick; O, replication origin. (a) A replication fork approaching a nick in the strand that is copied discontinuously. (b) The replication fork reaches the nick and that arm of the fork is broken, leaving a 3'-ended single strand at the tip. (c) The 3'-ended single strand at the tip of the broken arm invades the homologous region of an intact circle. Elongation at the tip generates a rolling circle. (a') A replication fork approaching a nick in the strand that is copied continuously. (b') The replication fork reaches the nick and that arm of the fork is broken and the replication apparatus falls off.
(c') The continued action of the other replication fork generates a rolling circle.

Enquist and Skalka (1973) came to the same conclusion from their study of λ DNA replication using various combinations of host *recA* and *recB* mutants and phage *red* and *gam* mutants.

Skalka (1974) postulated that during the early (circular) mode of replication of λ a nick is sometimes present in one of the parental strands. If the nicked strand is that with 5' to 3' polarity in the direction of movement of the replication apparatus (Fig. 39(a)), it will be copied discontinuously, that is, in the direction away from the growing point, so new synthesis is unlikely to occur near the 3' end caused by the nick (Fig. 39(b)). Skalka suggested that this simplex tip to the broken molecule could lead, through the action of the Red recombination system, to recombination with the homologous region of a duplex λ molecule. This might be the other arm arising from the same replication fork, or it might be a molecule of different parentage. She assumed that an as yet unidentified protein catalysed or stabilized an opening of the recipient duplex to admit the invading strand, which would thus be able to anneal with the complementary strand of the recipient molecule (Fig. 39(c)). Skalka supposed that the 3'-ended strand, after invading the other molecule, could act as a primer for replication, although an RNA segment is usually required for this purpose. Assuming, as is likely, that the recipient λ molecule had a closed-circular configuration, replication by addition of nucleotides to the 3' end that had invaded it would generate a rolling circle.

She pointed out that if the initial nick was in the other strand – that is, the one that was copied continuously (Fig. 39(a')) – replication at that fork would stop, because the replication apparatus would fall off the broken arm (Fig. 39(b')). Any gap resulting from the discontinuous synthesis in the other arm of the replication fork could be repaired by the sequential action of the host DNA polymerase I and ligase (Fig. 39(c')). The structure could then become a rolling circle through the continued action of the other replication fork of the bidirectional pair.

Skalka was led to suggest these two different ways of generating a rolling circle – via the host repair system, or via the λ Red system – because she had found that the Red-mediated events that affect phage growth do not involve the host DNA polymerase I or ligase (because in $polA^-$ or lig^- cells growth of λ is almost normal) and conversely the functioning of the polymerase-plus-ligase repair pathway seemed to be essential for growth of red^- phage. She pointed out, however, that her hypothesis did not clarify the function of λ exonuclease in red recombination.

She also suggested that the significance of the inhibition of the host $recBC$ DNase by the phage γ-protein was the potential threat posed by the host enzyme to the initiation of rolling circle replication, such replication being essential to λ since, as already implied, two cos sites in one molecule seem to be necessary for packaging the DNA into the phage coat. By using a temperature-sensitive gam mutant, Greenstein and Skalka (1975) were able to activate the $recBC$ nuclease at various times after infection. They concluded that the structure that was sensitive to this enzyme was probably not the tail of the rolling circle, but an intermediate in the transition from the early to the late mode of replication, such as the free ends to nucleotide chains in the hypothetical pathways outlined above.

Stahl and Stahl (1974), like Skalka (1974), also proposed that the rolling circle type of DNA replication, necessary for the production of mature virus particles, can be initiated by Red action. The Stahls suggested that the Red system generates a dimer of rolling circle structure from two circular monomers (Fig. 40). Approximately reciprocal exchange of 3'-ended strands takes place (Fig. 40(a)) and the junctions are sealed, giving a figure of eight configuration. The exchange point is free to move by reciprocal branch migration and is assumed to do so until it encounters a nick (Fig. 40(b)). This nick, like those in Skalka's model, is thought to be accidental in the sense that it is not caused by an identified recombination gene. The λ exonuclease – the product of one of the red genes – is then presumed to act at the site of the nick (though according to Cassuto $et\ al.$ (1971) a gap or a redundant joint is required, not a nick – see Section 10). This enzyme erodes 5' ends, so, depending on the strand orientation, the breakdown may take place in the same direction (Fig. 40(c)) as the previous branch migration. On approaching

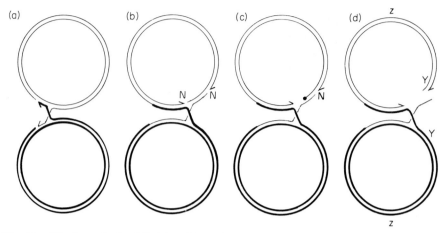

Fig. 40.   The hypothesis of Stahl and Stahl (1974) to explain the origin of a rolling circle by Red-mediated recombination of circular monomers of phage λ. The parental origin of the strands is indicated by the thickness of the lines. Half arrowheads show 3′ ends and a dot the site of action of λ exonuclease. N, nick; Y, Z, two of many alternative positions for *cos* sites. (a) Approximately reciprocal exchange of 3′-ended strands. (b) Reciprocal branch migration has reached a nick. (c) Erosion by λ exonuclease. (d) Opening of the circle when the exonuclease approaches a nick in the other strand.

a nick (again thought to be of 'accidental' origin) on the other strand, the exonuclease is expected to leave its substrate, but nevertheless the circle will open (Fig. 40(d)). Elongation by a polymerase at the 3′ end resulting from the first nick will give rise to a rolling circle configuration. A full replication complex could then take over, bringing about synthesis on the circle as well as its tail. The tail synthesis would be towards the replication fork, while that on the circle would be away from the fork and hence discontinuous. Stahl and Stahl (1974) called this an unrolling circle because the strand polarities are the opposite of those in the rolling circle model (Gilbert and Dressler, 1969) for DNA replication. This reversal is to accommodate the conclusion of White and Fox (1974) that in crosses in which appreciable DNA synthesis is prevented by a double block (resulting from the presence of mutants of the host *dnaB* gene and of the phage *O* and *P* genes) the progeny phage have, as expected, undergone recombination involving a heteroduplex segment formed by annealing parental molecules with staggered breaks in the two strands but, surprisingly, the longer strands are always those with 3′ ends. This unexpected discovery is discussed in Section 12.

Stahl and Stahl (1974) pointed out that if the reciprocal branch migration (Fig. 40(b)) happened to stop near the genome end on the linkage map (that is, *cos* sites at Y in Fig. 40(d)) the single-stranded DNA synthesis may reach the end. This would mean that a complete phage genome with a *cos* site at each end was present in the tail of the unrolling circle, and could be cut from it and incorporated in the phage coat. On the other hand, if the exchange point was more centrally placed in the genome (*cos* sites at Z in Fig. 40(d)), it would be necessary to invoke the full replication complex and appreciable DNA synthesis before the tail became long enough to include two *cos* sites. It is assumed, as already mentioned, that the packaging of the λ DNA into the phage head requires the presence of two such sites in the tail of the unrolling circle before a monomer can be cut from the concatemer. Thus the Stahls' hypothesis would explain why more DNA synthesis is required in mid-genome than near the ends for the production of Red-mediated recombinants. They pointed out that only those recombinants that met the requirements for packaging the DNA into the phage head would be detected, and so it was possible that other kinds of recombinant molecules were formed but did not appear in the progeny.

Stahl and Stahl (1974) attributed the difference between Rec- and Red-mediated recombination in the requirement for DNA synthesis to the existence in the Rec system (but not in the Red) of an endonuclease capable of converting the figure of eight structure (Fig. 40(b)) into a double-size circle by reciprocal break-reunion.

12. LONG HETERODUPLEX SEGMENTS

Russo (1973) took advantage of the double block to λ replication which McMilin and he had devised (see Section 11) to study the structure of recombinant DNA in λ. For this purpose it is essential to recover the molecules before any duplication has taken place. Their method allowed this, since there was less than 4% duplication. Crosses were made between two amber mutants of gene O, one parent also having a cI mutant. In these crosses each parent in turn was labelled with ^{15}N and ^{13}C, the other being of normal density (^{14}N and ^{12}C). The crosses were made in cells and culture medium of normal density. Recombinants in gene O could be selected by plating the progeny phage on a suppressor-negative host. Density-gradient centrifugation in caesium formate enabled the amount of DNA inherited from each parent to be measured. In the recombinant molecules, the peak density nearest to the light parent had incorporated about 10% of the DNA of the heavy parent, while the position of the peak nearest to the heavy parent corresponded to the incorporation of about 15% of light material, a third of which was believed to represent the limited DNA synthesis that took place despite the double block to replication. Thus, Russo concluded that the contribution to the recombinant DNA from each parent can be from 10 to 85% of the whole genome. Since nearly 20% of the genome lies to the right of gene O, Russo suggested that the minimum value of a 10% contribution from one parent might represent a single strand extending from the right-hand end of the genome, with recombinants produced largely by excision and repair of mismatched bases. Such repair would explain, at least qualitatively, the occurrence of localized negative interference (see Section 4).

Russo tested this hypothesis of long heteroduplex segments by repeating the crosses with a mutant of gene R present in one parent. The idea of the experiment was that $cI/+$ heterozygotes could be recognized by their mottled plaques (see Section 3), and the heteroduplex condition assumed to be responsible for the heterozygosity would often extend to gene *R* (at a distance of about one-fifth of the λ genome length) if the heteroduplex segments were long. It was found that only about 16% of the *c*I heterozygotes were also heterozygous at *R*. The crosses were made with the *c*I and *R* mutants in the *cis* and also in the *trans* configuration. The double heterozygotes had the corresponding structure, as expected if they arose from the annealing of single strands of DNA, one from each parent, over the *c*I–*R* interval. When this experiment was repeated in a *recA*-deficient host (Chapter VI, Section 4), about 35% of the *c*I heterozygotes were found also to be *R* heterozygotes. It was believed that in this experiment the recombination was largely *red*-mediated, where previously the host Rec system had probably predominated. Russo concluded that under both recombination systems heteroduplex segments were often extensive. He suggested that the comparatively low frequency of double heterozygotes for *c*I and *R*, particularly under the Rec system, might be a consequence of mismatch correction.

Russo's work gives no information about how the long heteroduplexes might arise, but their occurrence is consistent with Cassuto and Radding's strand assimilation model (Section 10).

Further evidence for long heteroduplex segments in recombinant λ DNA was obtained by White and Fox (1974). They used three amber mutants of λ, one (no. 29) in gene *O* and two (nos 3 and 80) in gene *P*. By crossing an *O P* double mutant (nos 29 and 80) with the other *P* mutant (no. 3) in a temperature-sensitive *dnaB* mutant of *E. coli*, λ DNA synthesis was doubly blocked, since it requires the active products of host gene *dnaB* and λ gene *P*. As in Russo's experiments, the cross was also heterozygous for a *c*I mutant. One parent was labelled with deuterium (^{2}H) and ^{15}N, the other parent and the culture medium being of normal density. The crosses were made on a suppressor-negative bacterial strain, with the result that only the recombinant progeny, wild type for the *O* and *P* genes, were able to grow. Centrifuging the recombinant molecules in a caesium formate gradient gave results similar to Russo's: the molecules ranged more or less uniformly from heavy ones into which about 15% of light DNA had been inserted to the converse structure, namely, the substitution of about 15% of heavy DNA in light molecules. The even density distribution pointed to long heteroduplex segments.

Unlike Russo's crosses, which involved selection for recombination between two closely linked mutants both in gene *O*, White and Fox were selecting double recombinants, since the *P* mutant (no. 3) in one parent mapped between the *O* and *P* mutants contributed by the other parent:

$$\frac{O29 \quad + \quad P80}{+ \quad P3 \quad +}$$

To derive a + + + molecule from these parents requires an exchange in both intervals. It was evident that these molecules were not arising simply by inserting DNA from the upper molecule into the lower one so as to replace the *P*3 nucleotide sequence by the wild-type one, because the interval spanned by the outer markers (*O*29 and *P*80) represents only 2.8% of the length of the λ genome. A segment inserted into the lower molecule could not therefore exceed this size if wild type was to be generated, yet the minimum contribution of one parent to the other in the recombinant molecules was, as already stated, 15%. White and Fox concluded, as Russo had done, that the formation of long heteroduplex segments, followed by mismatch repair or random excision and repair, was the likely source of the recombinant molecules.

To test this idea further they made crosses using an *R* mutant as an additional marker, just as Russo had done. They found from the analysis of the progeny of single infections that about 40% of the *O P* recombinants were heterozygous for the *c*I mutant, and of these about 15% were also heterozygous at the *R* site one-fifth of the genome length away. This frequency of double heterozygotes was similar to what Russo had found. The results, like his, implied that long segments of hybrid DNA were of frequent occurrence.

White and Fox (1974) made a further discovery that was utterly unexpected. From the results just described it seems that when one selects for recombination in a particular region of the genome, one is selecting for a heteroduplex segment, often of considerable length, in that region. This segment would link DNA derived wholly from one parent at the left-hand end of the genome and wholly from the other parent at the right-hand end. There are two possibilities for the relative positions of staggered breaks in a DNA molecule. It may be broken such that the longer strand in each fragment has a 5′ end, or alternatively such that each has a 3′ end. There are therefore four basically different structures for a heteroduplex spanning a particular region and joining homoduplexes one of each parentage, allowing for the fact that each end may be of either parentage (Fig. 41).

Since the *O* and *P* genes, where White and Fox selected for recombinants, are not centrally placed in the genome but are situated about one-fifth of its total length from the right-hand end, the recombinant molecules are expected to have a density that is dictated primarily by the density of the parent molecule that contributes the left-hand end. The cross was between the *P*3 mutant labelled with heavy atoms and the *O*29 *P*80 double mutant of normal density. A single burst analysis (that is, analysis of the progeny of single infections) of the heavy recombinant phage revealed that some cells gave only wild types and others gave a mixture of wild type and *P*3, implying a heteroduplex at this site in the original recombinant. On the other hand, a corresponding analysis of the light recombinants showed, in addition to bursts that gave only wild type, there were others that released a mixture of wild-type and *O*29 phage. Both heavy and light recombinants also produced small numbers of bursts containing other genotypes, but the surprising observation was the difference in genotype of the predominant kind of mixed burst when of different density. This was not expected because each heavy molecule (Fig. 41(a),(c)) has its light counterpart (Fig. 41(b),(d)). White and Fox concluded that only one of the two kinds of heteroduplex junction contributed to the recombinant progeny, that is, either a heteroduplex composed of strands from each parent with 3′ ends (Fig. 41(a),(b)) or those with 5′ ends (Fig. 41(c),(d)).

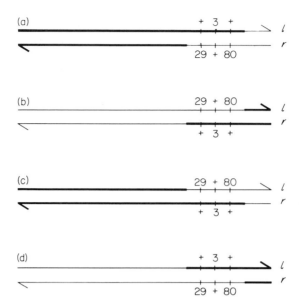

Fig. 41. Possible structures for recombinant DNA molecules in λ according to White and Fox (1974). The thick lines represent material derived from the heavy parent (*P*3) and the thin lines material from the light parent (*O*29 *P*80 double mutant). Half arrowheads represent 3′ ends. The strands of λ DNA are distinguished by the letters *l* (left) and *r* (right), indicating the direction of transcription from them.

In order to test this hypothesis, transfection experiments were carried out, that is to say, heteroduplex λ molecules were prepared *in vitro* and then allowed to infect host cells. DNA of λ of *P*3 genotype was denatured and the strands separated. The strand which is transcribed leftward on the standard map of the λ genome and which therefore has its 5′ end at the left-hand end is called the *l* strand. The other one, which is transcribed rightward and so has its 5′ end at the right-hand end of the map is called the *r* strand. The *l* strand is denser than *r* in alkaline caesium chloride. The technique which White and Fox used to separate the strands depends on the fact that the *r* strand binds guanine-rich ribopolymers about three times more than the *l* strand. On centrifuging in poly(U,G) caesium chloride the *r* strand is therefore denser than the *l* strand. DNA of the *O*29 *P*80 double mutant was likewise denatured and the strands separated. The *l* strands of one genotype were then mixed with the *r* strands of the other, and vice versa, under re-annealing conditions. In this way the two different kinds of heteroduplex were obtained *in vitro*. In separate experiments host cells were infected with each kind. Plaques containing wild-type phage were tested for the genotypes present.

It was found that the heteroduplexes containing the *l* strand of *P*3 and the *r* strand of *O*29 *P*80, in addition to bursts containing only wild-type phage, gave numerous mixed bursts containing wild-type phage and *P*3 phage, that is, they behaved like the mixed bursts from the heavy recombinants in the density-labelled cross. This similarity led White and Fox to conclude that the heavy bursts containing *P*3 arose from a recombination intermediate of the kind shown in Fig. 41(a). They supposed that mismatch repair (or random excision and repair) subsequently took place at sites 29 and 80 to give a wild-type homoduplex at these sites, so generating the wild-type recombinant strand that was being selected for.

The heteroduplexes in the transfection experiment that consisted of an *l* strand of *O*29 *P*80 genotype and an *r* strand of *P*3 gave rise, in addition to bursts that were wholly wild type, to others containing a mixture of wild-type phage and phage of *O*29 genotype. This was like the light recombinants in the density-labelled cross and led White and Fox to conclude that the latter arose from the structure shown in Fig. 41(b) followed by correction to wild type at sites 3 and 80.

Four conclusions can be drawn from the transfection experiments:

(1) Recombinants can be formed *in vivo* from the heteroduplexes prepared *in vitro*.

(2) Close double recombinants can arise from long heteroduplex segments, just as Russo and White and Fox had thought. The mechanism was likely to be mismatch repair, though neither random excision and repair nor limited replication were ruled out.

(3) This repair would explain the phenomenon of localized negative interference. The *P* mutants used by White and Fox were those with which Amati and Meselson (1965) had demonstrated the clustering of genetic exchanges (see Section 4).

(4) As predicted, only one of the two kinds of heteroduplex junction seems to occur in the formation of recombinant progeny when λ strains are crossed, and this was now shown to be the one with 3' ends (Fig. 41(a),(b)). The alternative with 5' ends was believed not to occur, as it would generate mixed bursts of kinds that were rarely found, namely, wild type and *P*3 of light density, and wild type and *O*29 of heavy density. This inference was based on the fact that, apart from the nature of the heteroduplex junction, the structures depicted in Fig. 41(a) and (d) differ only in density, and similarly with Fig. 41(b) and (c).

Several questions are left unanswered by White and Fox's experiments.

(1) Why do heteroduplex junctions with 5' ends (Fig. 41(c),(d)) not occur, or at least not give rise to progeny? The Stahls' hypothesis (Section 11) provides a possible answer. It is noteworthy that junctions with 3' ends, such as are found (Fig. 41(a),(b)), are those predicted by Cassuto and Radding (Fig. 37).

(2) Why are bursts heterozygous at *P*80 rarely found? In other words, according to the hypothesis of long heteroduplex segments and mismatch correction to explain the origin of recombinant progeny, why is correction to wild type at site 80 more frequent than at the sites of the other two amber mutants?

Further studies by White and Fox (1975*a*) approach this second question. They made crosses similar to those already described, λ DNA synthesis being doubly blocked and one parent being labelled with heavy atoms. But, whereas in the previous experiments the crosses were made on non-permissive bacteria, that is, amber suppressor-negative, so that only those phages that were wild type for the *O* and *P* genes yielded progeny, in the present experiments crosses were also made on permissive bacteria, that is, amber suppressor-positive. It was then necessary to identify plaques containing recombinant phage by screening.

The results obtained from single burst analyses of suppressor-positive infections differed markedly from the suppressor-negative results. Two crosses were used for this comparison: heavy *P3* x normal *P80*, and heavy *P3* x normal *O29 P80*. The genotypes of the bursts are given in Table 7 for a central fraction from the density gradient from each cross. The results from both crosses were similar and revealed a greater proportion of purely wild-type bursts with the suppressor-negative host. Moreover, the mixed bursts were not only relatively more frequent with the suppressor-positive host than the suppressor-negative, but differed markedly in the proportions of the different kinds. Examination of the products of transfection with artificially constructed heteroduplex molecules derived from the parents of the three-point cross, and using either suppressor-positive or supressor-negative cells, gave results similar to those shown in Table 7 (White and Fox, 1975*b*).

To account for the differences in relative frequency of the different kinds of recombinants on the two kinds of cells, White and Fox (1975*a*) suggested that the necessity, if progeny were to be obtained under suppressor-negative conditions, for a wild-type transcribed strand to arise in the *O* and *P* genes, might be responsible. They argued that in such cells hybrid DNA at the *O* and *P* mutant sites, consisting of an unaltered strand from each parent molecule, might persist for a substantial time, since it would be unable to replicate in the absence of wild-type *O* and *P* templates for messenger synthesis – the *r* strand in these genes. During this period there might be opportunity for mismatch correction. The greater proportion of unmixed wild-type bursts under suppressor-negative than suppressor-positive conditions might therefore be a consequence of more prolonged opportunity for the correction of mispairing at the mutant sites in hybrid DNA. The differences in the frequencies of the various kinds of mixed bursts between the suppressor-positive and suppressor-negative cells remain unexplained, but perhaps relate also to differences in opportunity for mismatch correction. Question (2) above thus remains unanswered.

When the initial infection was of suppressor-positive bacteria, the plaques selected for further examination, each containing only the progeny of a single infecting phage, were those shown by spot test to contain both wild-type alleles of the two *P* mutants in the two-point cross, or at least one wild-type allele from each parent in the three-point cross. Thus, on the suppressor-positive cells phage genotypes were recognized that could not be detected with a suppressor-negative host. Only recombinants with at least one wild-type strand can given progeny in suppressor-negative cells, so all the bursts include wild-type phage (Table 7).

It is of particular interest that among the mixed bursts lacking wild-type phage that could be recognized from the suppressor-positive crosses there were significant numbers corresponding to an unaltered heteroduplex at the mutant sites,

Table 7. Numbers of single bursts of various genotypes found by White and Fox (1975a) in a central fraction of the density gradient in crosses involving the *P3* mutant of λ carrying heavy atoms and the *P80* mutant or the *O29 P80* double mutant of normal density. The crosses were made on suppressor-positive and on suppressor-negative cells. A dash indicates that the genotype could not be recognized, that is, bursts lacking wild type on a suppressor-negative host.

Cross, and genotype of progeny	Host cells	
	Suppressor-positive	Suppressor-negative
Cross: $\dfrac{3\ +}{+\ 80}$		
$+\ +$	13	84
$\dfrac{3\ +}{+\ +}$	26	5
$\dfrac{+\ +}{+\ 80}$	15	0
$\dfrac{3\ +}{+\ 80}$	27	—
Cross: $\dfrac{+\ 3\ +}{29\ +\ 80}$		
$+\ +\ +$	11	30
$\dfrac{+\ +\ +}{29\ +\ +}$	1	9
$\dfrac{+\ +\ +}{29\ +\ 80}$	10	1
$\dfrac{+\ 3\ +}{+\ +\ +}$	0	7
$\dfrac{+\ +\ +}{+\ +\ 80}$	7	0
$\dfrac{+\ 3\ +}{+\ +\ 80}$	12	—
$\dfrac{+\ 3\ +}{29\ +\ +}$	17	—
$\dfrac{+\ 3\ +}{29\ +\ 80}$	13	—
$\dfrac{29\ 3\ +}{29\ +\ +}$	1	—
$29\ +\ +$	3	—
$\dfrac{29\ +\ +}{29\ +\ 80}$	1	—
$+\ +\ 80$	5	—
$\dfrac{+\ +\ 80}{29\ +\ 80}$	5	—

that is,

$$\frac{3 +}{+ 80} \quad \text{and} \quad \frac{+\ 3 +}{29 + 80}$$

in the two- and three-point crosses, respectively (Table 7). This provides direct support for the idea that such a molecule is of frequent occurrence in recombination in λ, the recombinants for mutants lying within the heteroduplex being presumed to arise by subsequent mismatch correction. The diversity of genotypes found among the single bursts (Table 7) and the high frequency of a number of them among the recombinants implies that mismatch correction — if that is the process resulting in homozygosity — is often confined to short regions of the hybrid DNA.

In the three-point cross *c*I and *R* mutants were also present and were found in the single bursts examined often to show both mutant and wild-type alleles. Moreover, most of the *R* heterozygotes were also heterozygous for *c*I and preserved the parental arrangements of the mutants, just as Russo (1973) had found. These observations agreed with his conclusion that the initial heteroduplex often extends throughout the *c*I–*R* interval. A non-permissive single burst analysis with an additional right-hand marker, in gene *Q* located between the *O P* region and *R*, revealed that 90% of the bursts that were heterozygous at *R* were also heterozygous at *Q*, but only 50% of those heterozygous at *Q* were also heterozygous at *R*. These results support the idea that the frequency of occurrence of hybrid DNA falls off with distance from the region where recombinants are selected (in this case, the *O P* region), and that the heteroduplex is continuous where it occurs.

White and Fox (1975*a*) emphasized that their results provided no test of the occurrence of single-strand insertion heterozygotes, that is, hybrid DNA associated with outside markers showing a parental genotype with one another (Fig. 32(b)). As already mentioned, it seems that, in order to produce mature phage, molecules longer than unit length are required from which unit lengths are cut for packaging. As pointed out in Section 11, Stahl and Stahl (1974) had suggested that when DNA synthesis is severely restricted, recombination between two λ genomes could provide an alternative to rolling circle replication for producing the molecule with two *cos* sites needed if packaging is to occur. Insertion heterozygotes, however, would not meet this requirement, and so would not be detected, as they would give no progeny.

In the replication-blocked λ crosses with one parent density-labelled there is a discrepancy between the inferences drawn from experiments with distant markers such as *A* and *R* at opposite ends of the λ genome (Section 11) and those with closely linked markers such as *O* and *P* mutants (this Section). With the distant markers the density distribution of the recombinants appears to reflect the distribution of exchanges along the genome (Fig. 38), with the implication that the position of the exchanges is well defined or, in other words, the heteroduplex segments are short relative to the total length of the λ genome. From the results with closely linked markers, however, it was concluded that the heteroduplex segments were long, occupying a substantial fraction of the genome. This conclusion was based not only on genetic data (heterozygosity at cI and R, with parental relationships between them, in single bursts containing recombinants at the intervening O and P sites) but also on molecular data (the even density distribution in the caesium chloride gradient of the O–P recombinants).

A possible explanation for the apparent discrepancy between the results with close and with distant markers is that the extent of the hybrid DNA segments in the recombinant molecules is highly variable. This possibility has already been raised in Section 11. In order to test this idea Stahl and Stahl (1976) made crosses involving close and distant markers simultaneously. They used the amber mutants 3 and 80 in gene P as close markers and temperature-sensitive mutants in genes A and R as distant markers, a cI mutant also being present. One parent was heavy ($^{13}C, ^{15}N$), the other and the culture being of normal density. The parents were

A	cI	+	$P80$	+	heavy
+	+	$P3$	+	R	light

The double block to replication was used, as before, that is, a temperature-sensitive host $dnaB$ mutant was present in addition to the $\lambda\, P$ mutants.

Stahl and Stahl (1976) confirmed the earlier results both for close and distant markers, and concluded that the occurrence of both short and long segments of hybrid DNA was a possible explanation. They found, however, that the recombinants that were wild type both for the close and distant markers included some that were nearly as heavy as the heavy parent. This parent had contributed the A and $P80$ mutants (see above). The simplest explanation of the origin of these recombinants was to suppose that the wild-type alleles of the widely spaced A and $P80$ mutants had been contributed by the light parent in separate events each involving only a short segment of DNA.

This possibility led the Stahls to make a triparental cross:

A	$O29$	+	$P80$	+	heavy
+	$O29$	+	$P80$	R	light
A	+	$P3$	+	R	light

such that although the selected markers (wild type for O and P) required recombination involving only two parents, to obtain in addition a wild type for the temperature-sensitive A and R mutants required the participation of a third parent. Such progeny were indeed found, and from their frequency Stahl and Stahl (1976) estimated that about half the close marker recombinants had undergone a second independent recombination event elsewhere in the genome. In the biparental cross about one-quarter of the close recombinants that were wild type for A and R would then owe this to a second event. Such a frequency would account for those that were nearly as heavy as the heavy parent. The Stahls concluded that likewise about one-quarter of the occurrences of simultaneous heterozygosity at cI and R in conjunction with recombination at O and P found by Russo (1973) and White and Fox (1974, 1975a) (see above) were not caused by a single stretch of hybrid DNA extending from cI to R but by two independent recombination events, one at R and the other in the cI–O–P region. Evidently the frequency of long heteroduplex segments was lower than these authors had inferred. The Stahls suggested that the double events might include insertion heterozygosity (Fig. 32(b)) at the site of the close markers and a crossover (Fig. 32(a)) elsewhere. As already pointed out, insertion heteroduplexes alone will not be detected as they will not give rise to mature phage in a cross where replication is blocked.

13. MISMATCH REPAIR

The first indication that mispaired bases in λ heteroduplex DNA might be removed by an enzyme system was obtained as an incidental consequence of an investigation with a different objective. Hogness et al. (1967) were trying to discover the orientation of gene N in the λ genome. Their method took advantage of the difference in molecular weight of the two strands of λ DNA. After denaturing the molecules, the heavy and light strands were separated by centrifuging to equilibrium in an alkaline caesium chloride gradient. The separated strands of wild-type λ and of an amber mutant in gene N were mixed in the appropriate pairs to give, under renaturing conditions (pH 10.5 and 37 °C), the two kinds of heteroduplexes, that is, the wild-type heavy (l) strand paired with the mutant light (r), and vice versa. It was anticipated that, on introducing the heteroduplex molecules into host cells, gene N activity would be manifest only if the transcribed strand (that is, the one that functions as template in the synthesis of messenger RNA) was of wild-type genotype. Since activity of gene N is essential for replication of λ DNA, only the appropriate kind of heteroduplex would produce progeny phage. It was found, however, that both heteroduplexes had about half the activity of the wild-type homoduplex.

Hogness et al. (1967) attributed this result to mismatch correction in the heteroduplex DNA, inferring that it occurred about equally often in each direction, that is, excision of the wild-type mismatch and its replacement by the mutant base sequence complementary to the other strand of the heteroduplex, or conversely excision of the mutant base sequence. In order to test this hypothesis, they irradiated the host cells with ultraviolet light immediately before introducing the heteroduplex λ. The assumption was that the ultraviolet radiation would cause such damage to the host DNA that the repair enzyme activity would be saturated and so would not be available to correct the mismatch in the λ heteroduplexes. In agreement with this hypothesis, Hogness et al. found that heteroduplexes with a wild-type l strand were active, but those with a wild-type r strand were not. They inferred that the l strand of gene N is the one that acts as template in transcription.

But how far is their conclusion justified that mismatch correction provides the explanation for the differential behaviour of the two heteroduplexes in irradiated cells? In an attempt to eliminate the possibility that slight activity of gene N in the mutant allowed sufficient replication to occur to generate a wild-type template from either heteroduplex, Doerfler and Hogness (1968) repeated the experiment using heteroduplexes in which two amber mutants were present simultaneously in one strand. Results similar to the earlier experiment were obtained, though the activity of each double mutant heteroduplex in irradiated cells was below the value of 25% of the wild-type homoduplex expected if the direction of mismatch correction (that is, to wild type or to mutant) at the one site were independent of that at the other. For activity, correction would have to be to wild type at both sites. In any case, to explain the results through leakiness of the block to gene N activity, this leakiness would have to be sensitive to ultraviolet irradiation of the host, since the wild-type r / mutant l heteroduplex is active only in unirradiated cells.

The results obtained by Russo (1973) and White and Fox (1974, 1975a) from crosses when one parent was density-labelled, and replication was blocked, pointed to mismatch correction in hybrid DNA segments of variable length as the recombination mechanism, but other means of deriving a homoduplex from a heteroduplex, such as random excision and repair or limited replication, were not ruled out (Section 12).

Direct evidence for the occurrence of mismatch repair was obtained by Wildenberg and Meselson (1975). Artificially constructed heteroduplex molecules of λ, heterozygous for amber mutants, were used to infect $E.$ $coli$ under conditions where conventional recombination was minimized by using $recA^-$ bacteria and int^- red^- phages. Eight amber mutants were studied: no. 63 in gene L, no. 2 in gene I, nos 53 and 96A in gene N, no. 8 in gene O, nos 3 and 80 in gene P and no. 73 in gene Q. Transfection was carried out with a randomly annealed mixture of equal quantities of DNA of amber mutants $L63$ and $I2$, and similarly with all the other pairwise combinations of the mutants. For each pair the infecting DNA would thus consist of equal amounts of four kinds of molecules, namely, the two parental homoduplexes and the two kinds of heteroduplexes (the l strand of one mutant with the r strand of the other, and vice versa).

The frequencies of wild-type phages in the transfection lysates ranged from 0.2% for $N53/N96A$ to 9.9% for $I2/P80$. Since half the infecting molecules were homoduplexes the actual frequencies of wild type resulting from heteroduplex transfection were twice these values. From the wild-type frequencies in the 28 different pairs, the repair frequencies at the eight individual mismatched sites were calculated on the following assumptions: (1) that the two mismatched sites in each heteroduplex are repaired independently; (2) that repair to wild-type and to amber are equally likely; (3) that descendants of both strands of a transfecting DNA molecule usually contribute to the progeny phage and, on average, to an equal degree. The calculated frequencies of repair to wild type ranged from 2.2% for $N53$ to 20.7% for $I2$, with no relationship apparent between the frequency and the gene in which the mutant was located or its position in the genome. The frequencies, therefore, seemed to be characteristic of the individual mutants.

For each pair of widely spaced mutants the frequency of wild types from the heteroduplex transfections was in good agreement with the sum of the two individual repair frequencies, as calculated. This agreement implied that the repair at the two sites was occurring independently. For mutants less than 4000 nucleotides apart, however, the observed frequency of wild types was less than the sum of the individual calculated frequencies, as expected if simultaneous repair sometimes occurred, that is, a strand was excised stretching from one site to the other. There would then be correction to wild-type at one site and to mutant at the other, and no overall wild-type would be generated.

In order to obtain more precise information about the length of DNA strand excised in mismatch repair, Wildenberg and Meselson (1975) prepared heteroduplex DNA with mismatches at $P3$ and $P80$ and also at the flanking sites cI and mi. If excision generally extends for 2000 nucleotides, which is the approximate distance from $P3$ to cI, transfection with these molecules and selection for wild type at P would increase the frequency of non-parental $cI\,mi$ combinations compared with unselected progeny. Only a small increase was found. It was concluded that repair tracts are usually shorter than 2000 nucleotides. Three-quarters of the wild types for P had a parental genotype for the flanking markers, and the two non-parental genotypes were about equally frequent in the remaining quarter. This is a pattern expected for mismatch repair, with the non-parentals attributed to repair at one or other flanking marker site in addition to that selected for at P.

Further information about mismatch repair in artificially prepared λ heteroduplexes was obtained by Wagner and Meselson (1976) by studying separately the two kinds of heteroduplex, that is, the l strand (L in their terminology) of one genotype annealed with the r strand (H in their terminology) of the other, and vice versa. Two genotypes were investigated:

$$\frac{+\ cI\,P\,mi}{N+\ +\ +} \quad \text{and} \quad \frac{h\,N\,O\,P}{+\ +\ +\ +},$$

where h is a host-range mutant, cI and mi are mutants with clear and minute plaques, respectively, and N, O and P are amber mutants nos 53, 8 and 80 in these genes, respectively. The approximate intervals in nucleotides between the mutant sites (from left to right) are 2000, 1800, and 8500 for the first cross, and 12 000, 3500, and 300 for the second. Recombination (other than mismatch repair) was prevented, as in the previous investigation, by using int-negative red-negative DNA strands and $recA$-negative host cells. With the first genotype, transfected cells were plated on a suppressor-positive $recA$-negative culture before lysis, giving rise to infective centres. Phages from individual centres were then tested for genotype. With the second genotype, the wild-type λ used as one parent overgrew the other in infective centres, so a single burst analysis was carried out in liquid culture.

Wagner and Meselson (1976) found that about 60% of the infective centres and single bursts yielded phages of only one genotype and the remainder yielded two genotypes. The relative frequencies of the various genotypes were the same among the single yielders as among the double yielders. Wagner and Meselson concluded that the single yielders arose, not from repair at every mismatch, but from failure of the second strand to produce progeny phage. They considered that this conclusion probably also applied to data of others, such as the pure bursts found by White and Fox (1975*b*) in their transfection studies, which therefore needed to be reinterpreted.

On the assumption that repair occurred independently at each site, the repair frequency at each mismatch was calculated separately for each kind of heteroduplex. The results are given in Fig. 42. It is evident that there is considerable variation in the frequency of repair to wild type and to mutant, not only between one mutant and another, but also between the two kinds of heteroduplex for each mutant. The latter variation was to be expected because with substitution mutants the nature of the mismatch will differ between the two: for example, a $G \cdot C$ to $A \cdot T$ transition will have G mispaired with T in one heteroduplex and A with C in the other.

The assumption of independent repair at each site was tested by comparing the observed frequency of each type of double yielder with that calculated from the product of the frequencies of repair or no repair at each site. It was found that this assumption was incorrect: between a half and three-quarters of all the double yielders in which there had been repair at N and P showed it also for the intervening site (cI or O) and with the repair to the same parental genotype at all three. This is the pattern expected if, in repair, there is extensive excision of one strand initiated at a mismatch.

Moreover, in both experiments the two kinds of heteroduplex showed consistent differences in the frequencies of different genotypes among the progeny phage. Using

$$\frac{A\ B\ C}{a\ b\ c}$$

to represent the parental genotypes for the most closely linked trio of mutants in each experiment, that is,

$$\frac{+\ cI\ P}{N\ +\ +} \quad \text{and} \quad \frac{N\ O\ P}{+\ +\ +},$$

respectively, it was found that the *A b c* and *A B c* genotypes were more frequent, both in single and double yielders, when *A B C* was the *l* strand than when it was the *r* strand, and conversely the *a b C* and *a B C* genotypes were more frequent when *A B C* was the *r* strand than when it was the *l* strand (Fig. 43). These results are explicable on the assumption that excision initiated at the central mismatch commonly extends to one of the neighbouring sites but not to both, and that the direction of this extension is non-random. If excision is in the 5′ to 3′ direction,

156

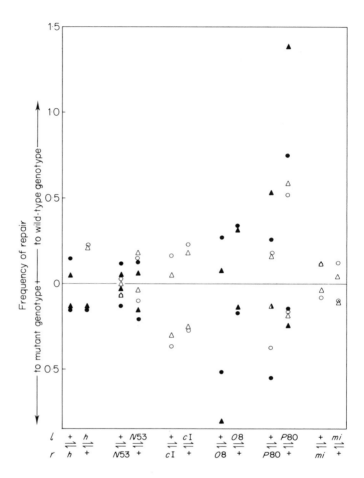

Fig. 42. Results obtained by Wagner and Meselson (1976) for mismatch repair follow-ing transfection of *E. coli* with heteroduplex molecules of phage λ prepared *in vitro*. In one experiment (○, △) the λ genotype was

$$\frac{+ \, cI \, P \, mi}{N + + +}$$

and in the other (●, ▲) it was

$$\frac{h \, N \, O \, P}{+ + + +}.$$

For each genotype the two different heteroduplexes (the *l* strand of one genotype annealed with the *r* strand of the other, and vice versa) were studied separately. Half arrowheads are at the 3′ termini. ○, ●, Repair frequency at each mismatch calculated on the assump-tion that repair occurs independently at each site. △, ▲, Repair initiation rates obtained from computer simulations on the assumption that excision is initiated at a mismatch and proceeds in the 5′ to 3′ direction.

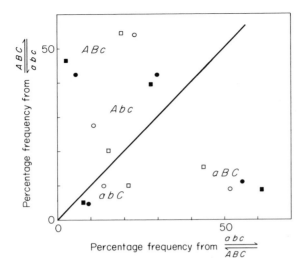

Fig. 43. Data of Wagner and Meselson (1976) for the frequencies of different genotypes in phages from infective centres or single bursts arising from transfection of *E. coli* with heteroduplex λ. Four genotypes, *A B c*, *A b c*, *a B C*, and *a b C* were studied from heteroduplexes of genotype

$$\frac{A\ B\ C}{a\ b\ c}.$$

The percentage frequency of each (out of the total for the four) derived from a heteroduplex with *A B C* in the *l* strand and *a b c* in the *r* strand is plotted against the corresponding frequency for the other heteroduplex (*a b c* in the *l* strand and *A B C* in the *r* strand). ○, □, *A* = +, *a* = *N*; *B* = *c*I, *b* = +; *C* = *P*, *c* = +. ●, ■, *A* = *N*, *a* = +; *B* = *O*, *b* = +; *C* = *P*, *c* = +. ○, ●, Single infections giving only one genotype. □, ■, Single infections giving two genotypes.

that is, the direction of the half arrowheads below, then in the

$$\frac{1}{r}\ \frac{A\ B\ C}{a\ b\ c}$$

heteroduplex, extensive excision initiated at *B* and followed by synthesis complementary to the other strand will give rise to a strand of *A b c* genotype. Similarly, excision at *b* will give rise to *A B c*. The corresponding events in the

$$\frac{1}{r}\ \frac{a\ b\ c}{A\ B\ C}$$

heteroduplex will generate *a b C* and *a B C*. These predictions correspond to the results observed. Excision in the opposite direction would give the converse bias. Wagner and Meselson (1976) concluded that excision in mismatch repair in phage λ takes place, at least predominantly, in the 5′ to 3′ direction.

On the assumption that excision begins at a mismatch and extends only in this one direction, computer simulation was used to estimate the average length of strand excised and the frequency with which repair was initiated at each site. Separate computations were made for each adjacent pair of markers, the period of time during which repair can occur being divided for the purpose of the calculations into 100 small intervals such that the occurrence of two repair events in one time interval could be neglected. Equations could then be given for the frequencies of every possible combination of repair to wild type, repair to mutant, or lack of repair at each of the two sites in each kind of heteroduplex in terms of the frequency of initiation of repair at each site, and of the fraction of events that extended (in the 3′ direction) at least as far as the other marker of the pair. For each pair, the value obtained for this fraction by the computer analysis was plotted against the distance between the sites. The average value for the excision length obtained in this way was about 3000 nucleotides – a larger figure than the earlier estimate. The initiation rates for repair to wild type and to mutant at each site in each kind of heteroduplex, obtained from the same analysis, are shown in Fig. 42. The values resemble the repair frequencies calculated previously on the assumption that repair occurred independently at each site. This agreement was to be expected since several of the site intervals were long.

Further transfection experiments with λ heteroduplexes are discussed in Chapter VIII.

14. SUMMARY

Study of the mechanisms of recombination in phage λ has proved difficult. This is partly because of the existence of the host Rec and the λ Red and Int systems all acting on λ DNA. A further complication in the analysis of Red-mediated recombination is the restriction imposed on the formation of progeny by the need for the λ DNA that is packaged in the protein head of the virus to be cut from a dimeric or polymeric molecule containing more than one genome in a linear array.

Recombination caused by the Red system has been shown to involve breakage and joining of parental molecules and to be non-reciprocal. The process involves λ exonuclease and its associated β-protein. The exonuclease erodes the 5′ ends of strands, both in duplex DNA and also where there is a 3′-ended redundant strand, which falls into place as the 5′ end is broken down, to leave only a nick. How far the recombination pathway suggested by this *in vitro* behaviour occurs *in vivo* is not known. There is evidence, however, for the occurrence of heteroduplex segments of variable length, sometimes of many thousands of nucleotides, and the longer strand from each parent is 3′-ended, as predicted if the other has been eroded by λ exonuclease. Mismatch correction at the site of mutations in the hybrid DNA has been demonstrated, and its frequency seems to be a characteristic of the individual mutants. For any particular mutant the frequencies of repair to wild type and to mutant are often unequal and, moreover, differ according to whether wild type or mutant is in the transcribed strand of the DNA. Excision in mismatch repair often extends for several thousand nucleotides and occurs, at least predominantly, in the 5' to 3' direction. Mismatch repair will explain, at least qualitatively, the occurrence of localized negative interference.

The relationship between Red-mediated recombination and replication is complex: at least in the middle of the genome recovery of progeny showing recombination by the Red pathway is dependent on DNA replication. Conversely, replication is slow in *red*-deficient mutants. It has been suggested that the switch in replication from simple circle to rolling circle, necessary for the production of mature virus particles, is triggered by a nick in one of the strands of the circle. Depending on which strand (in relation to the replication fork) has been nicked, the host DNA polymerase and ligase or the phage Red system may generate the rolling circle. The need for appreciable DNA replication when the recombination occurs in mid-genome has been attributed to the necessity for an entire λ genome, stretching from one *cos* site to the next, to lie within the tail of the rolling circle, before the DNA can be packaged in the phage head. Little is known about the occurrence of insertion heteroduplexes in λ since they do not produce a dimeric molecule and so cannot easily be detected.

The integration of the λ genome into that of *E. coli* involves site-specific recombination in which the product of the *int* gene plays a central part. There is a 15-nucleotide sequence common to the bacterial (B) and phage (P) attachment sites, with unique sequences on each side. In consequence, reciprocal exchange in the common region generates new combinations of the unique sequences, the left (L) and right (R) attachment sites of the prophage. Excision of the prophage from the host genome requires a small additional polypeptide, the product of the *xis* gene. In conjunction with the Int protein the L and R attachment sites are then believed to be recognized, where the Int protein alone fails to do so.

Study of integration and excision has been facilitated by the development of *in vitro* techniques that take advantage of special strains of λ containing the B and P and the L and R attachment sites, respectively. An unexpected outcome of this work has been the discovery of the need for supercoiled DNA if integration is to take place.

Recombination in λ is of particular interest because this phage possesses two such dissimilar mechanisms as Int and Red, the one reciprocal and site-specific, the other non-reciprocal and not restricted as to site. Nevertheless, the mechanisms share breakage and reunion. Many details of each process have yet to be resolved.

VI.   Recombination in *Escherichia coli*

1. INTRODUCTION

Genetic recombination was discovered in *Escherichia coli* by Lederberg and Tatum (1946). They used a strain called K-12 that had been isolated in the USA in 1922 from human faeces. Mutants lacking the ability to synthesize growth factors had been obtained by treatment with X-rays or ultraviolet light. By successive treatments, strains with several requirements had been obtained. Two triple-mutant strains were grown in mixed culture. One required threonine, leucine and thiamin, and the other required biotin, phenylalanine and cystine. Cells from the mixed culture were inoculated into synthetic medium to which various supplements had been added. About one cell in 10^6 was found to be a recombinant. The recombinants included wild-type strains with no growth factor deficiencies, single mutants requiring only thiamin or phenylalanine, and double mutants deficient in the synthesis of biotin and of one of the requirements of the other parent. Lederberg and Tatum concluded that a sexual process occurred in *E. coli*, since attempts to induce transformation, using sterile filtrates, had been unsuccessful. Because of the low frequency of recombination, use was made in later studies of selective techniques: wild-type recombinants of auxotrophic mutants could be selected by plating on minimal medium, and recombination frequencies observed of unselected markers such as sensitivity or resistance to specific drugs or bacteriophages. By plating on minimal medium containing appropriate supplements, auxotrophic mutants could be included in the unselected category.

It was established through the work of Lederberg, Cavalli and Lederberg (1952), Jacob and Wollman (1958), Hayes (reviews, 1966a,b, 1968) and others, that the transfer of DNA from one cell to another could occur by conjugation. The DNA is believed to pass from donor to recipient through filamentous appendages called pili. Agitating a culture during conjugation can cause separation of conjugating cells. By varying the time of agitation in relation to the start of conjugation, the time of transfer of specific genes can be determined. From recombination frequencies a single linear linkage group was established. A gene map based on the time of transfer in conjugation was found to agree with the map based on recombination frequencies. The conjugation is associated with the presence of a fertility (F) factor which confers donor properties on a cell, so F+ strains are donors and F– recipients. In F+ strains the F factor exists in the cytoplasm of the cell and is transferred to F− by infection. The F factor has been shown to be a DNA molecule long enough to carry 100 or more genes, and in strains showing high frequency recombination (Hfr) this DNA molecule is covalently joined to that which forms the *E. coli* genome. The F+ and Hfr states arise from one another with low frequency. Recombination frequencies using Hfr strains indicated a circular linkage map, but in these strains the genome is transferred as a linear oriented structure to F− cells by conjugation. The part of the DNA molecule which enters first, and the direction in which the molecule enters, differ from one Hfr strain to another. It appears that the behaviour in conjugation depends on where the F factor becomes integrated into the circular *E. coli* genome (Fig. 44). It appears, moreover, that there are several sites on the F factor where integration into the *E. coli* genome can take place (see Deonier and Davidson, 1977). As a result of integration the circular genome becomes linear, with part of the F factor covalently joined to each end. Through random breakage

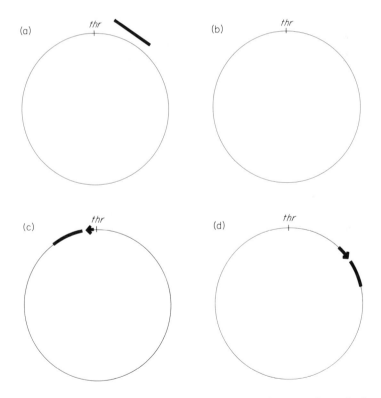

Fig. 44. Genomes of donor and recipient strains of *E. coli*. The F factor is shown by a thick line. The *threonine* (*thr*) locus is the mapping origin, being the first known gene transferred by Hfr H. (a) F+ strain, with the F factor in the cytoplasm. (b) Recipient strain, lacking an F factor. (c) Hfr Hayes and (d) Hfr Cavalli, showing the position at which the F factor is integrated in the circular genome, and the direction of transfer in conjugation (arrow).

during transfer, genes near the conjugation terminus, together with the tail end of the F factor are transferred only rarely.

Parts of the *E. coli* genome can be transferred from one cell to another by means of bacteriophages. In generalized transduction, a phage-sized piece of the *E. coli* DNA is incorporated in the protein envelope of the phage in place of the phage DNA. In this way any part of the host genome may be transferred. In specialized transduction host genes located close to the site of integration of the phage genome may be transferred, as a result of a reciprocal exchange between a site in the prophage and a nearby region of the host genome. Such an exchange would detach a circular DNA molecule containing part of the phage genome and neighbouring parts of the *E. coli* genome. In a similar way, when the F factor is detached from the *E. coli* genome of an Hfr strain as a circular molecule it can carry with it neighbouring bacterial genes as a result of a reciprocal exhange in an abnormal position. An F factor carrying such genes is called F-prime (F').

2. DEGREE OF RECIPROCITY IN RECOMBINATION

Scaife and Gross (1963) concluded that reciprocal exchange took place between an F' factor and the homologous region of the *E. coli* genome before the DNA was transferred in conjugation. This exchange would integrate the circular DNA molecule of the F' into the genome. Their evidence for this exchange was based on genetic studies with an F' carrying a mutation in the lactose operon of *E. coli*. The sequence of entry into the recipient cell of the lactose mutant and of the wild-type allele from the *E. coli* genome was studied. The sequence would depend on the position of the exchange relative to that of the mutant.

Support for Scaife and Gross's crossover model was obtained by Pittard and Adelberg (1964) using an F' that carried a number of genetic loci concerned with the synthesis of methionine, arginine and isoleucine–valine. Furthermore, from the recombination frequencies they inferred that the crossover between the F factor and the genome, necessary for the cell to function as a donor in conjugation, could occur with equal probability anywhere in the homologous region, that is, the segment of bacterial DNA present in the F' and also in the genome (though the two differed by several mutations). On the other hand, after conjugation, a short region at the leading end of the transferred material showed a 36-fold increase compared with the remainder in the frequency of recombination between donor and recipient. Pittard and Adelberg (1964) suggested that recombination was favoured at free ends of the DNA, but this end effect was not shown by the recombination between the F' and the genome because the F' was a closed circular DNA molecule with no free ends.

Evidence that recombination between an F' and the genome was reciprocal was obtained by Herman (1965) using an F' that carried the lactose (*lac*) operon with a mutation in the galactoside permease gene *lacY*. The *E. coli* genome had a polar mutant in the structural gene, *lacZ*, for β-galactosidase. This gene adjoins the promoter of the operon with the result that *lacY* was also inactivated. The cells were therefore unable to ferment lactose, but lac^+ segregants arose as a consequence of recombination between the sites of the two mutants:

$$\left.\begin{array}{ll} \text{F}' & Z^+\,Y^- \\ \text{Genome} & Z^-\,Y^+ \end{array}\right\} \text{ giving rise to } Z^+\,Y^+.$$

Altogether, 19 such lac^+ segregants were investigated. Each was treated with acridine orange, which is known to remove detached DNA molecules such as F′ factors: the process is called curing. Eleven of the 19 strains were found then to be lac^-. It was inferred that in these 11 the F′ had the genotype $Z^+\,Y^+$. The Y and Z genotypes of the *E. coli* genome were then tested and seven of the 11 proved to have the genotype $Z^-\,Y^-$, implying that reciprocal recombination had occurred between F′ and genome (Fig. 45). A complication arises, however, from the fact that if the F′ is circular, a reciprocal exchange between it and the genome will integrate the one into the DNA of the other. To explain, therefore, how the strain could nevertheless be cured by acridine, that is, the F′ be removed, Herman postulated that a second reciprocal exchange took place to detach the F′ again. The two reciprocal exchanges would need to be at different positions, only one of them being between the sites of Z and Y, to give the genotype observed in the seven lac^+ segregants in question.

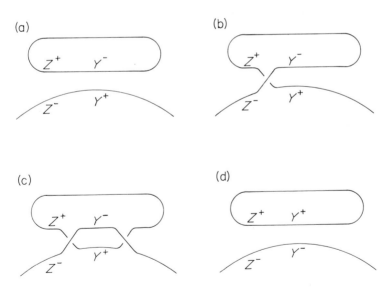

Fig. 45.   Diagram to illustrate the hypothesis of Herman (1965) of reciprocal recombination between an F′ and the *E. coli* genome. (a) The F′ carries a mutation in *lacY* and the genome a mutation in *lacZ*, so the cell is unable to utilize lactose. (b) A reciprocal exchange between the sites of the Y and Z mutations generates a wild-type lactose operon, but also integrates the F′ into the genome. (c), (d) A second reciprocal exchange, outside the Y–Z interval, detaches the F′ again.

Further evidence that recombination between an F′ and the *E. coli* genome is often reciprocal was obtained by Meselson (1967*b*). He used an F′ that carried a wild-type prophage λ and an *E. coli* genome that also carried prophage λ, but with three widely spaced mutants: *c* (clear plaque, due to a mutation in the *c*II gene), *mi* (minute plaque) and *h* (host range), mapping in the order quoted. Free phages derived from the prophages in both the F′ and the genome are spontaneously liberated at a low frequency during the growth of the bacteria. So the genotype of the two prophages could be determined simply by observing the plaque type produced by the phages associated with a culture. A total of 5735 unselected cells was analysed in this way. About 90% had the two parental prophage genotypes (*c mi h* and + + +). Among the recombinants, there were 36 with a reciprocal exchange between *c* and *mi* (prophage genotypes *c* + + and + *mi h*) and 68 with a reciprocal exchange between *mi* and *h* (prophage genotypes *c mi* + and + + *h*). If the reciprocal recombinants were arising by two independent non-reciprocal events that happened to occur in the same interval, one would expect similar numbers of occurrences where the two non-reciprocal events were in different intervals. This would generate the pairs *c* + + with + + *h* and *c mi* + with + *mi h*, but only eight of each of these were observed. Meselson concluded that recombination between genome and F′ of relatively well spaced mutants was often reciprocal.

Sarthy and Meselson (1976) investigated the extent to which recombination in *E. coli* was reciprocal by studying single bursts from three-point crosses of phage λ. The phage recombination systems (Red and Int) were blocked by making use of mutants defective in these pathways. The phage markers used were *h*, *c* and *mi*, as before, though the clear-plaque mutant used in this experiment was in the *c*I gene. The parental genotypes were

$$\frac{h\ c\ mi}{+\ +\ +}.$$

As indicated, the map sequence in the phage is a circular permutation of that in the prophage described above. The degree of reciprocity was found by analysing the correlation in the yield of reciprocal recombinant types from single bursts. A highly significant association was found between the occurrence of reciprocal recombinant types, that is, *h* + + with + *c mi*, and *h c* + with + + *mi*, indicating that recombination was largely or entirely reciprocal. Non-reciprocal pairs such as *h* + + with + + *mi* showed no such correlation. The double recombinants, *h* + *mi* and + *c* +, showed a pronounced excess, as measured by the coefficient of coincidence, which was much larger than the coefficient of heterogeneity. They inferred that the double recombinants were usually not the result of two independent events. With the double recombinants, the association between the reciprocal types, though significant, was not as strong as the association of reciprocal singles. Sarthy and Meselson favoured the model proposed by Meselson and Radding (1975) as an explanation, as this model predicts processes capable of producing not only reciprocal pairs of single and double recombinants but also an excess of double recombinants unaccompanied by their reciprocals (see Chapter VII, Section 14).

3.   RECOMBINATION FOLLOWING CONJUGATION

Evidence that the DNA of the recipient genome is directly incorporated into recombinant molecules, following conjugation, was obtained by Siddiqi (1963). The DNA of the recipient strain was labelled with tritiated thymidine. The donor DNA was unlabelled but carried a genetic marker: a mutation conferring resistance to phage T6, the recipient being sensitive to this phage. In addition, the donor was sensitive to streptomycin, the recipient being resistant. One hour, at 37 °C, after mixing the two strains, streptomycin was added to prevent multiplication of the donor strain. At various times up to 10 h after the initial mixing, samples of the culture were treated with phage T6. This caused breakdown of the T6-sensitive cells. It was found that the T6-resistant cells were radioactive. This showed that tritium atoms from the DNA of the recipient cells had entered the recombinants. Sensitivity to T6 is dominant over resistance, so for a cell to become resistant it is necessary for the sensitivity gene to be lost from the genome of the recipient cell and the resistance gene integrated into the genome in its place. The experiment therefore demonstrated the association of new genetic information with the old DNA. Furthermore, Siddiqi estimated that the T6-resistant recombinants constituted 5–6.5% of the recipient cells and contained 4–5% of the total tritium activity. These frequencies implied that in those cells that had become recombinant, more than 80% of the DNA of the recipient parent had been transmitted to it. From genetic evidence the donor contribution is thought often to be substantially less than half, in agreement with this molecular evidence.

Bresler and Lanzov (1967) designed an experiment complementary to Siddiqi's. They labelled the DNA of the donor with ^{32}P. The recipient was sensitive to streptomycin and the donor carried a recessive mutant conferring resistance to streptomycin. Resistant progeny, following conjugation, would therefore be true recombinants and not heterozygotes. Bresler and Lanzov were able to demonstrate the presence of parental donor DNA in the recombinants, selected by their resistance to streptomycin, by studying the kinetics of the decline in number of recombinants with time. This decline was exponential and its rate indicated that it was caused by cell inactivation through radioactive decay of ^{32}P.

Siddiqi's demonstration of parental recipient DNA and Bresler and Lanzov's of parental donor DNA, in the recombinant molecules, establishes that recombination following conjugation is by breakage and joining of parental contributions.

An investigation of recombination following conjugation was made by Verhoef and de Haan (1966) using a streptomycin-sensitive donor. The recipient strain was streptomycin-resistant and carried five auxotrophic mutants giving requirements for proline (*proA*), leucine (*leuA*), threonine (*thr*), adenine (*purA*), and arginine (*argH*), respectively. The mutations mapped in the order given (see Fig. 46), starting with the gene nearest to the conjugation origin in the donor strain used. One hour after mixing the two strains at 37 °C, the mating was

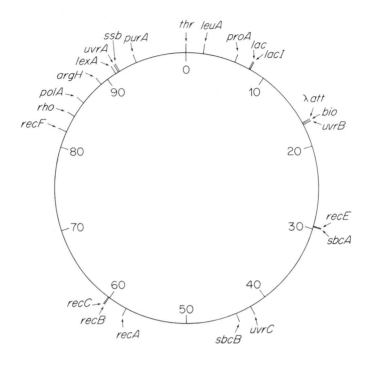

Fig. 46.   Genetic map of the *E. coli* genome, based on that of Bachmann and Low (1980). The map is derived from the time of transfer during conjugation at 37 °C shown by the original donor strain (HfrH) that was used for mapping purposes. The first marker to be transferred by this strain was the threonine (*thr*) locus, which is therefore allotted time zero on the map. The whole genome is transferred in 100 min.

interrupted by violently shaking the culture. Recombinants were selected for the most distal marker (arginine) by plating on minimal medium supplemented with streptomycin and with all the growth factors except arginine. All the auxotrophic markers other than arginine were therefore unselected, but nevertheless must have entered the recipient cell as their sites were nearer the conjugation origin than the selected marker. It was found that the frequency with which the *arg*$^+$ recombinants carried the unselected donor markers declined with the distance of the marker from *argH* (45% *purA*$^+$, 35% *thr*$^+$, 33% *leuA*$^+$, 27% *proA*$^+$). Using estimates of the distances between the markers based on their time of transfer (see Fig. 46), good quantitative agreement was obtained between the experimental results and those expected if there is random breakage of the DNA. Proximal and distal segments showed the same breakage probability per unit length. This implied that recombination did not begin until after DNA transfer had been completed, for otherwise proximal segments would show more breaks.

In contrast to this evidence for random breakage and joining of widely spaced markers, clusters of exchanges in short intervals were observed by Cavalli-Sforza and Jinks (1956) and Jacob and Wollman (1961) following conjugation, and by Gross and Englesberg (1959) following generalized transduction by phage P1. Maccacaro and Hayes (1961) found that the clustering of exchanges, that is, negative interference between exchanges, declined with distance between the mutant sites, but was still detectable when the separation was about 10^6 nucleotides. They suggested that this negative interference extending over long distances might be accounted for if, as soon as pairing between donor and recipient DNA has occurred in any region, the probability of pairing at any other region is no longer random but is increased in inverse proportion to its distance from the first region. Multiple events within a pairing region, postulated to account for negative interference over short intervals, in conjunction with the pairing interaction outlined above, would then account for negative interference over more extended intervals.

The molecular events following conjugation were studied by Oppenheim and Riley (1966) using density-labelled (^{2}H and ^{15}N) recipient cells and radioactive labelling (^{3}H-labelled thymine) of the donor DNA. After mating, DNA was isolated at successive times and fractionated by density-gradient centrifugation in caesium chloride solution. In this way a recombinant fraction was isolated which contained both donor and recipient labels. Samples of this fraction were investigated further, some by sonication to fragment the molecules and others by denaturation by heat or alkali to separate the strands. The products of these treatments were examined in caesium chloride or sucrose gradients. Oppenheim and Riley (1966) found that at least some of the recombinant molecules contained both strands of the recipient parent, both being replaced by donor material at intervals. These duplex donor insertions were revealed by the ^{3}H-labelling and occurred as several small segments near together. The evidence for duplex insertions of donor material was the density of sonicated fragments, and the inability of exonuclease I, which acts only on simplex DNA, to degrade them.

Genetical studies by Wollman, Jacob and Hayes (1957), Jacob and Wollman (1961) and Maccacaro and Hayes (1961) had indicated extensive donor material in recombinant progeny. Oppenheim and Riley (1966) pointed out that their technique would not have detected such extensive insertion. They suggested that the region of alternating donor and recipient material that they had detected might be near the end of an extended donor insert.

Oppenheim and Riley (1966, 1967) separated the recombinant DNA from parental DNA by repeated density-gradient centrifugation in caesium chloride. The ^{3}H-labelled and density-labelled DNA segments appeared to be joined mainly by hydrogen bonds. Dilution with fresh culture medium, however, after conjugation, triggered covalent union of DNA segments, as revealed by denaturation and further centrifugation. Oppenheim and Riley concluded that covalent joining takes place after the non-covalent bonding of the donor and recipient segments.

Results contradicting those of Oppenheim and Riley (1966) concerning the donor DNA in recombinant cells were obtained by Piekarowicz and Kunicki-Goldfinger (1968). They labelled the donor DNA radioactively, using [^{14}C]-

thymine. At 30 min and 60 min after mixing the parent strains, cells were harvested, the DNA extracted and fractionated in columns of methylated albumin-coated kieselguhr. A labelled DNA fraction was found that was not present in a control experiment using non-conjugating cells. This fraction was released from the columns only by elution with hot 1.5 M saline or by raising the pH to 10. These are results expected with high molecular weight single-stranded DNA. The simplex nature of this DNA, known from its label to be of donor origin, was confirmed by a control experiment with non-conjugating cells to which heat-denatured DNA was added. This added fraction behaved like that observed following conjugation. In conjugation experiments in which ^{14}C was present in the recipient DNA instead of the donor, no labelled single-stranded fraction was observed. Piekarowicz and Kunicki-Goldfinger (1968) concluded that the single-stranded DNA was derived exclusively from donor cells. At the same time as this fraction appeared, there also appeared a soluble fraction derived from donor DNA, pointing to degradation of the second strand of the donor DNA. Piekarowicz and Kunicki-Goldfinger suggested that Oppenheim and Riley's failure to find single-stranded DNA was because they examined the DNA at a later time: 2 h after mixing, when it might have become bound to the recipient DNA.

Support for the hypothesis of single-strand transfer was obtained from genetic experiments by Vielmetter, Bonhoeffer and Schütte (1968). They used a recipient strain that carried a deletion of the lactose operon and adjacent *lacI* gene. A histochemical staining technique, dependent on differential utilization of a galactoside, allowed recognition of mutants of the *lacI* gene (*lacI*$^+$ light pink; *lacI*$^-$ dark red). Nitrosoguanidine (NG) was known from previous studies to cause mutation in only one strand of the DNA. Mutants of the *lacI* gene were induced with NG at the time of conjugation. Recombinants were selected by use of a culture medium lacking proline and containing streptomycin, since proline independence was a character of the donor and streptomycin resistance of the recipient. When *lacI*$^-$ mutants were induced in the donor immediately before transfer of the lactose region to the recipient, the mutants in the recombinant colonies were always homozygous *lacI*$^-$. Exposure to NG soon after transfer to the recipient, however, gave *lacI* mutants that were always heterozygous, that is, they gave rise to sectoring colonies (light pink and dark red). Vielmetter *et al.* (1968) conlcuded from these results that only one strand of the donor DNA was transferred, but underwent replication immediately afterwards. With the late mutagenesis *lacI*$^-$ would then be induced in only one of these strands.

The strand that is transferred in conjugation is polarized, with the 5′ end always leading. This was shown independently by Ihler and Rupp (1969) and Ohki and Tomizawa (1969), using as donor an *E. coli* genome carrying prophage λ. The two strands of the λ genome can be separated through differential binding to poly(U,G) and hence one will have a heavier buoyant density in caesium chloride than the other. These authors found that radioactive label from the donor appeared in the progeny in the light strand of λ if conjugation was in one direction, and in the heavy strand if a different Hfr was used with transfer in the opposite direction on the genetic map. It was suggested that a site-specific single-chain scission is made in a particular strand of the F factor, the 5' end then becoming the origin of transfer. Ohki and Tomizawa (1969) found that the complementary strand to that transferred was synthesized quickly in the recipient, in agreement with the conclusions of Vielmetter *et al.* (1968) already described.

It was not clear from the experiments described above whether a single strand of donor DNA was integrated into the recipient genome, or whether the replication of the donor strand immediately after transfer to the recipient cell allowed duplex integration of DNA of donor parentage. In order to try and resolve this question, Siddiqi and Fox (1973) used radioactive and density labels. The donor (Hfr) bacteria were labelled with [^{3}H]thymine and the heavy isotopes ^{13}C and ^{15}N. The recipient (F–) bacteria also carried these density labels but were non-radioactive. By allowing both donor and recipient cells to undergo one replication in a culture medium with normal (light) atoms before mixing them, it was possible to have both parents with DNA of hybrid density, that is, one heavy (H) and one light (L) strand. Since both parents were of hybrid density, duplex integration of donor DNA into the recipient genome will not change the density. On the other hand, simplex integration will give rise to segments of DNA with heavy atoms in both strands when the heavy donor strand replaces the light recipient one. It is this heavy donor strand which carries the tritium label used to recognize recombinant molecules. After conjugation, the DNA of the recipient cells was examined in a caesium chloride density gradient in order to observe the position of the tritium atoms. Radioactive DNA of density greater than the hybrid (LH) molecules was found, implying single-strand integration of donor DNA. Denaturation by boiling gave single-stranded molecules of intermediate density, showing that the heavy donor segment was covalently joined to the light recipient strand. Sonication to fragment this strand produced a shift in the tritium-labelled material to a heavier density, thus showing that the donor label was not uniformly distributed but concentrated in discrete regions, as expected.

A further experiment was carried out by Siddiqi and Fox (1973). The donor cells carried the tritium label as before, but in this experiment they were of normal (light) density, while the recipient cells had heavy atoms in both strands of the DNA. Following conjugation in light medium the donor radioactivity appeared in molecules of hybrid density (LH). None was found in the unreplicated recipient molecules, which were at the fully heavy (HH) position in the gradient. Furthermore, when the molecules of intermediate density (LH) carrying the donor label were denatured, the tritium was found to be exclusively in the light strand. Siddiqi and Fox concluded that the insertion of donor material occurs only in the newly formed strand of the recipient DNA, and hence possibly at a replication fork. The absence of donor label covalently associated with the heavy recipient strand provided further evidence that double-strand insertions of donor material do not occur. This conclusion conflicts with that of Oppenheim and Riley (1966, 1967) on the basis of similar experiments (see above). Siddiqi and Fox also pointed out, just as Oppenheim and Riley had done, that their experiments might not have detected the insertion of long pieces of duplex donor DNA into recipient molecules.

4. RECOMBINATION-DEFICIENT MUTANTS

Since it seemed likely that enzymes take part in recombination in *E. coli*, Clark and Margulies (1965) isolated mutants unable to form recombinants by conjugation. A leucine-requiring recipient (F−) strain was treated with the mutagen 1-methyl-3-nitro-1-nitrosoguanidine and the survivors were tested by mating with an adenine-requiring donor (Hfr) strain and looking for the formation of recombinants lacking a requirement for leucine and adenine and hence able to grow on minimal medium. In this way two mutants were obtained that failed to give recombinants. These strains were found also to be killed by ultraviolet light much more readily than normal strains. Revertants of the mutants were obtained by ultraviolet irradiation and were found to have regained resistance to ultraviolet light and ability to form recombinants. Clark and Margulies concluded that in each mutant a single mutation was responsible for both characters. The gene in which the mutations had occurred was subsequently called *recA*. Further mutants with similar properties were isolated by Howard-Flanders and Theriot (1966) by testing the products of mutagen treatment for sensitivity to killing by X-rays. The five most sensitive mutants were found to be defective in ability to form recombinants. By growing cells in the presence of [³H]thymidine and then exposing to ultraviolet light, the release of radioactive nucleotides can be measured. In normal cells between 200 and 500 nucleotides are removed per pyrimidine dimer induced by the ultraviolet light, as part of a repair process, but in the *recA⁻* mutants it was estimated that 5000–20 000 nucleotides were removed per dimer.

Using one of these *recA⁻* mutants, Clowes and Moody (1966) discovered that it was defective in the recombination between an F′ (see Section 2) and the *E. coli* genome. The evidence for this was a greatly reduced frequency of genome transfer to the recipient cell when the donor carried the *recA⁻* mutation. As already described, Clark and Margulies (1965) had found that a *recA⁻* mutant in the recipient led to a deficiency in integrating donor material into the recipient genome after conjugation. Under these circumstances, genome transfer was normal since the donor did not carry the mutant.

Recombination-deficient mutants of *E. coli* that were less radiation-sensitive than the *recA* mutants were obtained by Howard-Flanders and Theriot (1966) and by Clark (1967) who found that, unlike *recA* mutants, they showed some residual recombination. The gene affected was subsequently called *recB*. It was mapped by Emmerson (1968) and found to be separated by several intervening loci from *recA* (see Fig. 46). Willetts and Mount (1969) found that mutants with this less severe recombination deficiency and ultraviolet sensitivity fell into two complementation groups. The second group they called *recC*. These mutants were found to map close to *recB* (Fig. 46).

The extent of DNA degradation in a *recB* mutant following exposure to

Table 8. Summary of results obtained by Clark and Margulies (1965) and Howard-Flanders and Theriot (1966) for *recA* mutants, Clark (1967) and Emmerson (1968) for *recB* and *recC* mutants, and Willetts and Clark (1969) for multiple mutants.

Genotype	Recombination frequency	Ultraviolet sensitivity	Ultraviolet-induced DNA breakdown
Wild type	High	Low	Normal
recA⁻	None	High	Much greater than normal
recB⁻ *recC⁻* *recB⁻ recC⁻*	Low	Moderate	Less than normal
recA⁻ recB⁻ *recA⁻ recC⁻* *recA⁻ recB⁻ recC⁻*	None	High	Less than normal

ultraviolet light was studied by Howard-Flanders and Boyce (1966). They labelled the DNA with tritiated thymidine and after irradiating the cells measured the fraction of the label that became acid-soluble. They found that the *recB* mutant showed less DNA degradation than wild type and called its behaviour 'cautious', the *recA* behaviour with its extensive breakdown being 'reckless'. Emmerson (1968) studied the 'cautious' breakdown in more detail and found that *recC* mutants showed it as well as *recB*. The degradation was only 20% of that shown by wild type.

Willetts and Clark (1969) prepared the three double mutants and the triple mutant of *recA⁻*, *recB⁻* and *recC⁻*. They measured the recombination deficiency, ultraviolet sensitivity, and rate of ultraviolet-induced DNA breakdown of these multiple mutants and made two important discoveries: the *recB recC* double mutant behaved like *recB* or *recC* mutants alone; and the double or triple mutants involving *recA* showed the low ultraviolet-induced DNA breakdown of *recB⁻* or *recC⁻*, but the other characteristics of *recA⁻*. These results are summarized in Table 8. Willetts and Clark concluded that the products of *recB* and *recC* act in the same pathway, both being necessary for normal levels of recombination, and that the product of *recA* also acts in the same pathway since the multiple mutants involving *recA⁻* have the same recombination deficiency as *recA* mutants alone. They further suggested that one or both products of *recB* and *recC* may be nucleases involved in the DNA breakdown after ultraviolet irradiation, and that the normal *recA* product inhibits this breakdown.

5. THE *recBC* DEOXYRIBONUCLEASE

In several laboratories it was discovered that lysates of *recB* and *recC* strains lacked a deoxyribonuclease activity found in lysates of wild-type strains. This activity was exonucleolytic on duplex DNA and was dependent on adenosine triphosphate (ATP).

During experiments with a restriction endonuclease using the single-stranded circular DNA of phage fd, Goldmark and Linn (1970) found a contaminating endonuclease activity that was greatly stimulated by nucleoside triphosphate and was lacking from extracts of *recB* and *recC* mutants. They purified the enzyme and found that it acted both as an endonuclease and as an exonuclease on simplex DNA, but only exonucleolytically on duplex DNA. Further purification enabled them (Goldmark and Linn, 1972) to obtain more detailed knowledge of this remarkable enzyme – the *recBC* DNase, also called exonuclease V. It was found to have a molecular weight of approximately 270 000 daltons and to comprise two non-identical polypeptides of similar size. Four reactions were catalysed: (1) endonucleolytic cleavage of single-stranded DNA; (2) exonucleolytic cleavage of single-stranded DNA; (3) exonucleolytic cleavage of duplex DNA; (4) hydrolysis of adenosine triphosphate (ATP). The exonucleolytic activities were dependent on the presence of ATP, with more than 20 ATP molecules hydrolysed to ADP and inorganic phosphate per phosphodiester bond of DNA broken; the end-products were 5'-phosphoryl-terminated oligonucleotides consisting of three to six or seven nucleotides. Conversely, the hydrolysis of ATP was dependent on the presence of a degradable DNA substrate. The endonucleolytic activity was found to be strictly confined to single-stranded DNA and to be stimulated sevenfold by ATP.

Temperature-sensitive *recB* and *recC* mutants were obtained by Tomizawa and Ogawa (1972) by treating lactose non-fermenting (*lac⁻*) streptomycin-resistant recipient cells with a mutagen, and incubating them at a high temperature (42 °C) with *lac⁺* streptomycin-sensitive donor cells in a culture medium containing streptomycin. Recombination-deficient mutants were revealed by a failure to find *lac⁺* cells – all were necessarily streptomycin-resistant. Some of the mutants showed the recombination deficiency (and associated characteristics such as ultraviolet sensitivity) only at the high temperature. The ATP-dependent deoxyribonuclease extracted from a temperature-sensitive *recB* mutant was more sensitive to heat than the wild-type enzyme. This indicated that *recB* is a structural gene of the enzyme. On the other hand, the enzyme extracted from a temperature-sensitive *recC* mutant showed normal heat sensitivity. Tomizawa and Ogawa (1972) concluded that the *recC* product (1) regulates the synthesis of the enzyme, or (2) modifies a precursor of it, or (3) forms a complex enzyme with the *recB* product. In the last-mentioned case, it would be necessary to postulate that whereas the *recC* product alone is heat-sensitive, the complex is heat-resistant. Hypothesis (1) was eliminated when Lieberman and Oishi (1973) demonstrated *in vitro* complementation between extracts of *recB* and *recC* mutants.

Lieberman and Oishi (1973) also found that the *recBC* DNase activity was rapidly lost when the purified enzyme preparation was exposed to a high salt concentration, but gradually restored on removal of the salt. They attributed this effect to dissociation and reassociation of subunits, and used the denaturation by salt to isolate the subunits by ion exchange chromatography (Lieberman and Oishi, 1974). In agreement with Goldmark and Linn's results, two subunits, which they called α and β, were found, and active enzyme molecules were reconstituted when they were mixed. They found that the β subunit on its own retained a small but significant ATP-dependent DNase and DNA-dependent ATPase activity, two of the characteristics of the enzyme. On the other hand, the α subunit on its own showed neither activity. They concluded that the β subunit is the core protein of the *recBC* DNase, and that α protein, by binding to β subunit, enables the β subunit to function more efficiently.

When they looked at the genetic origin of the subunits, Lieberman and Oishi (1974) made an unexpected discovery. An extract from a *recB* mutant was incubated with α subunit, and the *recBC* DNase activity produced by complementation was measured. The experiment was repeated with the β subunit in place of α, and also with an extract from a *recC* mutant in place of *recB*. They found that the α subunit did not produce any *recBC* DNase activity with either *recB* or *recC* extract. Conversely, the β subunit gave enzyme activity with either, and indeed with an extract of a *recB recC* double mutant, too. They concluded that the α subunit is present in both *recB⁻* and *recC⁻* strains, and the β subunit is absent from both. As regards the function of the *recC* product, they suggested that it was either a structural component of the β subunit, or an enzyme that modifies β subunit precursor (the *recB* product) to form active β subunit. To account for the absence of recombination-deficient mutations in the unidentified structural gene for the α subunit, Lieberman and Oishi suggested that such mutants were either (1) lethal, because the α subunit also catalysed some essential function such as replication, or (2) not recombination-deficient, because the β subunit alone was the recombination enzyme, the αβ complex having an additional function.

Kushner (1974) partially purified the *recBC* DNase from the temperature-sensitive *recB* and *recC* mutants obtained by Tomizawa and Ogawa (1972) and made a surprising discovery. He found that only the ATP-dependent exonucleolytic hydrolysis of duplex DNA was abnormally temperature-sensitive. The exo- and endonucleolytic degradation of single-stranded DNA was found to be no more temperature-sensitive than that catalysed by the wild-type enzyme. The significance of Kushner's discovery was revealed by Taylor and Smith (1980) when they found that under physiological conditions the initial action of the *recBC* DNase on duplex DNA was the unwinding of the strands. This was evident from electron micrographs prepared after linear duplex DNA had reacted with the enzyme. They concluded that the nucleolytic activity of the enzyme is all at the single-strand level, and that both of the temperature-sensitive mutants studied by Kushner were mutant for the unwinding activity but not for the nuclease. A further inference was that the unwinding reaction was involved in genetic recombination.

A temperature-sensitive *E. coli* mutant defective for growth above 32 °C was obtained by Das, Court and Adhya (1976) and found to have an altered rho factor. This protein functions in terminating transcription. The mutation was evidently in the structural gene of the protein. At 42 °C the mutant showed a 70-fold reduction in recombination frequency. Das *et al.* suggested that the α subunit of the *recBC* DNase might be controlled by the *rho* gene.

Infection of *E. coli* with phage λ gave rise to cells that resembled *recB* mutants. Unger and Clark (1972) studied this phenomenon. They tested the ATP-dependent duplex exonucleolytic activity and the ATP-stimulated simplex endonucleolytic activity of the *recBC* DNase and found that both were lacking in λ-infected cells. They identified the λ gene responsible for this effect as *gamma*, first described by Zissler, Signer and Schaefer (1971) (see Section 6). With infection by *gamma* mutants of λ the ATP-associated nuclease activity of the cells remained normal. The product of *gamma* appeared to be a specific inhibitor of the RecBC recombination pathway, because Unger, Echols and Clark (1972) found (1) that in the presence of a *gamma* mutation, but not otherwise, the host RecBC pathway could generate substantial phage λ recombination, and (2) that another host recombination pathway, RecE (see Section 6), was unaffected by infection with phage λ.

6. THE *sbcA* LOCUS AND THE RecE PATHWAY

Revertants of *recB* and *recC* mutants were studied by Barbour *et al.* (1970). The revertants were obtained by treatment with chemical mutagens and were found to fall into two classes. One appeared to be true back-mutants, with wild-type recombination proficiency and resistance to ultraviolet, and with the ATP-dependent nuclease activity found in the wild type. The other class of revertants resulted from indirect suppression of the *recB* or *recC* mutants as a result of mutation elsewhere in the genome leading to the production of a high level of another nuclease activity. Unlike the *recBC* nuclease this activity was independent of ATP. Barbour *et al.* called the new mutations *sbc* (suppressor of *recB* and *recC*) and favoured the idea that an alternative pathway of recombination was activated. This pathway would normally be of minor importance and would account for the low but significant level of recombination found in *recB* and *recC* mutants. They argued that the *recA* gene participated in both pathways since *recA* mutants formed no recombinants. Kushner *et al.* (1971) described these mutants that gave rise to the high level of ATP-independent nuclease activity as *sbcA* mutants, because they had discovered a second class of indirectly suppressed *recB* and *recC* mutants, which they called *sbcB* (see Section 7).

Kushner, Nagaishi and Clark (1974) reported genetic and enzymic tests on the *sbcA* mutations. They found that the ATP-independent exonuclease synthesized in these mutants did not correspond to known nucleases of *E. coli*, so they called it exonuclease VIII. They found that this enzyme showed a marked preference for duplex DNA over heat-denatured, that is, single-stranded, DNA, and it showed no endonucleolytic activity. It is presumed, therefore, that in the recombination process it substitutes for the exonucleolytic action on duplex DNA of the *recBC* nuclease, rather than for any of the other activities that the latter enzyme can catalyse. As the structural gene for exonuclease VIII plays a part in recombination, Clark (1973) gave it the name *recE*, since 'recD' had already been reserved for a mutant that appeared not to map at the *recA*, *recB* or *recC* loci, Kushner *et al.* (1974) favoured the idea that *sbcA* is not *recE* but is a control gene for exonuclease VIII, since it is easier then to understand how *sbcA* mutations lead to enzyme production. This conclusion was supported by Lloyd and Barbour (1974), who found that the genotype *recB⁻ sbcA⁻*, with *sbcA⁺* carried by an F′, was ultraviolet-sensitive, but became ultraviolet-resistant on curing, that is, when the F′ was removed by treatment with acridine orange. These results implied that *sbcA* $^+$ was dominant over *sbcA* $^-$, and were in agreement with the hypothesis that the *sbcA* $^+$ product represses *recE*. Lloyd and Barbour (1974) mapped the *sbcA* locus using donor strains that initiated conjugation at various positions on the linkage map, and concluded that the gene was at about minute 30 (Fig. 46).

In the course of genetic analysis of *sbc* mutations, Templin, Kushner and Clark (1972) made the surprising discovery that *sbcA* mutants could not be obtained in a particular genetic background, perhaps because it was deficient in *recE*. This would prevent production of exonuclease VIII even if *sbcA* $^-$. Templin *et al.* (1972) suggested that *sbcA* and *recE* might belong to a phage genome, which was integrated in the *E. coli* genome as a prophage in some host strains but absent from others. This hypothetical prophage was called Rac (recombination activation) by Low (1973).

Support for this idea has been obtained, unexpectedly, from studies with phage λ. In this virus there is a connection between recombination and growth. Mutants of λ defective in either of the Red recombination genes (coding, respectively, for λ exonuclease and β protein – see Chapter V, Section 7) will grow on a recombination-defective (*recA* $^-$) host, but Zissler, Signer and Schaefer (1971) discovered a third λ recombination gene, *gamma*, a defect of which prevented *red* $^-$ λ from growing on a *recA* $^-$ host (Chapter V, Section 11). When they selected for reverse mutants that, although *red* $^-$ and *gamma* $^-$ as a result of a deletion, were nevertheless able to grow on *recA* $^-$ cells, they found they had obtained λ phages in which recombination was restored to the wild-type level in a *recA* $^-$ host, perhaps as a result of the insertion of host recombination genes into the phage genome. These reverse mutants could not, however, simply have an insertion of the normal host recombination genes, because Zissler *et al.* (1971) found that reverse mutants could be obtained when the host was *recA* $^-$ *recB* $^-$. Signer (1971) suggested that the reverse mutants might carry the host genes that are expressed in *sbcA* $^-$ strains.

Gottesman *et al.* (1974) obtained further reverse mutants of phage λ and, from electron micrographs of heteroduplexes between their DNA and that of normal λ, showed that they indeed carried a segment of host DNA in place of the deleted λ recombination genes. Three reverse mutants appeared to carry identical inserts. Mapping studies, using phage λ that carried a reverse mutant and was integrated in the host genome as prophage, revealed that the prophage was not at the normal λ attachment site (17 min) but between 27 and 30 min, which corresponds to the position of *sbcA* (Fig. 46). Gottesman *et al.* (1974) therefore favoured the idea of Templin *et al.* (1972) that *sbcA* was on a previously undetected prophage. Reverse mutants of phage λ would then result from integration of this prophage into the λ genome. Gillen *et al.* (1977) purified the deoxyribonuclease produced by reverse mutants of phage λ and found it to be identical with exonuclease VIII by all the criteria they tested, which included the molecular weights of the active enzyme and of its subunits, and the effects of magnesium ion concentration and pH on enzyme activity. Their results support the hypothesis that the DNA of host origin in reverse mutants of phage λ includes *recE*, the structural gene for exonuclease VIII.

Kaiser and Murray (1979) investigated the physical structure of the Rac prophage that carries the *sbcA* and *recE* loci. They used restriction endonucleases to fragment the DNA of *E. coli*. The DNA of a reverse mutant of phage λ was labelled with ^{32}P and used as a probe to find where homologous nucleotide sequences occurred. The *E. coli* DNA fragments were separated by electrophoresis in agarose gels, transferred to nitrocellulose filters and hybridized with the ^{32}P-labelled DNA. The fragments containing homologous DNA were identified by autoradiography. Kaiser and Murray's results confirmed the existence of the Rac prophage near minute 30, and supported the origin of reverse mutants of λ by recombination between λ and Rac phages. The latter phage was thought to be defective in view of the absence of plaques attributable to it.

7. THE *sbcB* LOCUS AND THE RecF PATHWAY

In the course of surveying the enzymic activities of mutants that indirectly suppressed *recB* and *recC* mutants, Kushner *et al.* (1971) discovered that, unlike the *sbcA* mutants, some of the suppressors lacked an enzyme that degraded single-stranded DNA. They called these mutants *sbcB*. Enzyme purification and studies with a specific antiserum indicated that exonuclease I was lacking in the *sbcB* mutants. Lehman and Nussbaum (1964) had found that *in vitro* exonuclease I degrades single-stranded DNA from a 3'-OH terminus, but has no effect on duplex DNA. Kushner *et al.* (1971) argued that the *sbcB* mutations could either be in a structural gene for exonuclease I or result in the production of an inhibitor of the enzyme. No sharp increase in exonuclease I activity was found during attempts to purify the enzyme from an *sbcB* mutant. They concluded that no inhibitor was being removed from the enzyme and that *sbcB* was therefore probably a structural gene for exonuclease I. Mapping experiments by Templin, Kushner and Clark (1972) placed the *sbcB* locus near 40 min on the standard linkage map of *E. coli* (Fig. 46). This position is near the histidine operon and not close to any of the other genes known to affect recombination.

The discovery of the *sbcB* mutants with their exonuclease I deficiency led Kushner *et al.* (1974) to propose the existence of an alternative pathway for recombination to that involving the *recBC* nuclease. They called this pathway that was activated by exonuclease I deficiency the RecF pathway, *recF* being one of the hypothetical genes necessary for it to function. The idea was that the RecBC pathway, using the products of the *recB* and *recC* genes, accounted for 99% of wild-type recombination and the RecF pathway for the remaining 1%. The RecF pathway was a minor one in wild-type cells because of inhibition or repression of one or more of its enzymes, or because of degradation of an essential intermediate. It was assumed that exonuclease I was responsible for shunting an intermediate in the RecF pathway into the RecBC pathway (Fig. 47). In the absence of exonuclease I, that is, in an *sbcB* mutant, the drain on the intermediate would be relieved and the RecF pathway would become much more effective.

In order to test these ideas, Horii and Clark (1973) obtained recombination-deficient mutants in a *recB⁻ recC⁻ sbcB⁻* (and therefore recombination-proficient) strain. They expected some of the mutations to be in genes of the RecF pathway and not to affect the RecBC pathway. Among the mutants isolated, following nitrosoguanidine treatment, one was *sbcB⁺*, three were *recA⁻*, and 11 were

Parental DNA

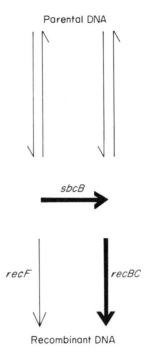

sbcB

recF   recBC

Recombinant DNA

Fig. 47. Hypothetical relationship proposed by Horii and Clark (1973) between the RecF and RecBC pathways of recombination. The RecF pathway is important for ultraviolet repair but is only a minor route (thin arrow) for recombination unless both *sbcB* and *recBC* are mutant.

mutants in genes not previously recognized, one of which was named *recF*. Tests with one of the *recF* mutants showed that it conferred ultraviolet light sensitivity as well as recombination deficiency on the $recB^-$ $recC^-$ $sbcB^-$ genotype. This *recF* mutant was transferred to a $recB^+$ $recC^+$ background and found to be recombination-proficient. This showed that *recF* did not act in the RecBC recombination pathway, in agreement with expectation if RecF and RecBC are alternative pathways. A surprising discovery, however, was that the *recF* mutant in a $recB^+$ $recC^+$ background remained ultraviolet-sensitive (see Table 9). Horii and Clark (1973) concluded that, as predicted, the *recF* gene was indeed independent of the RecBC pathway for recombination (Fig. 47). Unexpectedly, however, the RecF pathway functions in the repair of ultraviolet-induced damage even in the presence of a normal *sbcB* gene. In other words, exonuclease I (the *sbcB* product) inhibits repair by RecF less than it inhibits recombination by RecF.

A further and unexpected discovery was made about the genetic basis of exonuclease I. The *sbcB* mutants, in a $recBC^-$ but otherwise wild-type background, had recovered normal recombination proficiency and normal resistance to ultraviolet light, and had also recovered normal resistance to mitomycin C, to which unsuppressed *recB* or *recC* mutants are sensitive. A mutant had been

Table 9. Results obtained by Horii and Clark (1973) on the behaviour of a *recF* mutant in various genetic backgrounds. The behaviour of *recF+* strains is also shown.

Genotype				
recF	*recB* and *recC*	*sbcB*	Recombination proficiency	Ultraviolet sensitivity
+	+	+		
+	+	−	Proficient	Resistant
+	−	−		
+	−	+	Deficient	Sensitive
−	+	+		
−	+	−	Proficient	Sensitive
−	−	+		
−	−	−	Deficient	Very sensitive

obtained, however, by B. Weiss by screening directly for loss of exonuclease I activity in survivors of mutagen treatment, and Kushner, Nagaishi and Clark (1972) found that this mutant, besides being deficient in exonuclease I, was resistant to ultraviolet light and to mitomycin but, unexpectedly, was recombination-deficient. They called the gene that had undergone mutation *xonA*, because of the exonuclease deficiency. Thus, in the presence of the *xonA* mutant, the mitomycin and ultraviolet sensitivity conferred by *recB* or *recC* mutations was indirectly suppressed, presumably through the loss of exonuclease I activity, but this loss did not suppress the recombination deficiency. The *xonA* mutant showed close linkage to *sbcB* mutants. Kushner *et al.* (1972) obtained further mutants by treating a *recB recC* double mutant with nitrosoguanidine and plating on a medium containing mitomycin C. Of the resistant colonies obtained, 84% were *sbcB* mutants, that is, recombination-proficient, and 16% were *xonA* mutants. Two *sbcB* mutants that had been the subject of earlier investigations were crossed with a *xonA+ sbcB+* strain, but no *xonA−* progeny were obtained in nearly 8000 tested. It was concluded that these *sbcB* mutants did not carry undetected *xonA* mutants.

Kushner *et al.* (1972) put forward several alternative explanations for these remarkable results:

(1) The *sbcB* and *xonA* mutations are in the same gene, but *xonA* mutants, unlike *sbcB*, leave sufficient residual exonuclease I activity to prevent recombination, but insufficient to permit repair of ultraviolet- and mitomycin-induced damage to the DNA.

(2) The *sbcB* mutations are in a regulatory gene, and prevent the synthesis both of exonuclease I, coded by *xonA*, and also of an inhibitor of recombination.

(3) The *xonA* and *sbcB* mutations are in different genes, both of which are needed for the 3′ to 5′ exonucleolytic activity of exonuclease I on single-stranded DNA that prevents repair, but one of which, *sbcB*, is also required for an unknown activity that prevents recombination. Loss of the *sbcB* subunit of the enzyme would then allow both repair and recombination, whereas loss of the *xonA* subunit would allow repair only, in agreement with the observed behaviour of the mutants.

(4) The *xonA* and *sbcB* mutations are in the same gene, but their polypeptide product possesses two types of activity corresponding to those in (3) above.

Of these alternatives, no. (2) was not favoured, for reasons given earlier. There was insufficient evidence to discriminate between the others.

Further insight into the genetic basis of exonuclease I was obtained by Yajko, Valentine and Weiss (1974). They attempted to isolate mutants with a temperature-sensitive exonuclease I by the tedious procedure of testing 10 000 isolates, following mutagenesis, for exonuclease I activity at 42 °C. In this way six mutants were obtained. The enzyme was isolated from one of them, *xon*101, and found to be temperature-sensitive. Yajko *et al.* (1974) concluded that this muta-tion was in a structural gene for exonuclease I. The mutant was transferred to an F′, which was introduced into cells carrying in turn, an *sbcB* mutant, a *xonA* mutant, and each of the other new exonuclease I mutants. In every case the cells retained their low exonuclease I activity at 42 °C. This lack of complementation was also evident when an *sbcB* mutant was used in the F′, and the use of *xon*⁺ in the F′ revealed that all the exonuclease I mutants were recessive. *In vitro* complementation tests were also carried out, by mixing cell-free extracts and assaying for exonuclease I activity at 42 °C, and again no complementation was observed. Furthermore, all the mutants were found to map in the *sbcB* region of the genome. With the exception of *xon*101, all the new exonuclease I mutants, when grown at 42 °C in a *recB⁻ recC⁻* strain, proved to be resistant to mitomycin C and to ultraviolet light and to be recombination-deficient. These mutants were assigned to the *xonA* gene, as they behaved like the *xonA* mutants studied by Kushner *et al.* (1972). At 42 °C in a *recB⁻ recC⁻* background, the *xon*101 mutant, however, was mitomycin- and ultraviolet-sensitive, as well as being recombination-deficient, or in other words, like *xon*⁺ or *sbcB*⁺. This unexpected result extended still further the phenotypic diversity of the exonuclease I mutants.

To account for these results, Yajko *et al.* (1974) favoured the first hypothesis of Kushner *et al.* (1972), namely, that all the mutations are within a single gene, a structural gene for exonuclease I, and that their different phenotypes result from differing levels of residual exonuclease I activity. These levels would be too low to have been detected. Yajko *et al.* pointed out that the *sbcB* mutations might be polar mutations or promoter mutations that affected *xonA* and a gene involved in the expression of recombination deficiency by *recBC*-deficient strains. In other words, the *xonA* gene was not necessarily altered in *sbcB* mutants, and the indirect suppression of *recBC* deficiency by *sbcB* mutants was not necessarily caused by lack of exonuclease I.

Yajko *et al.* (1974) found that *sbcB* mutants survived mitomycin treatment better than most *xonA* mutants in a *recB⁻ recC⁻* background. This might explain why Kushner *et al.* (1972) found so many *sbcB* mutants using their technique of selection for mitomycin resistance, whereas Yajko *et al.* found none among their non-selected exonuclease I mutants.

8. LOCALIZED REGIONS OF HIGH ACTIVITY OF THE RecBC PATHWAY

McMilin, Stahl and Stahl (1974) studied the distribution of recombination events in the genome of phage λ. DNA synthesis was blocked using the technique developed by McMilin and Russo (1972) – the combined effects of mutations in the host *dnaB* gene and either gene *O* or gene *P* of the phage (see Chapter V, Section 11). One parent was density-labelled (^{15}N and ^{13}C) and carried a marker in gene *A*. The other parent was unlabelled and carried a marker in gene *R* at the other end of the genome. The wild-type recombinants, lacking both markers, will appear equally often at all positions in a caesium chloride density gradient, if there is a uniform distribution of recombination events along the genome. On the other hand, if there are favoured sites for recombination there will be favoured densities for the recombinants. It was found that when both parents carried deletion b1319, or both carried deletion b1453, most of the wild-type recombinants had a nearly fully light density, implying a favoured site ('hot spot') for recombination near gene *R*. A similar localization of events was observed with these deletions in the presence of DNA synthesis, indicating that the non-random distribution was not a consequence of blocking the DNA synthesis. The deletions investigated are extensive and remove all the phage genes concerned with recombination, so the hot spot appeared to be the consequence of localized activity of a recombination pathway of the host. Lam *et al.* (1974) identified mutations responsible for this recombinational hot spot activity and called them 'crossover hot-spot instigators' (Chi). They found that the activity is lost in a *recB⁻* host and concluded that the RecBC pathway was involved in hot spot activity. The Chi mutations were found to map near gene *R* and could therefore be at the site where recombination was stimulated. Crosses in which one parent only had a Chi mutation revealed that the mutation was dominant.

A further experiment was carried out by McMilin *et al.* (1974). They investigated a λ cross in which both parents carried substitution *bio1*. This strain showed a recombinational hot spot near the middle of the λ genone. The *bio1* substitution involves the deletion of the phage recombination genes, which are in this middle region of the genome, and the insertion in their place of the host *biotin* operon. The hot spot activity of the *bio1* genotype was not shown in a *recB⁻* host, again implicating the RecBC pathway. McMilin *et al.* suggested that the host DNA inserted into the λ genome in the *bio1* strain might carry a Chi mutation.

In order to investigate further which recombination pathways in *E. coli* are affected by Chi mutants, Stahl and Stahl (1977) made λ crosses in various genetic backgrounds. The crosses involved three markers, A, cI and R, thus:

$$\frac{A \quad cI \quad +}{+ \quad + \quad R}$$

Recombinants that were wild type for A and R would be of cI genotype if the exchange occurred in the left-hand interval, but not if it occurred in the right-hand interval. In one experiment a Chi mutant was present in the left-hand interval and in a second experiment a Chi mutant in the right-hand interval was used instead. Thus, in a genetic background in which these recombination hot spots are active, there will be a marked difference between the two crosses in the proportion of the recombinants $(A^+ R^+)$ that were of cI genotype. No such difference would be expected if the Chi mutants were inactive. Using this test of Chi activity, they found that the triple mutants *recB recC sbcA* and *recB recC sbcB* showed no Chi activity, implying that the Chi sites did not function in the RecE or RecF pathways. In these tests the phage recombination systems (Red and Int) were inactivated by mutants. When the host recombination pathways were not functioning (as a result of a mutation in *recA*), but the phage general recombination pathway (Red) was active, there was again no Chi activity. On the other hand, in the presence of a functional host RecBC pathway, as previously shown, the hot spots were effective. If the Chi sites are initiators of recombination, they would evidently be active at the start of the RecBC pathway (Fig. 47, top right) but not at the start of either the RecE or the RecF (Fig. 47, top left) pathways.

The idea that a Chi mutation represents the creation of a recombinational hot spot at its site was confirmed by Stahl, Crasemann and Stahl (1975). They isolated three additional Chi mutations by taking advantage of the fact that these mutants suppress the small-plaque phenotype shown by λ mutants defective in the λ recombination genes *red* and *gam*. They found that the new Chi mutants mapped at widely spaced sites in the λ genome and that each was associated with a recombination hot spot at or near its site. It was evident tht the Chi mutants were not associated with gain or loss of a gene function but rather with a nucleotide sequence that provokes the host RecBC recombination system. This conclusion was supported by the fact that one of the new Chi mutants was within the cII gene (*clear* plaque gene no. 2), which is not a recombination gene.

The effect of the recombination hot spot that had arisen spontaneously within the cII gene of phage λ was investigated by Stahl and Stahl (1975). They found that in its presence recombination between mutants in genes J and N to the left of cII on the map was increased to a greater extent than recombination between mutants in genes P and R to the right of cII. This discovery was the more surprising because the J–N region is further from cII than the P–R region. It was evident that stimulation of recombination could occur at a distance from the mutant site and with an asymmetrical distribution about it. Stahl and Stahl (1975) also investigated the effect of structural heterozygosity on the stimulation of recombination. They found that when one parent carried the normal λ immunity region and the other the immunity region of phage 434 in place of that of λ, the stimulation of recombination by the Chi mutation in gene cII was unaffected. The region of structural heterozygosity was between cII and the J–N region. It was evident that the promotion of recombination at a distance from the Chi site was not occurring through heteroduplex formation, as the lack of homology between the nucleotide sequences of the λ and 434 immunity regions would prevent hybrid DNA formation. Stahl and Stahl suggested that the stimulation of recombination at a distance from the mutant site might occur through exonucleolytic breakdown of one strand of the duplex from a nick at the Chi site or, alternatively, that Chi was a binding site for a nicking enzyme that often moved along the DNA before acting. In either case, the free end to a nucleotide chain would trigger recombination. Structural heterozygosity would not interfere with the promotion of recombination at a distance from the Chi site, because the distant effect would depend on movement of an enzyme (either an exonuclease or an endonuclease) along one of the DNA molecules before the other became associated with it in the recombination event.

As already described, McMilin et $al.$ (1974) had found a recombinational hot spot that seemed to be associated with the host DNA of the $bioI$ substitution in phage λ. Malone et $al.$ (1978) isolated a new bio substitution and found that this also contained a Chi site. Furthermore, this site was indistinguishable in its properties from the Chi mutations known to arise in λ by mutation. Thus, it required functional $recA$ and $recB$ genes for the stimulation of recombination to take place. By sampling the $E.$ $coli$ genome through the inclusion of portions of it in phage λ, the discovery was made that Chi sites are frequent. The estimated number was one per 10 000 nucleotide pairs of the duplex. This estimated frequency was subsequently doubled: see below. This high frequency in $E.$ $coli$ DNA contrasts with wild-type λ, where there are believed to be none.

The molecular nature of four Chi mutations located in the cII gene of phage λ was investigated by Sprague, Faulds and Smith (1978). The mutations were mapped with the aid of a series of deletions that extended for various distances into the cII gene. All four mutants were found to map between the end-points of the same two deletions. These end-points were then located physically by determining which of the known endonuclease cleavage sites of λ DNA were removed by the deletions. Restriction endonucleases $Hind$II, MboII and TaqI, derived respectively from $Haemophilus$ $influenzae$ strain d, $Moraxella$ $bovis$ and $Thermus$ $aquaticus$, cut λ DNA at one, two and two sites, respectively, within the cII gene. The nucleotide sequence of the DNA of the endonuclease fragment spanning the end-points of the two deletions was determined for two of the Chi mutants and compared with the known sequence for the normal cII gene. Both Chi mutants were found to have an $A \cdot T$ to $T \cdot A$ transversion at identical positions. The nucleotide sequence of the relevant part of the l chain of the DNA is as follows:

wild type $5'$ -C-A-G-A-T-C-A-G-$C$$\left\{\begin{matrix}\text{-A-}\\\text{-}T\text{-}\end{matrix}\right\}$$G$-$G$-$T$-$G$-$G$-A-A-G-A- $3'$
Chi mutants

It is evident that these two Chi mutants arose by a single base-pair change. The other two Chi mutants mapping in *c*II appeared to be similar. Wild-type λ, though showing some RecBC recombination, is devoid of Chi sites, so Sprague *et al.* (1978) suggested that the RecBC pathway, besides acting at a high rate in the vicinity of the Chi nucleotide sequence, also acted at a low rate with other sequences. Smith, Schultz and Crasemann (1980) examined the nucleotide sequence of a second Chi site in λ, located between *xis* and *red*. They found that this site shared with the *c*II site the eight bases italicized above, it having arisen by a $C \cdot G$ to $G \cdot C$ transversion at position 1 of these eight. Confirmation that these eight base pairs constitute the Chi sequence was obtained by Schultz, Swindle and Smith (1981) by inactivating the Chi site in *c*II by mutations induced with nitrous acid. The mutations, which were all $G \cdot C$ to $A \cdot T$ transitions, occurred at positions 1, 2, 4 and 7 of the italicized sequence.

Faulds *et al.* (1979) investigated the asymmetry, already referred to, in recombination stimulation by Chi sites. Using restriction endonucleases it was possible to fragment the DNA of phage λ carrying host Chi sites. Subsequent reannealing of ends and sealing by DNA ligase would give some molecules in which a fragment containing a Chi site was reversed. The surprising discovery was made that Chi sites become inactive when reversed. Some of the fragments did remain active, but these were believed to carry Chi sites in both orientations. Some inactive segments of DNA became hot spots when reversed. Thus, the previous estimate of Chi site frequency in *E. coli* DNA is doubled to one per 5000 nucleotide pairs to allow for the cryptic sites with reverse orientation. A further peculiarity of the results is that the Chi sites in active orientation always gave more recombination to their left on the map than to their right. It is not yet clear whether these remarkable results have a relatively trivial explanation in terms of the entry of λ DNA into the head of the phage, which is believed to occur only in the left to right direction in terms of the map, that is, gene *A* first and gene *R* last, or whether the data reflect basic features of the recombination process. Concerning the latter, one of the possibilities considered by Faulds *et al.* (1979) was that strand exchange in the RecBC recombination pathway might always involve one particular strand (say, the *l* strand – defined in Chapter V, Section 12) of phage λ (Fig. 48). They argued that exonuclease V, the *recBC* nuclease, might enter the recombinant structure at the site of the strand exchange, move along the DNA from it, and bind to a Chi site if the latter 'faced' the exchange point (Fig. 48(a)), but would not recognize a Chi site with the reverse orientation (Fig. 48(b)). Furthermore, the asymmetry of the Chi nucleotide sequence would mean that the enzyme would cut only one strand – always the same one in terms of Chi nucleotide sequence. If the orientation was such that the *l* strand was cut, resolution of the strand exchange (by cutting the other *l* strand and exchanging them) would not give overall crossing over, since the same strands would have been cut twice (Fig. 48(c)). A dimer – two phage genomes joined end to end – is necessary for packaging the λ DNA in the head of the phage, so only the orientation of Chi resulting in cutting of the *r* strand (the other *r* strand being subsequently cut, and

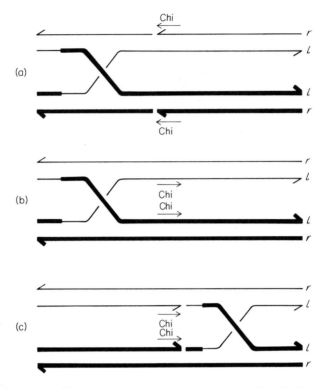

Fig. 48.   Diagram to illustrate one of the hypotheses proposed by Faulds *et al.* (1979) to explain orientation-dependent hot spot activity. The parental contributions are distinguished by the thickness of the lines, and have undergone strand exchange in the *l* strands: *l* and *r* identify the strands of λ DNA. (a) Active Chi site because exonuclease V, moving from the site of the strand exchange, recognizes the Chi site (arrow facing it) and cuts the strand at this site. This leads to cutting at the corresponding position in the other molecule, as shown, followed by strand exchange. (b) Reversed Chi site, inactive because the exonuclease V, moving from the site of the strand exchange, does not recognize it (arrow pointing away from it). (c) Reversed Chi site, inactive because the *l* strands are cut (see text).

the two exchanged) would give viable progeny (Fig. 48(a)). The need for recombination to give rise to a genome dimer is a consequence of the experimental conditions used to study Chi sites, with DNA replication blocked. Under normal conditions the end-to-end joining of genomes results from replication by rolling circle (Chapter V, Section 11).

Further investigation of Chi sites was made by Stahl *et al.* (1980). They made λ crosses under conditions when DNA replication was blocked, with one parent density-labelled with ^{15}N and ^{13}C. With every Chi site that they tested they found that recombination by the RecBC pathway was stimulated to its left on the map, and furthermore the two complementary recombinant genotypes were of unequal frequency. This was shown by the relative size of the peaks for light and heavy recombinants in a caesium formate density gradient. In each instance the recombinant carrying the Chi element was produced less often than the complementary one. Stahl *et al.* (1980) concluded that the stimulated exchange was frequently non-reciprocal. They favoured the idea that both this and the orientation of the stimulation related to the oriented packaging of λ DNA in the head of the phage particle. If the Chi site functions after packaging has begun, the failure to stimulate recombination to the right might be because the Chi site had already been packaged, while the bias against the recombinant carrying Chi might relate to the gap in one strand where the *recBC* nuclease had acted.

9. DNA REPAIR

The main products of exposure of DNA to ultraviolet light are pyrimidine dimers, formed by the covalent joining of adjacent pyrimidines in the same strand to give a cyclobutane ring. The dimers are a substrate for photoreactivating enzyme which, in the presence of visible light, reverses the action of ultraviolet light. A second repair mechanism involves excision of dimers by a specific endonuclease that cuts a phosphodiester bond near the dimer on its 5′ side to give a 5′-hydroxyl end on the dimer side and a 3′-phosphoryl end on the other side of the cut. The latter end is converted to a 3′-hydroxyl group by a 3′-phosphatase. The 5′-exonuclease activity of DNA polymerase I removes the dimer in an oligonucleotide of six or seven nucleotides, and the gap is filled by the same enzyme adding nucleotides to the 3′-hydroxyl end, using the intact complementary strand as template. When the gap has been filled, the junction is sealed by DNA ligase. Mutants of *E. coli* defective in the excision of pyrimidine dimers have been found at three genetic loci: *uvrA*, *uvrB* and *uvrC* (Fig. 46). All have a similar phenotype. The *A* and *B* genes are believed to code for the endonuclease, and the product of *C* to be responsible for the stability of the cut. Howard-Flanders and Boyce (1966) discovered that a *uvrA recA* double mutant, in the absence of photoreactivation, was killed by very low doses of ultraviolet, whereas over one-third of cells of a *uvrA* mutant with a normal *recA* gene survived an ultraviolet dose that put 50 dimers into the genome. They suggested that normal cells contain a mechanism that allows survival with unexcised dimers, but cells carrying a *recA* mutant do not.

This idea was investigated by Rupp and Howard-Flanders (1968). They examined the size of newly synthesized denatured DNA from cells of a *uvrA* mutant treated with a dose of ultraviolet light sufficient to produce about 360 pyrimidine dimers in the genome. The size was measured by sedimentation in alkaline sucrose gradients, particular care being taken to minimize breakage as a result of shearing. It was found that the newly synthesized strands had a molecular weight of about 14×10^6 daltons, whereas unirradiated cells gave a value of 100×10^6 daltons. During incubation of cells after ultraviolet treatment, the sedimentation rate of the DNA synthesized immediately after irradiation (identified by a ^{3}H label introduced from [^{3}H]thymidine) increased and approached that of normal DNA. Rupp and Howard-Flanders (1968) concluded

186

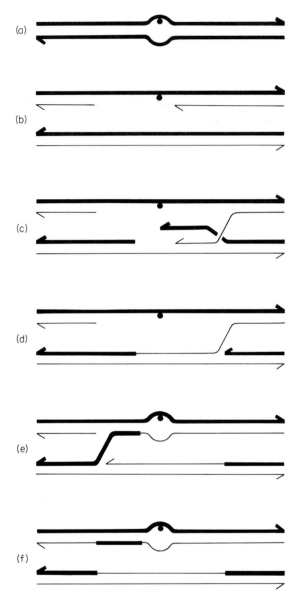

Fig. 49. Hypothesis favoured by Rupp *et al.* (1971) to accommodate their evidence for exchanges in post-replication repair. (a) A DNA duplex containing a pyrimidine dimer, shown as a dot. (b) The duplex is replicated (thin lines) leaving a gap opposite the dimer. (c) The strand at the gap initiates an exchange with the sister duplex, the corresponding strand of the latter being cut. (d) New synthesis fills the gap in the sister duplex and the junction is sealed, while the displaced strand is broken down. (e) The gap opposite the dimer is filled by a change of pairing partner. New synthesis fills the resulting gap in the sister duplex. (f) The strand linking the two duplexes is cut and gaps are filled and sealed.

that the daughter-strand DNA synthesized after ultraviolet irradiation contained gaps, but that these disappeared during further incubation. Estimates of the number of gaps indicated a value similar to the number of pyrimidine dimers, as though a gap appeared in the new chain opposite each dimer. It was not clear whether these gaps were a primary feature, as a result of interrupted synthesis, or arose secondarily through enzyme action following recognition of abnormal linkages in the DNA strand opposite the dimer. As excision-defective cells surviving ultraviolet irradiation usually produce normal daughter cells, Rupp and Howard-Flanders thought that the gaps opposite the dimers might be filled by making use of the base sequence in the sister duplex, which would be intact at this point (Fig. 49(a),(b)). In other words, genetic exchanges might occur between the sister duplexes. The *recA* gene was believed to participate in this post-replication mechanism for the reconstruction of the correct base sequence.

In order to see whether exchanges could be detected between sister duplexes after ultraviolet irradiation of a *uvrA* mutant kept in the dark, Rupp *et al.* (1971) grew cells for several generations in medium containing ^{13}C and ^{15}N. The cells were then exposed to various doses of ultraviolet and grown for less than a generation (30 min at 37 °C) in light medium containing [^{3}H]thymidine. As a result, part of the DNA will be hybrid in density (heavy old strand, light new one) with the ^{3}H label in the light strand. If exchanges occur between sister duplexes, light ^{3}H-labelled strands will become covalently joined to heavy strands. The DNA was extracted from the cells, denatured by heating to 98 °C for 5 min, and centrifuged in neutral caesium chloride gradients. Strands of intermediate density were indeed found. This was unlike the results obtained with unirradiated cells, which gave discrete heavy and light bands in the gradient, after denaturation. Furthermore, the strands of intermediate density from the irradiated cells separated into heavy and light components only after shearing to 5×10^5 daltons molecular weight. This implied that segments of greater molecular weight than this had been exchanged. From the proportion of the ^{3}H-labelled molecules that were of intermediate density, Rupp *et al.* concluded that one exchange occurred for every one to two pyrimidine dimers, implying that dimers and post-replication gaps cause genetic exchanges with high efficiency. They favoured the idea that the gap in the new strand opposite a dimer is filled from the sister duplex, the resulting gap in the latter being filled by new synthesis using the intact complementary strand (Fig. 49(c)–(e)).

In theory, the recombination process outlined by Rupp *et al.* (1971) would give efficient repair unless there were pyrimidine dimers in both strands of the DNA in proximity to one another, following the exposure to ultraviolet light. After replication had occurred, the gaps in the newly synthesized strands might then overlap. Evidence for inefficient repair in this pathway involving recombination was obtained independently by Miura and Tomizawa (1968) and Witkin (1969), who independently discovered that ultraviolet light will not induce mutations in *recA* mutants. The inference was that the repair process involving the normal *recA* gene was prone to errors, and these were the source of ultraviolet-induced mutations.

For progress in understanding the function of the *recA* product, and to assist in isolating it, conditional mutants are required. Mount (1971) obtained an amber *recA* mutant. He used a strain that did not carry an amber suppressor, treated the cells with the mutagen nitrosoguanidine, and then plated them on a culture medium containing methylmethane suphonate (MMS). Small colonies indicated sensitivity to MMS and these were tested for ultraviolet sensitivity and ability to undergo recombination. Four *recA* mutants were obtained, one of which, no. 99, was evidently an amber mutant as it was suppressed by amber suppressors.

To find a conditional mutant of a different kind, namely, showing sensitivity to temperature, Lloyd *et al.* (1974) took advantage of the discovery, made by Gross, Grunstein and Witkin (1971) that *recA* mutants cannot survive in a strain defective in DNA polymerase I (*polA⁻*). In agreement with this, a cold-sensitive *pol⁻* mutant as the double mutant (*pol⁻ recA⁻*) was found to grow only poorly below 30 °C. Lloyd *et al.* (1974) treated a *polA* mutant with a mutagen and looked for survivors unable to grow at a high temperature (42 °C). In this way they isolated a temperature-sensitive *recA* mutant, no. 200. At 42 °C it was recombination-deficient and ultraviolet-sensitive, but not at 35 °C or below.

Another temperature-sensitive *recA* mutant, no. 44, was investigated by Hall and Howard-Flanders (1975). They found that the short DNA strands synthesized after ultraviolet irradiation increased in molecular weight at 32 °C but not at 45 °C. This implied that the product of the *recA* gene was required for the process that leads to the filling of the gap between the short strands, as implied by the earlier studies on the mechanism of post-replication repair. Moreover, incubation at 32 °C for 40 min after the irradiation allowed a shift towards higher molecular weights, but this proceeded no further on transfer to 45 °C. It was inferred that there was a continuing need for the normal *recA* product in the subsequent phases of the post-replication repair process.

Cole (1973) studied the repair of DNA containing cross-links between the strands. These covalent links had been induced by treatment with 4,5′,8-trimethylpsoralen followed by exposure of light of wavelength 360 nm for 10 min. The DNA had been labelled with [³H]thymidine. After various periods of incubation, following the cross-linking treatment, samples of the DNA were taken and sedimented in an alkaline sucrose gradient. It was found that after 30 min incubation the DNA, denatured by the alkali, sedimented more slowly than at 0 min. This implied that the single strands had been cut into discrete pieces. Their molecular weight was equivalent to a single strand of length equal to about twice the estimated average distance between cross-links. Cole (1973) suggested that these results could be accounted for if one of the strands of the DNA was cut on each side of each cross-link. Since either strand might be cut, the single-stranded fragments released in the alkaline sucrose would, on average, measure twice the cross-link separation (Fig. 50). Later work supported this idea, with the cut on the 5′ side of the cross-link made by the endonuclease coded by *uvrA* and *uvrB*, and that on the 3′ side by the nuclease activity associated with DNA polymerase I. After 90 min incubation before extracting the DNA, Cole found that a wild-type strain gave fast-sedimenting DNA, indicating high molecular weight, but in a *recA*

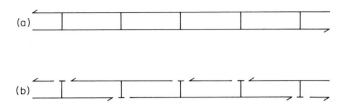

Fig. 50. Initial step in the repair of DNA containing cross-links, according to Cole (1973). (a) DNA duplex with cross-links. (b) Two cuts at each cross-link, both in the same strand and one on each side of the cross-link.

mutant (no. 1) the DNA fragments had been further degraded into smaller pieces. It was evident that the normal *recA* gene was involved in the pathway that led to the covalent joining of the fragments found at 30 min to give the high molecular weight DNA found at 90 min.

In a second experiment Cole (1973) used heavy isotopes to label the DNA and then allowed one replication in light medium containing [³H]thymidine before giving the psoralen-plus-light treatment. In this way sister duplexes would each have one heavy non-radioactive and one light ³H-labelled strand, and the strands of the same polarity in the two molecules would be one of each kind, prior to the cross-linking and its repair. After the psoralen plus light, the cells were incubated in light non-radioactive medium for 60 or 90 min before DNA extraction. Centrifuging in alkaline caesium chloride density gradients would allow the complementary strands to separate and band at their respective densities. It was found that in wild-type cells some of the ³H label occurred at intermediate and heavy densities, but no such shift of the radioactivity from the light density was observed in *recA⁻* cells. It was inferred that in the presence of *recA⁺*, but not otherwise, recombination occurred between sister duplexes in the repair of DNA containing cross-links.

Genetic evidence for recombination in the repair of DNA containing cross-links was obtained by Lin, Bardwell and Howard-Flanders (1977). The bacteria carried prophage λ with a mutation in gene *P* and were infected with λ with a different mutation in the same gene. Wild-type (*P⁺*) recombinant prophage arose with low frequency (less than 0.1%), but increased 50- to 100-fold to frequencies above 1% when the phage was treated with psoralen plus light before infection. Few if any exchanges were induced, however, if the bacteria carried a *uvrA*, *uvrB*, or *recA* mutant (no. 1). Exchanges were frequent in *recB* or *recC* mutants, implying that the *recBC* nuclease, exonuclease V, is not needed for recombination initiated by incisions in DNA containing cross-links. The same conclusion can be drawn for recombination initiated by post-replication gaps at pyrimidine dimers, since post-replication repair in ultraviolet-irradiated *recB* or *recC* mutants is normal. Further evidence is given in Section 10 that the RecBC pathway is not needed for ultraviolet repair.

The mechanism of genetic exchange that is apparently initiated by a nick or a gap in one strand of a DNA molecule, and which requires the *recA* gene product, is discussed in Section 11.

10. THE *recA* GENE PRODUCT

From the experiments described in Sections 4 and 9 it is evident that knowledge of the function of the *recA* gene product is central to an understanding of recombination in *E. coli*. The product is inferred to be a protein because conditional mutants, both amber and temperature-sensitive, have been isolated (see Section 9).

The *recA* protein was identified by McEntee, Hesse and Epstein (1976) by making use of a mutant strain of phage λ. The genome of this strain became integrated as prophage at an abnormal position in the host genome near the *recA* gene. It was possible then to isolate phages in which excision had led to the inclusion of neighbouring host genes, including *recA*. These phages could thus transfer the *recA* gene to a host cell by specialized transduction. A *uvrA* mutant of *E. coli* was given a heavy dose of ultraviolet, with the result that the cells lost the capacity to incorporate amino acids into proteins, owing to reduced transcription of the damaged genome. The λ phage carrying the *recA* gene was then introduced into the cells, with the result that amino acids were preferentially incorporated in proteins coded by the phage. Uptake of isoleucine labelled with ^{14}C was studied by polyacrylamide gel analysis of cell extracts. It was found that a significant fraction of the label was incorporated in a single protein band that was lacking when a phage carrying a *recA* mutant was substituted for that carrying wild-type *recA*. These results suggested that the protein in question was the product of the *recA* gene. The mutant *recA* used was the amber mutant (no. 99) isolated by Mount (1971). When an amber suppressor host strain was substituted for the suppressor-free strain previously used, and the experiment repeated, infection with λ carrying the amber *recA* mutant led to the synthesis of the protein, as expected if it was the *recA* product. Furthermore, this protein band was the only detectable difference between the non-suppressing and the suppressing cell extracts. McEntee *et al* concluded that this protein was the product of the *recA* gene.

By comparing the mobility on polyacrylamide gels of the *recA* protein and known λ proteins. McEntee *et al.* (1976) estimated the molecular weight of the *recA* product to be 43 000 daltons. The protein was extracted and purified by following the distribution of radioactivity during fractionation. Centrifugation left the radioactivity in the supernatant, and chromatography resulted in a peak of radioactivity that eluted as a protein with a molecular weight of about 150 000 daltons, possibly a tetrameric form of the *recA* protein.

Inouye and Pardee (1970) had discovered that when DNA synthesis was inhibited, either by thymidine starvation or by treatment with nalidixic acid, a specific protein was produced at a greater rate than normal. This protein gave a peak at a position they called X in the acrylamide gel when the membrane fraction of the cells was subjected to electrophoresis. From its position the protein was estimated to have a molecular weight of about 39 000 daltons. Later studies by Gudas and Pardee (1976) showed that this protein X, as it came to be called, was found in the cytoplasm to a much greater extent than on the membrane, and that *in vitro* in the presence of magnesium ions it had an affinity for single-stranded DNA.

Emmerson and West (1977) separated protein X from other proteins of *E. coli* by a two-dimensional gel electrophoretic technique which separates proteins according to isoelectric point in one dimension and according to molecular weight in the second. When protein X is induced in wild-type cells it has an isoelectric point of about 6.0, but protein X from a mutant called 'tif' was found to have an isoelectric point of about 6.2. Complementation studies by Castellazzi *et al.* (1977) indicated that tif is a mutation of the *recA* gene. It follows that protein X is the *recA* gene product. Similar results to those of Emmerson and West were obtained by McEntee (1977) and Gudas and Mount (1977). Protein X and the *recA* protein were found to have identical molecular weights and isoelectric points, and this agreement was maintained in a mutant (no. 12) with a smaller than normal product, and in the tif mutant with its abnormally basic product.

These results also confirmed that tif is a *recA* mutant. It had earlier been thought to belong to another gene because, although known to be closely linked to *recA*, it did not cause recombination deficiency. It had been isolated by Goldthwait and Jacob (1964) through its sensitivity to high temperature, and had been called 'tif', standing for 'thermal induction and filamentation', by Castellazzi, George and Buttin (1972) because, when lysogenic for λ, phage production is induced at 42 °C; and, when non-lysogenic, filaments are formed at this temperature due to a defect in the formation of cross septa.

Howard-Flanders and Boyce (1966) had described a mutation which they called *lex*, standing for 'locus for X-ray sensitivity'. It was located at about 90 min on the genetic map (Fig. 46) and was recognized by its increased sensitivity to ultraviolet and ionizing radiation compared with wild type. The mutant was studied in more detail by Mount, Low and Edmiston (1972), who also described two new mutants (nos 2 and 3) at this locus which they had isolated through their X-ray sensitivity. All three *lex* mutants were found to be dominant and to have no effect on recombination. Their genetic locus was subsequently called *lexA* when Blanco, Levine and Devoret (1975) isolated a mutant with the 'lex' phenotype that mapped near *recA* and to which they gave the name 'lexB'. Morand, Blanco and Devoret (1977) studied three mutants with the lexB phenotype, one being one of the mutants defective in recombination isolated by Van de Putte, Zwenk and Rörsch (1966). Morand *et al.* (1977) found that these mutants were less sensitive to ultraviolet than *recA* mutants, showed some recombination proficiency (ranging from 1% to 30% of wild type depending on the mutant), and when carrying prophage λ the phage was not released on exposure to ultraviolet. An early discovery with *recA* mutants was that ultraviolet failed to release the phage, so in this respect the lexB mutants resembled *recA* mutants. Moreover, Morand *et al.* found that the three lexB mutants altered unequally and independently several of the functions affected. On account of this diversity and the overlap with *recA* mutant characters, they favoured the idea that lexB mutations were located within the *recA* gene. Support for this idea was given by the discovery by Castellazzi *et al.* (1977) of a mutant (called zab-4) with lexB phenotype at 37 °C and *recA* mutant phenotype at 30 °C. These authors studied dominance and complementation between tif, lexB and *recA* mutants, using partial diploids. They found that lexB mutants, unlike *lexA*, were recessive, and they found no complementation, or only partial complementation, between *recA* and lexB mutants as regards resistance to ultraviolet light, as if the same product were affected and these mutations were all in the same gene. They suggested that the *recA* product had two active sites, one (site no. 1) needed for recombination and one (site no. 2) for induction of phage λ and for inhibition of cell division. The part of the *recA* protein involved in recombination would be permanently active, whereas the other part of the molecule needed to be activated before it could function. This activation would occur in response to an interruption of DNA synthesis, or in tif cells as a result of exposure to high temperature. Castellazzi *et al.* (1977) further suggested (1) that tif mutants affected that part of the *recA* product where interaction occurred with an effector molecule responsible for activating site no. 2, and (2) that lexB mutants modified this active site (no. 2). This hypothesis would account for the recombination proficiency of lexB mutants, if site no. 1 was unaffected (or only slightly altered) in these mutants. The hypothesis would also explain why tif mutants affected only the inducible functions of the *recA* product, such as the release of prophage λ and the control of cell division.

Stacey and Lloyd (1976) isolated recombination-deficient mutants by a new method. They used a stable F′ merodiploid heterozygous for two non-complementing mutants in the galactose operon and, following mutagen treatment, the cultures were screened for those that gave a reduced number of wild-type progeny, implying a lower frequency than normal of recombination between the mutant sites. One of the recombination-deficient mutants isolated in this way was an unusual *recA* mutant, less severe in its defects than previous *recA* mutants. Further studies by Lloyd and Low (1976) showed that this mutant (*recA*255) retained a residual capacity for recombination and, unlike other *recA* mutants, it did not prevent induction of prophage λ after ultraviolet irradiation. Morand *et al.* (1977) pointed out that this mutant appeared to be the converse of lexB mutants, affecting only the constitutive functions (active site 1) and not the inducible ones (active site 2).

Having established that protein X is the *recA* product, Emmerson and West (1977), McEntee (1977) and Gudas and Mount (1977) considered the question of how the synthesis of this protein is regulated. From their results they suggested that the synthesis of *recA* protein is repressed by the product of the *lexA* gene. They also proposed that DNA damage leads to an inducer molecule combining with the *recA* protein, such that it is activated to destroy the *lexA* product. In this way, DNA damage would lead the *recA* protein to derepress its own synthesis, and so the synthesis of *recA* protein would be self-regulating. The tif mutant would produce a temperature-sensitive form of *recA* protein which, at 42 °C, could spontaneously remove the *lexA* product from the operator of *recA* without the need for inducer.

It was pointed out above that *recA* protein is synthesized in response to (1) inhibition of DNA synthesis by nalidixic acid, and (2) DNA damage by ultraviolet light. McPartland, Green and Echols (1980) made the surprising discovery that these two responses involve different pathways. They found that both nalidixic acid and ultraviolet light increased the synthesis of *recA* messenger RNA. Furthermore, *recA* messenger induction by these agents was abolished in a *lexA* mutant (no. 3) and also in a *recA* mis-sense mutant (no. 1). As the latter mutation causes an amino acid substitution in the *recA* protein rather than premature termination of polypeptide synthesis, or deletion of part or all of the gene, it was evident that the failure to synthesize *recA* messenger could not be attributed directly to the genetic defect of the *recA* mutation. These results thus showed that the *lexA* and *recA* products regulate transcription of *recA*. McPartland *et al.* (1980) found that mutations in *recB* and *recC* prevent the induction of *recA* by nalidixic acid but not by ultraviolet light, while a *recF* mutation had the converse effect. They concluded that induction by nalidixic acid involves the RecBC pathway and by ultraviolet light the RecF pathway. That the RecBC nuclease, exonuclease V, was not needed for recombination initiated by ultraviolet-induced damage was already evident from the normal post-replication repair in *recB* or *recC* mutants, as pointed out in Section 9.

As already pointed out, recombination-deficient *recA* mutants (other than *recA*255), when carrying prophage λ, do not release the phage on exposure to ultraviolet light. Induction of the prophage is controlled by the product of λ gene *c*I, which is a repressor molecule. Roberts and Roberts (1975) investigated what happened to the repressor when λ was induced, that is, released. The proteins were labelled with ^{35}S and the repressor purified from cell extracts by precipitation with specific antibody. The precipitate was analysed by electrophoresis in polyacrylamide gel. It was found that the intact repressor disappeared from the culture during the half hour after induction. Instead of the repressor polypeptide at the 27 000 dalton molecular weight position in the gel, at least part of the molecule was found as a fragment of molecular weight 14 000 daltons. It was evident that induction of λ prophage by ultraviolet light resulted in cleavage of the repressor molecules. In a *recA* mutant, however, no breakdown of the repressor was found. It was clear from these experiments that proteolytic cleavage might be the primary mechanism of repressor inactivation following exposure to ultraviolet light, although the possibility that the breakdown was secondary could not be ruled out. It was also evident that *recA* activity was involved, conceivably itself being the protease in question.

In a further investigation of the proteolytic cleavage of the λ repressor during induction, Roberts, Roberts and Mount (1977) achieved the breakdown *in vitro*. They used an *E. coli* strain in which induction occurred constitutively. This was a triple mutant comprising mutant no. 3 at the *lexA* locus and the tif mutant at the *recA* locus, together with an additional mutation, 'spr' (standing for spontaneous repressor inactivation) that resulted in increased induction. This spr mutant had been isolated by Mount (1977), who found that it mapped at the *lexA* locus.

Roberts *et al.* (1977) found that λ repressor added to an extract of the triple mutant was inactivated by incubation in the presence of ATP and magnesium ions. The repressor was assayed by its DNA binding activity. Using the same method as in the previous investigation, it was found that cleavage accompanied the inactivation of repressor *in vitro*. The demonstration of specific proteolytic cleavage of λ repressor *in vitro* in an ATP-dependent reaction gave support to the idea that this is the inactivation mechanism and not a secondary process. Roberts, Roberts and Craig (1978) fractionated the crude extract of the *lexA*3 tif spr triple mutant and tested the components for their ability to destroy the DNA-binding activity of the λ repressor. With this assay it was found that a single predominant polypeptide of molecular weight 40 000 daltons was selected. This purified repressor-inactivating factor and protein X were analysed in parallel by poly-acrylamide gel electrophoresis in sodium dodecyl sulphate. The protein X was induced by growing *E. coli* in the presence of nalidixic acid to inhibit DNA synthesis, and also by growing the tif mutant at 40 °C. It was found that the inactivating factor is a protein of the same molecular weight as protein X. Roberts *et al.* (1978) concluded that the inactivating factor is protein X and, therefore, is the *recA* protein. This was confirmed by electrophoresis of protein precipitated with antibody to purified inactivating factor. A *recA* mutant (no. 12) that encodes a *recA* protein with a slightly greater mobility than wild type gave an inactivating factor that likewise had slightly greater mobility than normal. Roberts *et al.* (1978) pointed out that their work did not constitute a rigorous demonstration that *recA* protein is a protease: it was still possible that the *recA* protein modifies the repressor so that it is recognized by a separate protease.

Little *et al.* (1980) found that the *recA* protein can cleave the product of the *lexA* gene. This was demonstrated *in vitro* under the same conditions as for cleavage of the λ repressor. No such cleavage was found with *lexA* mutant no. 3. These results support the hypothesis that inactivation of *lexA* protein by the *recA* protease is necessary for derepression of the *recA* gene. Little *et al.* (1980) suggested that λ repressor had evolved sensitivity to the protease because release of phage would be advantageous in a cell with damaged DNA that might die.

Recombination was studied by Lloyd (1978) in *E. coli* strains in which *recA* protein was synthesized constitutively. The raised level of *recA* protein in the *lexA*3 tif spr triple mutant was found to be associated with an increased frequency of multiple exchanges between Hfr and F– DNA. A similar conclusion had been reached by Lloyd and Low (1976), who found that two leaky *recA*⁻ mutants that allowed a low residual level of recombination markedly reduced the proportion of recombinants with multiple exchanges. As indicated above, Castellazzi *et al.* (1977) favoured the idea that the part of the *recA* protein involved in recombination was permanently active, unlike the part of the molecule needed for prophage induction and the other induced activities. Lloyd (1978) suggested, however, that the *recA* activity which promoted increased multiple exchanges was the same as that necessary for the induced functions. This was because he had found that the tif mutation was necessary in conjunction with spr to give high levels of multiple exchanges, although spr without tif was known to suffice to promote constitutive synthesis of *recA* protein. Oliver (1982) also found that, following conjugation, selection for recombinants at a particular locus was associated in tif cells with an increase, compared with wild type, in multiple exchanges involving widely spaced unselected markers.

The nucleotide sequence of the *recA* gene was determined by Horii, Ogawa and Ogawa (1980) and Sancar *et al.* (1980). The gene was propagated in a plasmid, cleaved by restriction enzymes, and the fragments sequenced. The gene was found to contain 1059 nucleotide residues, corresponding to 353 amino acid residues in the polypeptide. From the amino acid composition the molecular weight of the *recA* polypeptide was calculated to be 37 842. The site of initiation of transcription *in vitro* was determined to be 50 base pairs upstream of polypeptide initiation. Several reverse repeats were found in the neighbourhood of the transcription origin. These could include the site where the *lexA* protein interacts with the DNA.

11. THE ACTION OF THE *recA* PRODUCT IN RECOMBINATION

As indicated in Section 9, there is evidence that cross-linked DNA, after the action of the endonuclease coded by the *uvrA* and *uvrB* genes, initiates exchange with a sister duplex in the presence of the *recA* gene product. This process of exchange was investigated by Ross and Howard-Flanders (1977*a*). *E. coli* lysogenic for phage λ, that is, carrying the λ genome integrated as prophage, was infected with ^{32}P-labelled λ and incubated. The infecting λ formed covalently closed, circular molecules by the annealing and sealing of its cohesive ends, as is normal following infection. The cells were then further infected with unlabelled λ that had been treated with psoralen plus light to induce interstrand links. After incubation the cell contents were centrifuged in alkaline sucrose. A reduction in the proportion of labelled, rapidly sedimenting molecules was found, and the size of the reduction depended on the dose of 360 nm light given to the unlabelled phage after the psoralen treatment. The rapidly sedimenting molecules are those that are covalently closed; those that sediment more slowly contain nicks. The cutting in *trans*, as Ross and Howard-Flanders called it, was evidently caused by the psoralen plus light treatment in view of the dose dependence. Furthermore, there was no reduction in sedimentation rate in *uvrA*⁻ or *recA*⁻ cells, showing that the normal alleles of these genes were required. Likewise, there was no reduction when the super-infecting phage was untreated λ. It was evident that λ DNA that has been cut by the *uvrA*-coded endonuclease can bring abouts cuts in *trans* in another λ genome in the presence of the *recA* gene product.

Ross and Howard-Flanders (1977*b*) went on to investigate whether the cutting in *recA*⁺ cells of an undamaged, covalently closed, circular λ genome in the presence of a λ genome cut by the *uvrA* endonuclease depended on the homology of nucleotide sequences in the two molecules. They used an *E. coli* strain lysogenic for phage λ and also for a phage called '186', which has no detectable base homology with λ. The bacterial cells were infected with ^{32}P-labelled λ and ^{3}H-labelled 186. Super-infection with unlabelled λ that had been exposed to psoralen plus light caused a marked decrease in the fraction of ^{32}P radioactivity sedimenting rapidly, but no decrease in the fraction of ^{3}H radioactivity sedimenting rapidly. Thus, the fraction of 186 DNA in the covalent circular form remained undiminished despite the presence of damaged λ DNA capable of inducing cuts in λ covalent circles. Conversely, super-infection with unlabelled 186 that had been exposed to psoralen plus light led to cutting of the undamaged ^{3}H-labelled 186 DNA, while the ^{32}P-labelled λ DNA remained intact. It was evident that the cutting in *trans* was specific to homologous DNA.

Lin *et al.* (1977) had found that recombination between genetic markers in prophage λ and infecting λ was much increased if the infecting phage had been treated with psoralen plus light (see Section 9). Ross and Howard-Flanders (1977*b*) repeated this experiment using the double lysogen-carrying prophages λ and 186. When the cells were infected with psoralen-damaged λ, the frequency of recombinants among the λ prophages was increased, as Lin *et al.* had found. When the cells were infected with undamaged λ and with psoralen-damaged 186, however, the frequency of λ recombinants was unaffected. The converse experiment, using genetic markers in prophage 186 and the infecting 186 gave comparable results: the stimulation of recombination was restricted to DNA homologous to that which was damaged.

To account for their results, both physical and genetic, Ross and Howard-Flanders (1977*b*) suggested that base pairing occurred between damaged molecule and undamaged homologue before cutting in *trans* took place. The data are not consistent with the idea that psoralen-damaged DNA activates a nuclease that cuts all the DNA molecules in the cell. The need for the *uvrAB* endonuclease showed that strands need to be cut to initiate the *recA*-dependent recombination. Ross and Howard-Flanders therefore inferred that the initial step in recombination was the pairing of the broken molecule with its homologue. Either the cut strand or the uncut strand of the damaged molecule might take part in the pairing (Fig. 51). There would follow a *recA*-dependent reaction by an endonuclease that cuts the undamaged member of the paired homologues. Thus, the sequence of events would be: cut, pair, cut.

Cassuto, Mursalim and Howard-Flanders (1978) described an *in vitro* system in which the *uvrAB*-determined cutting of DNA containing cross-links and the induced *recA*-dependent cutting of undamaged homologous DNA could be studied without purification of the enzymes involved. Cross-links were induced in ^{3}H-labelled covalent circular duplex DNA molecules of phage φX174 by psoralen plus light treatment. Production of the *recA* protein in cell extracts was amplified by using *E. coli* lysogenic for the λ phage carrying the *recA* gene. The phage was induced by heat treatment. Cutting in *trans* was observed by incubating the ^{3}H-labelled, cross-linked molecules with ^{32}P-labelled, undamaged, but otherwise similar φX174 DNA molecules in the presence of the *E. coli* extract, and following the sedimentation rate of ^{32}P-labelled molecules in alkaline sucrose. A reduction in the proportion of rapidly sedimenting molecules implied shorter strands as a result of cutting. Some of the results were puzzling, but evidence was obtained for

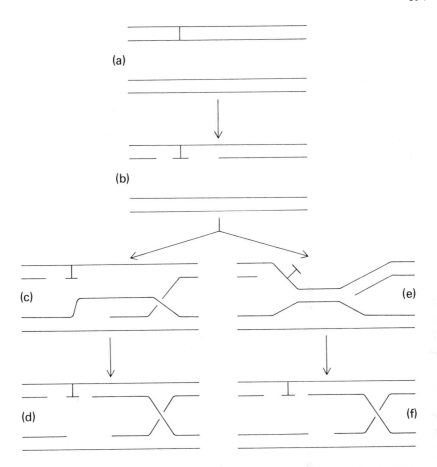

Fig. 51. Mechanisms proposed by Ross and Howard-Flanders (1977*b*) for the initiation of recombination following the incision of a DNA molecule containing a cross-link. Pairs of horizontal lines represent DNA duplexes. (a) Two homologous duplexes, one with a cross-link. The strands may be numbered 1–4 reading downwards. (b) A cut has occurred in strand no. 2 on each side of the cross-link, followed by some exonucleolytic breakdown. (c) The broken strand (no. 2) has paired with the homologue (strand no. 4). (d) The displaced strand (no. 3) of the undamaged molecule has been cut and has paired with no. 1. (e) The unbroken strand (no. 1) of the cross-linked molecule has paired with the homologue (strand no. 3). (f) Strand no. 3 has been cut and strand no. 2 has paired with no. 4.

cutting in *trans* in *uvr*$^+$ *recA*$^+$ extracts and its absence in *uvrB*$^-$ or *recA*$^-$ extracts.

Closed circular duplex DNA molecules, such as those of φX174 used by Cassuto *et al.* (1978) are superhelical, that is, the molecule has a locked-in twist that is lost if one of the strands is cut (cf. Chapter V, Section 9). It seems that in the formation of closed circular molecules the last closure of chain ends occurs before all of the winding of the two DNA strands into the double helix structure is completed. In consequence, closed circular DNA molecules are underwound, that is, they have negative superhelical turns, and owing to their compact form sediment much faster than nicked circular molecules which have a simple circular structure. In closed circular DNA molecules, the topological winding number, that is, the number of times one strand winds about the other when the molecule lies in one plane, is smaller than the number required for the B form of the Watson–Crick double helix. Free energy is required to form the superhelix turns, with the result that the strands will start to separate at a lower temperature than in a nicked circular molecule and free energy is released from the superhelical molecule when a strand is nicked.

In view of the free energy of superhelical DNA, Holloman et al. (1975) were led to test the hypothesis that superhelicity promotes the uptake of homologous single strands, as a first step in recombination. They used closed circular duplex DNA of ϕX174 as a source of superhelical DNA and found that stable complexes were formed spontaneously in vitro when ^{32}P-labelled single strands 25–80 nucleotides long of ϕX174 DNA were incubated with the superhelical molecules. The complexes were revealed by their sedimentation rate in neutral sucrose gradients. No such association was detected when single strands from other organisms were substituted, so base pairing was evidently involved. If the closed circular ϕX174 DNA was nicked by S1 endonuclease of Aspergillus oryzae, the DNA molecules failed to take up homologous single strands. It was evident that the superhelicity of the circular molecules, which is lost on nicking, was necessary for the formation of the complex.

Holloman and Radding (1976) related these in vitro observations to recombination by showing that superhelicity can strongly promote recombination that depends upon the recA gene. A mixture of ϕX174 superhelical DNA and single-stranded fragments of ϕX174 DNA differing by a genetic marker was added to E. coli spheroplasts. Recombinant ϕX174 progeny, incorporating the marker from the single-stranded fragments, were obtained with wild-type E. coli, but with a recA mutant (no. 13) the frequency was 10–20 times lower. When relaxed ϕX174 molecules, that is, without supercoiling, were substituted for the superhelical ones, the recombination frequency with wild-type E. coli was 20–100 times lower. Holloman and Radding pointed out that the evidence that superhelical DNA promoted recombination was of wide significance, because superhelicity was not confined to organisms with small circular genomes but appeared to occur in E. coli and in eukaryotes, through topological constraints on the DNA through folding.

Purified recA protein was obtained by Weinstock, McEntee and Lehman (1979) using cells containing high levels of it, as a consequence of the presence of the spr mutation at the lexA locus, and also through nalidixic acid treatment to inhibit DNA synthesis and so induce recA protein synthesis. The protein was assayed by polyacrylamide gel electrophoresis. It was found that denatured DNA incubated with recA protein, magnesium ions and ATP led to the rapid formation of large DNA aggregates containing many branched structures that included duplex regions as well as single strands. This was revealed by digestion with S1 nuclease and by electron micrographs. In the absence of the *recA* protein the DNA remained single-stranded. It was evident that the *recA* protein catalyses renaturation of single-stranded DNA. This annealing of complementary strands was coupled to the hydrolysis of ATP. Weinstock *et al.* (1979) also found that the *recA* protein from a cold-sensitive *recA* mutant (no. 629) did not catalyse renaturation at 28 °C but did so at 37 °C. This mutant is recombination-deficient at the lower temperature but not at the higher one, strongly suggesting that *in vivo* the *recA* protein catalyses the annealing of a single-stranded region from one DNA duplex to a complementary simplex region of another molecule.

The *recA* protein was also purified by Shibata *et al.* (1979*a*) from a strain carrying the *recA* gene on a plasmid, and using as an assay the hydrolysis of ATP, which the *recA* protein catalyses in the presence of DNA. They found that the *recA* protein catalysed ATP-dependent pairing of superhelical DNA and homologous single-stranded fragments. This was in direct agreement with the evidence obtained by Holloman and Radding (1976) implicating the *recA* gene in the interaction of these DNA molecules. The discovery that *recA* protein cata-lysed strand uptake by a duplex molecule gave support to the hypothesis of Hollo-man *et al.* (1975) that this reaction is involved in the initiation of recombination.

The interaction of *recA* protein with duplex DNA was also investigated by McEntee, Weinstock and Lehman (1979), who showed that the duplex molecule, even when relaxed, that is, without superhelical turns, can assimilate single-stranded DNA into a homologous region in the presence of the *recA* protein. They used phage P22 as a source of superhelical DNA, linear duplex DNA, and linear single-stranded DNA. They found that in the presence of ATP, magnesium ions and homologous simplex DNA, the *recA* protein converts ^{3}H-labelled duplex DNA, whether superhelical or linear, into a form that was retained on nitrocellulose filters. Enzyme treatment suggested that assimilation of single strands into the labelled duplex had occurred, displacing one strand of the duplex as a loop (Fig. 52(i)). The complex would be retained on the filter because of the single-stranded displacement loop (D-loop). Such a loop is defined as a triple-stranded region in DNA containing two paired strands and an unpaired strand, the latter covalently linked at both ends to the rest of the duplex DNA. The pre-sence of the loops was confirmed by electron microscopy. McEntee *et al.* (1979) found that the assimilation reaction could be separated into two steps. In the first, *recA* protein bound to duplex DNA. This binding required ATP or GTP and mag-nesium ions, but hydrolysis of the nucleoside triphosphate was not required. In the second step, a single-stranded region of the duplex DNA where the *recA* protein has bound to it hybridized with a complementary single strand, the other strand of the duplex forming the D-loop. This reaction was coupled to ATP hydrolysis, which appeared to be required to dissociate the *recA* protein from the DNA. Cun-ningham *et al.* (1979) independently found that superhelicity was not essential for single-strand assimilation by *recA* protein. They used nicked circular duplex DNA of phage fd, and found that complexes retained by nitrocellulose filters were one-half to two-thirds as frequent as with superhelical DNA, on addition of homologous single strands in the presence of *recA* protein and ATP.

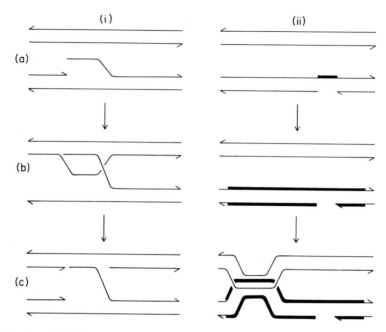

Fig. 52. Hypothetical sequences of events by which *recA* protein promotes association between two duplex DNA molecules.

(i) Strand uptake by an intact duplex to form a D-loop, as discussed by McEntee *et al.* (1979) and Cunningham *et al.* (1979). (a) A broken strand in one duplex uncoils. (b) The broken strand anneals with the complementary strand of the intact duplex, displacing the other strand to form a D-loop. (c) Digestion of the D-loop by *recBC* DNase takes place, as observed by Wiegand *et al.* (1977).

(ii) Four-strand homologous pairing before strand separation, as favoured by Cassuto *et al.* (1980) and West *et al.* (1981a). The thick lines show *recA* protein bound to the DNA strands. (a) An intact and a gapped DNA duplex, with *recA* protein bound to the single-stranded DNA. (b) The *recA* protein has bound to neighbouring duplex regions. (c) The *recA* protein has led to an association of the gapped molecule with the intact one. A search for homology has been successful, and has given rise to base pairing between the two duplexes.

The action of the *recBC* DNase on D-loops was investigated by Wiegand *et al.* (1977), who found that it digested the displaced strand (Fig. 52(i)). Indeed, their data suggested that this enzyme cleaves D-loops more specifically than it cleaves single-stranded DNA. They pointed out that the sequence *in vitro* of strand uptake, loop cleavage and covalent union of the assimilated strand with the recipient duplex could represent steps in recombination *in vivo*.

Cunningham *et al.* (1979) discovered that single-stranded DNA, whether homologous or not, stimulated *recA* protein to unwind duplex DNA. Even circular single-stranded DNA, which does not form D-loops, promoted unwinding. This was shown by agarose gel electrophoresis of nicked circular duplex DNA of phage fd incubated with *recA* protein and ATP, with the addition of simplex ϕX174 DNA. Ligase was also present, resulting in the sealing of some of the duplex molecules. The presence of the non-homologous single strands altered the migration rate in the gel. The products were centrifuged in a caesium chloride gradient to establish their molecular structure. Cunningham *et al.* (1979) suggested that the non-specific stimulation of unwinding by a single strand explained in part how *recA* protein promotes a search for homology. Their finding illuminates one of the most obscure aspects of recombination.

Further investigation of the *in vitro* properties of the *recA* protein by Shibata *et al.* (1979*b*) revealed that the amount of *recA* protein required for the homologous pairing of single-stranded DNA with duplex DNA was directly proportional to the amount of single-stranded DNA. They used phage fd DNA in these experiments and measured the formation of loops, revealed by retention by nitrocellulose. They found that the minimal amount of *recA* protein needed was one molecule per 10–18 bases of single-stranded DNA, and the optimum amount agreed with a previous estimate of one molecule per five to 10 bases. Using γ-S-ATP, an ATP analogue (adenosine 5'-γ-thiotriphosphate) that was not rapidly hydrolysed, they obtained stable complexes of *recA* protein and single-stranded DNA, and these complexes were found partially to unwind duplex DNA, whether homologous or not. They concluded that if the complexes formed with γ-S-ATP correspond to natural intermediates, *recA* protein must promote homologous pairing either by moving simplex and duplex DNA relative to one another, or by repeatedly forming complexes and then dissociating them again if there was no homology. The hydrolysis of ATP might drive either kind of motion.

Craig and Roberts (1980) showed that the proteolytic cleavage of λ repressor directed by *recA* protein required polynucleotide and ATP. They suggested that a complex of *recA* protein and single-stranded DNA was active both in destroying repressors and in initiating homologous pairing, through its ability to invade duplex DNA.

An *E. coli* protein that binds to single-stranded DNA was described by Sigal *et al.* (1972). It was subsequently found to be essential for DNA replication *in vitro*. A mutation causing temperature-sensitive DNA replication *in vivo* and *in vitro* was found by Meyer, Glassberg and Kornberg (1979) to have mutant single-strand DNA-binding protein (SSB). This established that the gene, *ssb*, in which the mutation mapped was the structural gene for the protein. They went on to investigate the effects of the mutation and found that the mutant was more sensitive to ultraviolet irradiation than wild type and about one-fifth as active in recombination (Glassberg, Meyer and Kornberg, 1979). It was evident that SSB played a part in DNA repair and recombination as well as in replication.

McEntee, Weinstock and Lehman (1980) studied the effect of the binding protein on the strand assimilation reaction catalysed by *recA* protein. They found that SSB increased the rate and extent of strand assimilation into homologous duplex DNA *in vitro*. It was also found that SSB greatly reduced the extensive ATP hydrolysis that accompanies strand assimilation. A mutant form of SSB was less effective in stimulating strand assimilation. McEntee *et al.* (1980) concluded that SSB participates in general recombination *in vivo*, since they considered that strand assimilation was likely to be a central feature of recombination and post-replication repair. Similar results concerning the effect of SSB in reducing the amount of *recA* protein required to catalyse the formation of D-loops from duplex DNA and homologous single-stranded fragments were obtained independently by Shibata *et al.* (1980).

The discovery by Cunningham *et al.* (1979) that *recA* protein was stimulated by single-stranded DNA, even when circular, to unwind duplex DNA led them (Cunningham *et al.*, 1980) to investigate the interaction of closed circular duplex DNA and homologous circular duplex DNA with a gap about 1200 nucleotides long in one of the strands. They used phage DNA, either fd or ϕX174. They found that in the presence of ATP and magnesium ions, *recA* protein promoted the pairing *in vitro* of the homologous intact and gapped duplex molecules. These molecules carried ^{3}H and ^{32}P labels, respectively, and the pairing was demonstrated by the retention of |^{3}H]DNA on nitrocellulose filters. The pairing was confirmed by electron micrographs, which showed that the molecules were sometimes held together in regions remote from the site of the gap. Cunningham *et al.* (1980) inferred that a free end of the interrupted strand in the gapped duplex paired with the complementary strand of the intact duplex to form a structure like a D-loop, which moved away from the gap by branch migration.

Similar experiments were made by West *et al.* (1980). They found that duplex DNA containing a single-strand gap of about 30 nucleotides was as effective in promoting association of *recA* protein and DNA as wholly single-stranded DNA. Their measure of this interaction was the associated ATPase activity. Sedimentation studies showed cooperative binding of *recA* protein to simplex or gapped duplex DNA, and this was confirmed by electron microscopy, following incubation in the presence of the non-hydrolysable γ-S-ATP: segments of the DNA up to 10^4 bases long had more than twice the normal thickness, with a straight rod-like appearance. No evidence was found from the electron micrographs that the *recA* protein was separating the strands of the gapped molecules. West *et al.* (1980) favoured the idea that *recA* protein binds initially at the gaps and then to neighbouring duplex regions (Fig. 52(ii)). Cassuto *et al.* (1980) went on to investigate the interaction between intact and gapped duplex DNA molecules derived from a plasmid. They found that incubation of these molecules in the presence of *recA* protein and ATP led to homologous pairing. This was detected by centrifugation in neutral sucrose gradients, the intact molecules carrying a ^{32}P label and the gapped ones a ^{3}H label, or vice versa. The pairing was also revealed by electron microscopy. If the gapped molecule contained only one gap, dimeric structures were seen, but with several gaps per molecule large complexes were formed involving more than 1000 molecules. Intact DNA of phage ϕX174 mixed with the gapped plasmid DNA showed no pairing. This indicated that the pairing was dependent on genetic homology. The *recA* product and ATP were also found to be essential for the pairing to occur. Single-stranded DNA-binding protein was found to influence the pairing, which was inhibited by SSB added at the start of incubation but considerably enhanced by SSB acting after *recA* protein. When Cassuto *et al.* (1980) incubated intact plasmid duplex DNA with *recA* protein,

ATP and singly gapped homologous DNA containing numerous cross-links induced with psoralen plus light, they found that pairing was unaffected by the cross-links. This implied that little or no unwinding of the gapped duplex took place. Cassuto *et al.* (1980) therefore favoured duplex–duplex pairing (Fig. 52(ii)), as proposed by McGavin (1971) and Wilson (1979).

The idea that *recA* protein can promote homologous association of DNA molecules without the need for a free end was also reached by DasGupta *et al.* (1980). They mixed phage G4 linear duplex DNA and *recA* protein with single-stranded circular phage M13 DNA containing a 274-nucleotide segment of G4 DNA from near the middle of the G4 genome (Fig. 53(a)). In consequence the region of homology in the duplex molecule was flanked on both sides by several thousand base pairs of non-homologous DNA. The location of the junctions of the two molecules relative to the ends of the linear one was measured on electron micrographs and found nearly always to be at or close to the region of homology. Heat treatment insufficient to separate the strands of duplex DNA nevertheless caused the pairing of the two molecules to disappear, as expected if the association was side by side and not in the normal interlocked double helix. Moreover, the single-stranded circular M13 DNA carrying the G4 fragment showed pairing with closed circular superhelical G4 DNA with the same frequency as with the linear duplex DNA, despite the total absence of free ends. These observations confirmed that *recA* protein can promote homologous pairing by a mechanism in which complementarity of nucleotide sequence is recognized by laterally associated strands instead of intertwined ones.

Further evidence that pairing can occur without prior unwinding of a free end of a DNA strand was obtained by West, Cassuto and Howard-Flanders (1981*a*) by incubating *recA* protein, ATP and closed circular duplex DNA of phage ϕX174 or a plasmid with homologous single-stranded circular DNA with an annealed fragment, that is, a short duplex region of about 400 base pairs (Fig. 53(b)). The duplex circles were labelled with ^{32}P. Pairing was revealed by the retention of ^{32}P-labelled DNA on nitrocellulose filters, which trap single-stranded DNA and molecules attached to it. The presence of the annealed fragment was necessary for pairing to take place with the duplex circle, and yet the fragment did not show single-strand assimilation into the duplex circle. The evidence for this was provided by labelling the fragment with ^{3}H, stopping the pairing reaction at the end of incubation by adding γ-S-ATP, and then cleaving the ^{32}P-labelled duplex circle with a restriction enzyme. Sedimentation analysis revealed no association of the ^{3}H label with the ^{32}P-labelled fragments. Furthermore, incubation with *recA* protein did not cause unwinding of a 200-nucleotide ^{32}P-labelled fragment annealed to a ^{3}H-labelled single-stranded circle of phage ϕX174 DNA. Sedimentation analysis showed no separation of the labels, such as was found when helicase I, a DNA-unwinding enzyme described by Abdel-Monem and Hoffmann-Berling (1976), was substituted for the *recA* protein.

From their experiment on the association of the two kinds of circular DNA molecules in the presence of *recA* protein, West *et al.* (1981*a*) concluded that duplex–duplex pairing was occurring (Fig. 52(ii)), the fragment annealed to the

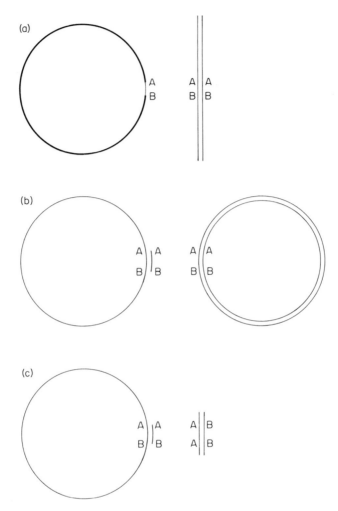

Fig. 53. Substrates for investigating the action of the *recA* protein that were used by (a) DasGupta *et al.* (1980) and (b), (c) West *et al.* (1981a, b), respectively. The letters A and B indicate homologous regions. (a) Single-stranded circular phage M13 DNA (thick line) containing a 274-nucleotide segment of phage G4 DNA (thin line) from near the middle of the G4 genome; and linear duplex G4 DNA. (b) Single-stranded circular DNA of phage ϕX174 or a plasmid, with an annealed fragment about 400 nucleotides long; and homologous closed circular duplex DNA. (c) Single-stranded circular ϕX174 DNA with a 392-base annealed fragment; and a linear duplex fragment 872 base pairs long of ϕX174 DNA homologous with the region of the circle containing the annealed fragment.

single-stranded circle being necessary for this. The single-stranded DNA would provide an initial binding site for the *recA* protein. As they had previously suggested, however, through cooperative binding the *recA* protein spreads along the DNA from simplex to neighbouring duplex regions. Following non-specific contact with the intact duplex, there would be a search for homology and then

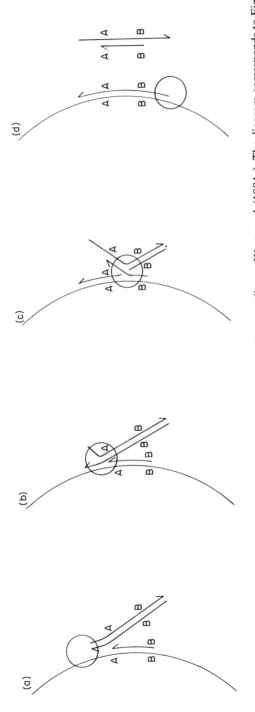

Fig. 54. Steps in strand exchange resulting from the action of *recA* protein, according to West *et al.* (1981c). The diagram corresponds to Fig. 53(c). The *recA* protein is shown as a circle moving in the 5′ to 3′ direction along the single-stranded DNA. Half arrowheads indicate 3′ ends.

base pairing between the two duplexes. Strand cutting, separation and exchange would occur subsequently.

Later the same authors experimented (West, Cassuto and Howard-Flanders, 1981*b*) with a ^{32}P-labelled linear duplex fragment, 872 base pairs long, of ϕX174 DNA in place of the circular duplex DNA (Fig. 53(c)). The other DNA molecule was the same as before, that is, a single-stranded circle of ϕX174 DNA to which a ^{3}H-labelled complementary fragment 392 base pairs long had been annealed. The fragments had been obtained with restriction endonucleases. The linear duplex fragment was homologous with the region of the circle containing the annealed simplex fragment. After incubation in the presence of *recA* protein, ATP and magnesium ions, strand exchange was detected by observing the sedimentation rate of the radioactive molecules in neutral sucrose. It was found that about 50% of the ^{3}H sedimented at the position of the linear molecule and 15% of the ^{32}P at the position of the circular molecule. Subsidiary experiments confirmed that the radioactive molecules in the new positions were held by hydrogen bonds. Reciprocal strand exchange had evidently taken place between the two molecules. When a linear duplex was substituted that was homologous with a part of the circle that did not include the annealed fragement, 15% of the ^{32}P was transferred to the circular molecule as before, but there was no transfer of the ^{3}H to the linear molecule. Strand exchange was blocked by the addition of γ-S-ATP, showing that hydrolysis of ATP was required. Strand exchange also failed to occur when the linear duplex consisted merely of segment AB of Fig. 53(c), that is, the 392 base pairs corresponding to the annealed fragment. It appeared as if annealing of one strand of the linear duplex to the circular molecule in proximity to the annealed fragment was required before reciprocal transfer of the latter could occur. In agreement with this, cross-links in the duplex fragment were found to prevent strand exchange. West *et al.* (1981*b*) concluded that, following homologous pairing, strand transfer is initiated by *recA* protein at the end of the linear duplex and progresses along the molecule, reciprocal exchange occurring when the annealed fragment is reached. Later studies of the same kind (West, Cassuto and Howard-Flanders, 1981*c*) indicated that strand transfer and exchanges are initiated by 3′ but not by 5′ ends. It seems as if *recA* protein binds to a single strand and moves in a 5′ to 3′ direction until a 3′ end on a complementary strand is encountered. The steps in strand exchange in the situation described above might then be as outlined in Fig. 54, with an association of all four nucleotide chains arising (Fig. 54(b),(c)) after an association of three (Fig. 54(a)) and being of transitory duration at any one position.

The manifold activities of the *recA* protein, recently revealed, are astonishing. Knowledge of its behaviour has greatly enlarged understanding of the initial steps in recombination. .

12. CONCLUSIONS

It seems that recombination in *Escherichia coli* is initiated through the actions of the *recA* gene product. This protein evidently functions in bringing about homologous pairing of DNA molecules. After binding to a single-stranded region in one of the molecules, further *recA* protein molecules bind cooperatively to neighbouring duplex regions, if present. Following association with a second duplex molecule, a search for homology takes place. It is not known whether this occurs by moving one molecule relative to the other, or by repeatedly associating and dissociating them. A free end is not required. The process is apparently driven by the hydrolysis of ATP. Strand transfer is promoted by the *recA* protein in the presence of ATP when a 3′ end from one molecule anneals with the complementary strand of another. If the recipient molecule is duplex, a D-loop will form. The cutting of one strand of the recipient duplex, following such invasion by a homologous donor strand, is also a *recA*-dependent reaction. Strand exchange is promoted if both molecules have free 3′ ends, but one of these strands must be longer than the other to allow the initial one-way transfer. A close association of all four strands will occur at the point of strand exchange between the two duplexes, but will have only a fleeting existence at any one point as strand exchange proceeds.

The part played by the *recBC* nuclease (exonuclease V) is not clear. Once a strand of the recipient duplex has been cut, the *recBC* nuclease could untwist the duplex and break down the single strand with a free end, allowing the heteroduplex formed by a donor strand to be extended. If DNA synthesis took place on the donor molecule to fill the gap left by the strand transfer, the situation would resemble that postulated by Meselson and Radding (1975).

As an alternative to the RecBC pathway, about 1% of recombination in wild-type *E. coli* is believed to occur by the RecF pathway. Exonuclease I is thought to shunt an intermediate in the latter pathway into the RecBC pathway. In the absence of exonuclease I the RecF route becomes more effective. The normal function of the RecF pathway may be in DNA repair where exonuclease I has less of an inhibitory effect. Conversely, there is evidence that the RecBC pathway is not needed for DNA repair.

A defective prophage, Rac, that occurs in the *E. coli* genome, carries genes that can play a part in recombination. The RecE pathway, involving exonuclease VIII, arises in this way and provides another alternative to the RecBC route.

Particular nucleotide sequences are favoured sites of action for the *recBC* nuclease, but do not function in the RecE or RecF pathways. The enzyme can move along the DNA from these sites that it recognizes, before promoting recombination with another molecule. This promotion requires the *recA* product as well as the *recBC* nuclease and is thought to involve the formation of a free end to a nucleotide chain, which then penetrates the other molecule and gives rise to a heteroduplex segment.

Mismatch repair in *E. coli*, studied by transfection with λ heteroduplexes, is discussed in Chapter V, Section 13, and Chapter VIII.

Despite much effort, the mechanism of recombination in *E. coli* is not yet understood in detail.

VII.   Recombination in eukaryotes

1. CROSSING OVER

Recombination of linked characters was discovered by Bateson, Saunders and Punnett (1905) with *Lathyrus odoratus* (Sweet Pea) when studying the inheritance of purple versus red petal colour and elongated versus disc-shaped pollen grains. They found that the recombination frequency of these character differences was about 12%, with the two reciprocal recombinants equally frequent. Subsequently they discovered other examples in *L. odoratus* of such partial linkage. Exchanges between homologous chromosomes had been predicted by De Vries (1903), but Bateson *et al.* (1905) thought exchanges conflicted with the evidence for the permanence of the chromosomes, and so they did not accept the chromosome theory of heredity.

When Morgan (1911*a*,*b*) discovered partial linkage of the white-eye and miniature-wing mutants of *Drosophila melanogaster*, the position was quite different, because both characters were sex-linked and appeared to follow the X-chromosome in inheritance. Morgan therefore favoured the idea of exchanges between homologous chromosomes and introduced the term *crossing over* (Morgan and Cattell, 1912) for the process of interchange by which new combinations of linked factors arose. De Vries (1903) had postulated, on purely theoretical grounds, that exchanges involved individual pairs of alleles, but Morgan and associates soon found that the exchanges in *Drosophila* involved extensive chromosome segments. The tendency for some genes to stay together in inheritance enabled Sturtevant (1912) to use the frequency of crossover progeny as a measure of their separation and obtain a linear linkage map. According to De Vries (1903), writing at a time when linkage had not been discovered, exchanges of alleles were so frequent that genes at different loci close together on the same chromosome would show independent inheritance. With the discovery of differing frequencies of crossing over fitting a linear linkage map, Morgan and associates favoured the hypothesis of Janssens (1909). Janssens had observed the nodes, or chiasmata as he called them, between loops in the paired chromosomes in late prophase of the first division of meiosis, and had expressed the belief that these nodes represented points of exchange between corresponding paternal and maternal chromosomes. Furthermore, he postulated that, of the two chromatids making up each chromosome at this stage (Fig. 55(a)), an exchange involved only one chromatid from each parental chromosome (Fig. 55(b)). This would explain why

(a)   (b)   (c)

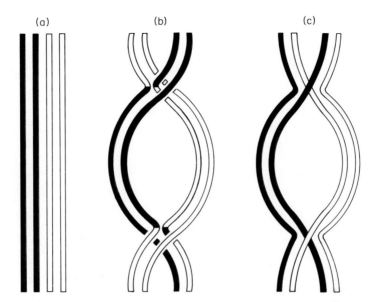

Fig. 55.   Diagram to illustrate the hypothesis of Janssens (1909) in which he attributed the chiasmata seen at diplotene of meiosis to crossing over at the four-chromatid stage. (a) A chromosome pair at pachytene. Each chromosome consists of two chromatids. (b) Diplotene, with crossovers between chromatids revealed as chiasmata. (c) The alternative explanation for chiasmata: a change of pairing partner without genetic exchange. The colours indicate parental origin.

at each chiasma two chromatids crossed one another and two did not. An alternative explanation of chiasmata, however, was possible, namely, that they merely represented the sites of a change of pairing partner (Fig. 55(c)) without any genetic exchange.

After protracted debate (review, Whitehouse, 1973a), Janssens's theory was accepted. A first step in establishing its truth was to demonstrate that exchange took place between chromatids, not whole chromosomes. This was first shown with abnormal strains of *D. melanogaster* with two X chromosomes and a Y chromosome, and later using a strain with the two X chromosomes joined together, and also with triploid flies having three X chromosomes. The first demonstration in an organism with a normal chromosome complement was obtained by Lindegren (1933) using *Neurospora crassa*. In this fungus the four nuclei resulting from meiosis undergo mitosis and then a spore is formed round each of the eight nuclei. The spores mature within a single cell, the ascus, forming a row in which their position reflects the preceding nuclear divisions (Fig. 56). Lindegren found 22.5% crossover progeny for two character differences, namely, mating type and pale versus normal orange conidial colour. When he dissected individual asci and isolated and germinated the spores, he found that whenever a

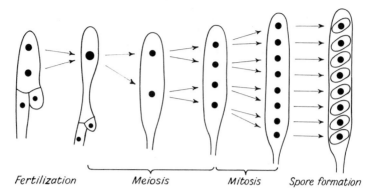

Fertilization   Meiosis   Mitosis   Spore formation

Fig. 56. Stages in the development of the ascospores of *Neurospora crassa*. (Reproduced from H. L. K. Whitehouse, *Towards an Understanding of the Mechanism of Heredity*, 3rd edn, Edward Arnold, London, 1973.)

crossover took place, two of the four pairs of spores had parental genotypes and the other two had crossover genotypes, as predicted with crossing over at the chromatid level. A visual demonstration that crossing over involves only two of the four chromatids is now possible using linked spore-colour mutants in *Sordaria brevicollis* (Whitehouse, 1973*a*): when a *buff* spore-colour mutant is crossed with a *yellow* mutant, about 25% of the resulting asci have two *buff* spores, two *yellow* spores, two black (wild-type) spores and two colourless (double mutant) spores (Plate 3).

A consequence of crossing over occurring at the chromatid level is that, as indicated above, when an exchange takes place only two of the four products of meiosis have a recombinant genotype. Since Sturtevant (1912) used frequency of crossover progeny as his measure of distance on the linkage map, it follows that the frequency of exchanges is double the map interval. In other words, in the *Neurospora* example quoted above, an exchange evidently takes place between the loci for mating type and conidial colour in 45% of meioses and, conversely, in the *Sordaria* example, the map distance between the spore-colour loci is 12.5 units.

Another early discovery about the process of exchange was that successive crossovers along a chromosome do not necessarily involve the same two chromatids. Again, this was first shown with abnormal strains of *Drosophila*, but the information about chromatids was necessarily incomplete. The first complete set of data on the chromatid relationship of adjacent crossovers was obtained by Lindegren by genetic analysis of spores from individual asci from crosses heterozygous for linked mutants. Subsequently, extensive results were obtained in this way by several authors, particularly with *Neurospora* and the liverwort *Sphaerocarpos donnellii* (see Table 10). It appears that the chromatids involved in one crossover are usually independent, or nearly so, of those taking part in a neighbouring exchange.

The fact that either chromatid of a chromosome could take part in exchange had an important influence on ideas about the mechanism of exchange. So long as

(a)

(b)

(c)

Plate 4. (a) Part of a pale magenta flower of *Antirrhinum majus* heterozygous at the *incolorata* locus showing a twin spot with ivory (*inc/inc*) and dark magenta (*Inc/Inc*) components. (b) Three such twin spots following caffeine treatment of an inflorescence. (c) A twin spot on an apple (*Malus*) variety Signe Tillisch. (Courtesy of B. J. Harrison and (a, b) Genetical Society of Great Britain from *Heredity*, **38**, 176 (1977), and (b, c) John Innes Institute from *John Innes Annual Report* No. 67, p. 55, 1976.)

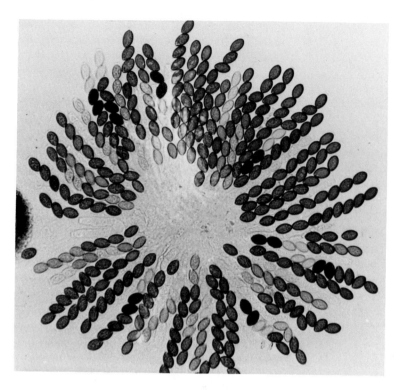

Plate 3. Cluster of asci of *Sordaria brevicollis* from a cross between *buff* and *yellow* spore-colour mutants. Whenever a crossover occurs between these linked loci, two black (wild-type) and two colourless (double mutant) spores are produced in the ascus, as well as two *buff* and two *yellow* ones.

Table 10. Chromatids involved in adjacent crossovers. The table shows the numbers of tetrads observed with two, three or four chromatids participating in a pair of adjacent crossovers. Figures significantly in excess of the 1:2:1 ratio expected with no chromatid interference are underlined (with a broken line if bordering on significance).

Organism and author	Total number of chromatids involved in the two crossovers					
	Within chromosome arms			Across the centromere		
	2	3	4	2	3	4
Neurospora crassa						
Lindegren and Lindegren (1942)	1	2	8	17	11	12
Strickland (1961)	42	64	29	–	–	–
Perkins (1962)	_191_	371	153	–	–	–
Bole-Gowda, Perkins and Strickland (1962)	-73-	132	51	88	168	77
(totals)	_307_	569	241	105	179	89
Aspergillus nidulans						
Strickland (1958b)	25	40	17	2	11	2
Chlamydomonas reinhardii						
Ebersold and Levine (1959)	43	76	46	–	–	–
Sphaerocarpos donnellii						
E. Knapp, E. Möller and W. O. Abel in Abel (1967) 10–20 °C	860	1785	995	_60_	43	15
22–28 °C	905	1722	824			

there was ignorance about how chromosomes were duplicated before nuclear division, it was possible to imagine that one chromatid was the daughter and the other the parent. If all crossovers had involved the same two chromatids, one from each parent, these could be supposed to be the daughter chromatids. The mechanism of crossing over need not then involve physical breakage and joining but merely the copying, first of a segment of the parental chromatid of one parent and then switching template and copying a segment from the chromatid of the other parent. This copy-choice hypothesis was dismissed when it was discovered that either chromatid may take part in crossing over. About 30 years later, Taylor, Woods and Hughes (1957) showed by tritium labelling of the nuclei of root tips of the Broad Bean, *Vicia faba*, followed by autoradiography two nuclear divisions later, that each chromatid contains two longitudinally arranged subunits, one old and one new (Fig. 57). This implied that the two chromatids could not be regarded as parent and daughter, but were equivalent. Replication was evidently occurring at the level of the half-chromatids, whereas crossing over was known, from evidence such as that given above, to occur at the level of the whole chromatid. Attempts

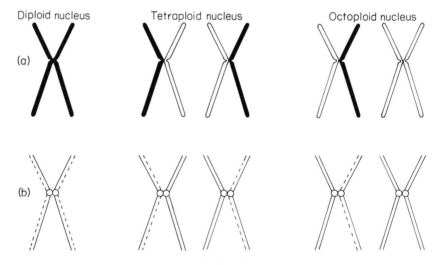

Fig. 57. Diagram to illustrate the experimental results obtained by Taylor, Woods and Hughes (1957) by labelling the DNA of *Vicia faba* root tips with tritium. After 8 h in [³H]thymidine solution, the roots were washed and placed in colchicine solution for either 10 h (nuclei diploid) or 34 h (nuclei tetraploid or octoploid). They were then fixed, stained, squashed and autoradiographed. The colchicine inhibits anaphase and allows chromatids to separate except at the centromere. (a) Appearance of metaphase chromosomes. The presence of tritium, revealed as silver grains in the autoradiograph, is shown black. (b) Interpretation of (a), with each chromatid containing two longitudinally arranged subunits. The presence of tritium in a subunit is indicated by a broken line. Newly synthesized subunits are shown on the outside of the pre-existing ones. The DNA replication preceding the diploid metaphase was in the presence of the [³H]thymidine, but subsequent replications to give the tetraploid and octoploid nuclei were in its absence.

were made about this time to resurrect the copy-choice idea, for reasons referred to in Section 7, but any such hypothesis would necessarily have to include breakage and joining also, in order to accommodate the differences of level (chromatid and sub-chromatid) of the exchange and replication processes.

Direct evidence that crossing over involves breakage and joining and not copy-choice was obtained by Taylor (1965) when he applied his tritium-labelling technique to meiosis. The technical difficulties here are considerable because the exposure to [^{3}H]thymidine has to be made at the penultimate replication before meiosis. Taylor used males of the grasshopper *Romalea microptera*. By this technique each meiotic chromosome had one chromatid labelled with tritium and the other not. Since crossovers occur between chromatids chosen at random, half of them will be between a labelled chromatid from each of the paired chromosomes, or an unlabelled chromatid from each. These crossovers will not be detected on the autoradiographs. The other half of the crossovers will be between a labelled chromatid in one chromosome and an unlabelled chromatid in its homologue. These will be visible as a label exchange in autoradiographs of anaphase I, metaphase II or anaphase II. Comparison of label exchange frequency with chiasma frequency showed that the former was indeed half the latter, as expected if all crossovers involved breakage and joining, with the chromatids chosen at random.

A complication in determining label exchange frequency is the possible occurrence of sister-chromatid exchange (see Section 18). This difficulty was avoided in experiments similar to Taylor's made by Peacock (1970) and Jones (1971) using other species of grasshopper, where the counts could be restricted to cells with only one chiasma per chromosome arm. Their results again indicated a label exchange frequency of half the chiasma frequency (20 out of 46, and 32 out of 60, respectively). Furthermore, in the grasshopper (*Stethophyma grossum*) used by Jones, the chiasmata are localized near the centromere. He found that the label exchanges showed similar proximal localization, in direct agreement with Janssens's hypothesis. Tease (1978) obtained similar results to Jones using the locust *Locusta migratoria* and applying a new technique for differential recognition of the two chromatids. This method, due to Latt (1973) and Perry and Wolff (1974), takes advantage of the fact that incorporation of 5-bromodeoxyuridine (BrdUrd) into DNA in place of thymidine alters its response to photosensitive dyes. After two rounds of replication in the presence of BrdUrd, each chromosome will have bromine atoms in one strand (polynucleotide) of one chromatid and in both strands of the other chromatid. After treatment with the fluorescent dye Hoechst fluorochrome, followed by exposure to light and then Giemsa staining, the chromatid with a bromine-free strand stains deeply whereas the other does not. Tease (1978) found that half the chiasmata (21 out of 46) were associated with a label exchange, the two coinciding in position. Polani *et al.* (1979), using the same method, obtained correspondence between chiasmata and label exchange in the mouse, *Mus musculus*. They used an ingenious technique, exposing the differentiating germ cells of embryonic ovaries to BrdUrd *in vitro*, before transplanting into young adults. The mature oocytes were subsequently removed and stained.

Another early discovery about crossing over was made by Muller (1916) from analysis of the progeny of *Drosophila* crosses involving three or more linked character differences. He found that the occurrence of crossing over reduced the likelihood of a second crossover in proximity to it. This *interference*, as he called it, was strong in the immediate neighbourhood of a crossover but decreased with distance. The degree of interference is measured by the *coincidence*, which is given by

$$\frac{n \times n_{pq}}{n_p \times n_q},$$

where n_p is the number of individuals showing crossing over in region p, irrespective of its occurrence in region q or elsewhere; n_q is the corresponding number for region q; n_{pq} is the number of individuals with crossing over in both p and q; and n is the total number of progeny. Regions p and q need not be adjacent on the linkage map. Strong interference is indicated by coincidence values close to zero and lack of interference by a value of 1 (cf. Chapter V, Section 4). When coincidence values were calculated for regions in opposite arms of a chromosome, no interference was found. In the fungus *Aspergillus nidulans* interference is lacking even within chromosome arms (Strickland, 1958b).

2. MITOTIC CROSSING OVER

Crossing over was thought to be restricted to meiosis until Stern (1936), from an extensive genetic analysis of single and twin spots in *Drosophila melanogaster*, established that it also occurred in somatic cells. Flies were made heterozygous for recessive genes whose effects were visible in very small areas of the insect's body, even single hairs. The genes mainly used were *yellow* (*y*) body and hair colour at the left-hand end (distal with respect to the centromere) of the X-chromosome linkage map, and *singed* (*sn*), 21 units away in the same chromosome, producing thickened and curved hairs. In heterozygous

$$\frac{y\ +}{+\ sn}$$

females he discovered twin spots formed by a yellow not-singed area adjacent to a singed not-yellow area. He inferred that a crossover had taken place in the interval between *sn* and the centromere leading to segregation whereby one daughter cell obtained the *yellow* gene (Fig. 58(a)) carried originally by one of the X chromosomes while the other cell obtained the *singed* gene carried originally by the other X (Fig. 58(b)). The further division and normal somatic differentiation of the two daughter cells gave rise to the mosaic twin areas.

Stern (1936) also observed single spots on the doubly heterozygous insects'

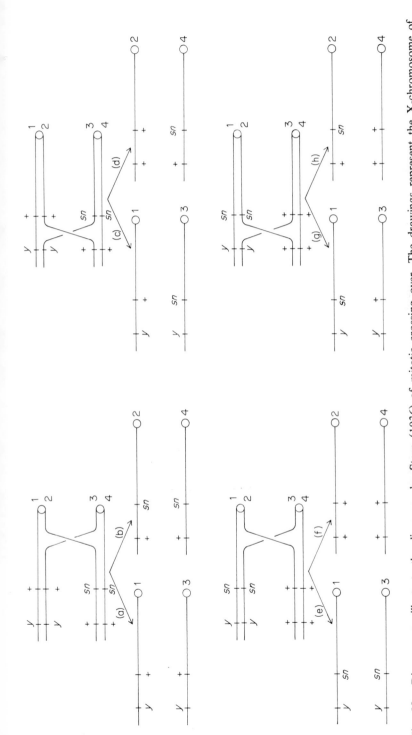

Fig. 58. Diagrams to illustrate the discovery by Stern (1936) of mitotic crossing over. The drawings represent the X-chromosome of *Drosophila melanogaster*, the circles indicating the centromere. *y*, a recessive mutation giving *yellow* body and hair colour; *sn*, a recessive mutation giving *singed* (that is, thickened and curved) hairs. The derivation of spots through mitotic crossing over at the four-chromatid stage is shown for the double heterozygote in repulsion ((a)–(d)) and in coupling ((e)–(h)). Crossing over may occur proximal to both loci ((a), (b), (e), (f)) or between them ((c), (d), (g), (h)). The chromatids are numbered, with the crossovers between nos 2 and 3 throughout. The diagrams all show chromatids 1 and 3 passing to one daughter cell and 2 and 4 to the other. The other possibility (1 and 4 separating from 2 and 3 at anaphase) is not shown because it leaves both daughter cells heterozygous at both loci and hence will not generate spots.

bodies. He attributed some of these to twin spots in which one of the two spots was invisible, either because it was so small that it failed to include a hair or because one of the segregating daughter cells failed to survive or failed to multiply. He suggested, however, that some of the yellow single spots might have arisen as a result of a crossover between the *yellow* and *singed* loci, the crossover having occurred at the four-chromatid stage (Fig. 58(c)). In 50% of cases (see legend to Fig. 58) mitotic crossing over is expected to lead to homozygosity for markers distal to the crossover.

Stern (1936) went on to investigate the double heterozygote with the genes in coupling. He failed to find twin spots but observed double-mutant single spots (Fig. 58(e)) and also yellow single spots (Fig. 58(g)). These are the counterparts of the twin spots and yellow single spots observed with the repulsion heterozygote (Fig. 58(a)–(c)), and strengthened the evidence that crossing over was occurring at the four-chromatid stage. Indeed, the yellow single spots in the coupling heterozygote could not be explained away as twin spots in which the singed spot had failed to develop, because no twin spots were found with this genotype. Stern also observed small numbers of singed single spots with the coupling heterozygote. He attributed these to double crossovers, one occurring between *y* and *sn* and the other between *sn* and the centromere.

Stern carried out many further experiments and these gave additional support for the hypothesis that mitotic crossing over was taking place and occurring, as at meiosis, at the four-chromatid stage. An interesting additional discovery was that the distribution of crossovers along the X chromosome at mitosis was not the same as at meiosis. At mitosis a larger proportion than at meiosis occurred to the right of *sn*, that is, nearer the centromere.

Mitotic recombination was discovered in diploid strains of *Aspergillus nidulans* by Pontecorvo and Roper (1953). To obtain the heterozygous diploid needed to detect the recombination, a balanced heterokaryon, that is, containing nuclei of two kinds, each with a different growth requirement, was grown on minimal medium. The conidia in *A. nidulans* arise in chains from uninucleate sterigmata. In consequence, all the conidia in a chain will be genetically alike and not heterokaryotic. Diploidy was induced by treatment with camphor vapour, but subsequently found also to occur spontaneously. Diploid conidia were selected by their ability to grow on minimal medium, where the normal haploid ones could not. The use of colour markers made the isolation of diploids even easier. The wild-type conidia are green. If the balanced heterokaryon was between a white and a yellow strain, the diploid was green. Mitotic recombination was revealed by the presence of white or yellow regions in such a green diploid, and was confirmed by scoring for auxotrophic mutants also present in heterozygous form in the diploid. Some of the mutant characters were linked. The first evidence of mitotic recombination between linked loci was obtained from a green diploid of genotype

$$\frac{pabaA \quad yA}{+ \qquad +},$$

where *pabaA* is a mutation giving a requirement for *p*-aminobenzoic acid and *yA*

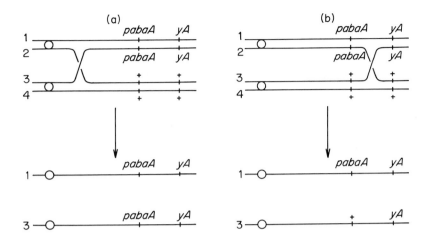

Fig. 59. Diagrams to illustrate the discovery of mitotic recombination of linked mutants in a heterozygous diploid strain of *Aspergillus nidulans* by Pontecorvo and Roper (1953): *pabaA*, a mutation giving a requirement for *p*-aminobenzoic acid; *yA*, a mutation giving yellow conidia. In (a) a crossover proximal to both loci has given rise to a diploid homozygous for both mutants, as a result of the passage of chromatids 1 and 3 to the same pole at anaphase. In (b) a crossover between the loci has given homozygosity only for the more distal mutant.

gives yellow conidia. Out of 51 yellow regions studied, 39 were *paba⁻* and 12 were *paba⁺*, suggesting that the centromere was to the left of the *pabaA* locus at a distance greater than the distance apart of the *pabaA* and *yA* loci (Fig. 59).

Evidence that mitotic crossing over in *Aspergillus* occurs between chromatids rather than between unreplicated chromosomes, just as in meiosis, was obtained by Pontecorvo, Gloor and Forbes (1954). They used diploids of genotype

$$\frac{pabaA \quad yA \quad +}{+ \quad + \quad biA} \quad \text{and} \quad \frac{pabaA \quad yA \quad biA}{+ \quad + \quad +} ,$$

where *biA* is a biotin-requiring mutation. They selected recombinants either by their yellow conidial colour or by their biotin requirement. All the detectable genotypes resulting from crossing over between the centromere and the *pabaA* locus, or between the *pabaA* and *yA* loci, were found, on the basis of crossing over at the four-chromatid stage. These genotypes and their postulated mode of origin are shown in Fig. 60. The genotype for characters showing the wild-type phenotype was determined by obtaining haploids from the diploid recombinants and observing their conidial colour and growth requirements. The haploidization occurs regularly with low frequency. Haploids can be selected by conidial size and by the expression of recessive genes in other linkage groups for which the diploid was heterozygous. The recombinant diploids at Fig. 60(c), (d) and (f), in particular, require crossing over to have occurred at the four-chromatid stage.

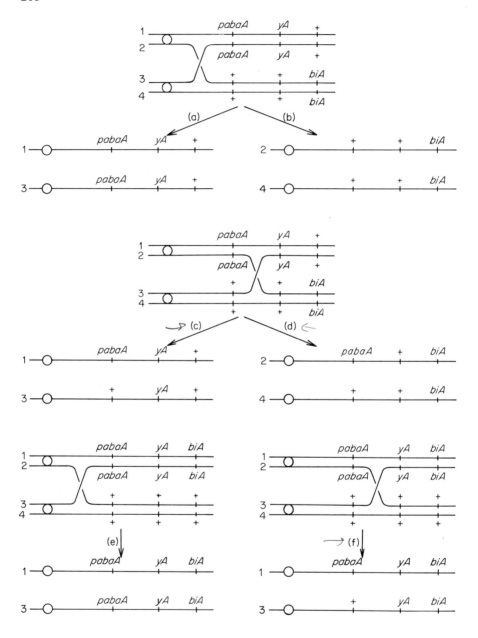

Fig. 60. Results obtained by Pontecorvo, Gloor and Forbes (1954) indicating that mitotic crossing over in *Aspergillus nidulans* occurs at the four-chromatid stage: *biA*, a mutation giving a requirement for biotin; other symbols as in Fig. 59. The recombinant diploids were obtained by selecting for yellow conidia ((a), (c)), biotin requirement ((b), (d)), or either of these characters ((e), (f)), in two heterozygous diploids, one shown in ((a)–(d)) and the other in (e) and (f).

Furthermore, genotypes

$$\frac{+\,yA\,+}{+\,yA\,+}\,, \quad \frac{pabaA\,+\,biA}{pabaA\,+\,biA} \quad \text{and} \quad \frac{+\,yA\,\,biA}{+\,yA\,\,biA}\,,$$

which would have arisen in place of (c), (d) and (f), respectively, if crossing over had taken place before chromosome duplication, were not found. The presumption was that all crossing over was occurring at the four-chromatid stage.

Mitotic crossing over was confirmed by the recovery by Roper and Pritchard (1955) of the two complementary products in single nuclei, analogous to the twin spots in *Drosophila* observed by Stern. Their heterozygous diploid had the genotype

$$\frac{pabaA\,\,yA\,+\,8\,+}{+\quad+\,\,16\,+.biA}\,,$$

where 8 and 16 are allelic adenine-requiring (*adE*) mutants. Conidia were plated on medium lacking adenine in order to select adenine-independent diploids. The majority were found to have the genotypes shown in Fig. 61(a),(b), that in Fig. 61(a) having the reciprocal products of mitotic crossing over included in the same nucleus. The genotype of the recombinant diploids was determined by obtaining haploids from them and testing their growth on minimal medium: mutant 16 showed reduced growth, while mutant 8 and the double mutant (16 8) showed none. The double mutant was recognized by the recovery of mutant 16 among its progeny.

The parasexual cycle, that is, recombination of hereditary determinants outside the sexual cycle, has subsequently been found to be of widespread occurrence in fungi, including species in which the sexual cycle is unknown. Gingold and Ashworth (1974) and Katz and Kao (1974) discovered mitotic crossing over in the cellular slime mould *Dictyostelium discoideum*. Diploids heterozygous for three markers in linkage group II, (a) white spore colour, (b) temperature-sensitive growth, and (c) methanol resistance, gave segregants homozygous for (a), (b) and (c), or for (b) and (c), or for (c) alone. The authors concluded that the markers mapped in the following sequence: centromere, (a), (b), (c). The occurrence of recombinants heterozygous at some of the linked loci and homozygous at others indicated that, just as in *Drosophila* and *Aspergillus*, mitotic crossing over was occurring at the four-chromatid stage.

Later studies of mitotic crossing over in *Aspergillus*, like Stern's with *Drosophila*, revealed striking differences between meiosis and mitosis in the relative frequencies of crossing over in different regions of the chromosomes (Pontecorvo and Käfer, 1958). These differences were most marked near the centromeres, but the direction of the difference was variable. In other words, some centromere regions were expanded and others contracted on mitotic linkage maps relative to meiotic maps.

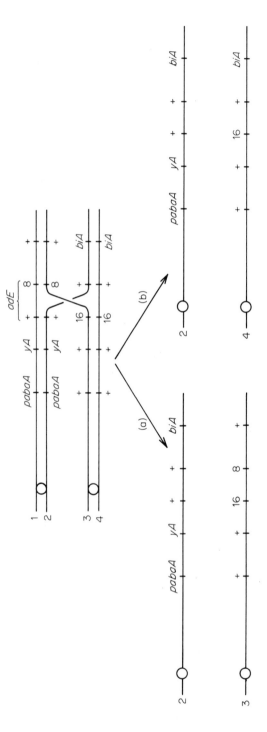

Fig. 61. Adenine-independent diploids in *Aspergillus nidulans* obtained by Roper and Pritchard (1955) as a result of mitotic crossing over. The numbers 8 and 16 denote alleles at the *adenineE* (*adE*) locus. Other symbols as in Figs 59 and 60. The linkage map is not to scale. The intervals, as measured by meiotic crossover frequency, were as follows: centromere to *pabaA*, 27; *pabaA* to *yA*, 16; *yA* to *adE*, 0.1; *adE* to *biA*, 6.

James (1955) discovered that ultraviolet irradiation induced sectoring in diploid *Saccharomyces cerevisiae* heterozygous for a recessive mutation (*gal1*) causing an inability to utilize galactose. Analysis showed that the sectors were homozygous for *gal1* and that sectors homozygous for *gal*$^+$ were also arising, both with high frequency. James (1955) suggested ultraviolet-induced somatic crossing over as an explanation, and this was confirmed by James and Lee-Whiting (1955), who extended the investigation to several other loci. Ultraviolet irradiation of cells heterozygous for two closely linked mutations

$$\frac{gal3 \quad +}{+ \quad trp1}$$

gave seven sectors homozygous for *gal3* +, as expected from mitotic exchange proximal to both loci, and one sector homozygous for *gal3* but not for *trp1*, as expected from an exchange between the loci, if *gal3* was distal to *trp1*. Furthermore, the frequency of homozygosis was greater for mutants distant from their centromere than for those near the centromere, as expected if homozygosis resulted from mitotic crossing over proximal to the mutant site. The centromere distances were obtained by tetrad analysis of crosses heterozygous for the mutants: the frequency of asci with all four products of meiosis of a different genotype (tetratype tetrad) from a cross between two unlinked mutants depends on their centromere distances.

Holliday (1961) exposed diploid sporidia of *Ustilago maydis*, heterozygous for four or five recessive auxotrophic mutants, to ultraviolet irradiation. Colonies formed on complete medium form the irradiated sporidia were replicated to minimal medium. All the markers were recovered among the segregants. Some mosaic colonies were found containing the reciprocal products of mitotic exchange. For example, a heterozygote for three linked but widely spaced auxotrophic mutants, of genotype

$$\frac{+ \quad ad\text{-}1 \quad me\text{-}1}{leu\text{-}1 \quad \text{I} \quad + \quad \text{II} \quad + \quad \text{III}} \;,$$

gave rise to several *leu/ad* mosaic colonies and also to *leu/ad me* mosaics. These are the results expected from mitotic crossing over in regions II and III, respectively, the two daughter cells and the half-colonies derived from them being homozygous for markers distal to the point of exchange. Holliday (1964a) obtained similar results with mitomycin C in place of ultraviolet irradiation. He found that this antibiotic also induced mitotic crossing over in *Saccharomyces cerevisiae*. A diploid was used that was heterozygous for a recessive mutation conferring resistance to actidione. Mitotic crossing over was demonstrated by the occurrence of colonies resistant to actidione, homozygous as a result of an exchange proximal to the site of the actidione-resistance mutation.

Morpurgo (1963) discovered that the bifunctional alkylating agents diepoxybutane and methyl bis (β-chloroethyl) amine induced mitotic crossing over in *Aspergillus* with high frequency. He suggested that the ability of these compounds to form intra- or intermolecular cross-links might lead to a bridge linking non-sister chromatids, and so favouring crossing over.

Indications of mitotic crossing over in a flowering plant were obtained by Ross and Holm (1960). In the tomato, *Lycopersicon esculentum*, a mutant called *aurea* gives rise to pale green plants when heterozygous and to yellow plants without chlorophyll when homozygous. Ross and Holm observed twin spots (one normal green and the other yellow) on the leaves of the heterozygote, as expected if mitotic crossing over was taking place in the chromosome region between the locus of *aurea* and the centromere. Single spots were also seen and could have been potential twins in which one of the segregants failed to multiply. Twin spots were subsequently seen by several authors in various flowering plants, in each case in heterozygotes with incomplete dominance as in the *aurea* tomato example.

Direct evidence for mitotic crossing over in a flowering plant was obtained by Carlson (1974). He used *Nicotiana tabacum* heterozygous for two linked mutants. The plants were of genotype

$$\frac{Su\ +}{+\ cl}\ ,$$

where *Su* (*Sulphur*) gives yellow-green leaves when heterozygous and yellow when homozygous, and *cl* (*chimeral*) is a recessive mutation causing regions of chlorosis in leaf tissue when homozygous, and mapping 38 units from *Su*. Twin spots (one yellow, the other green) were isolated by establishing *in vitro* cell cultures from each spot and regenerating plants from the tissue. Plants regenerated from the green tissue of 11 twin spots were backcrossed to the + *cl* homozygote. All gave a 1 : 1 ratio for *chimeral* and wild type. The yellow homozygote does not survive as an independent plant, but the yellow tissue of each twin spot could be induced to regenerate a shoot: none had the chimeral phenotype. It was evident that the green part of each twin spot had the genotype

$$\frac{+\ +}{+\ cl}$$

and the yellow part

$$\frac{Su\ +}{Su\ cl}\ .$$

Mitotic crossing over had evidently occurred between the loci at the four-chromatid stage, and the centromere was evidently near the *cl* locus, as homozygosity for *cl* was not found.

Harrison and Carpenter (1977) observed twin spots on the corollas of *Antirrhinum majus* heterozygotes. For example, a heterozygote (*inc/Inc*) at the *incolorata* locus had magenta flowers with a twin spot comprising white (*inc/inc*) and dark magenta (*Inc/Inc*) components (Plate 4), and a similar example was found at the *pallida* locus. In the latter example, the recessive allele was *pallida-recurrens* which shows flakes of pigment when homozygous as a result of muta-tion to the dominant allele (see Chapter IX). In these and other examples of twin

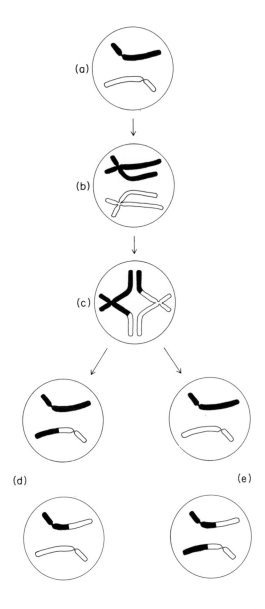

Fig. 62. Diagram to show the origin, according to German (1964), of a quadriradial configuration at metaphase as a consequence of mitotic crossing over. Two homologous chromosomes are shown (a) before and (b) after replication. In (c) the quadriradial configuration at metaphase is shown, following an exchange between chromatids. In (d) and (e) are shown the two alternatives for chromatid segregation, (d), unlike (e), being homozygous for regions distal to the crossover. (Reproduced by permission from J. German, *Science, N.Y.* **144**, 300 (1964). Copyright 1964 by the American Association for the Advancement of Science.)

spots, Harrison and Carpenter were able to test genotypes by administering specific anthocyanidin precursors to the acyanic component of twin spots, and observing whether or not pigment was synthesized. This allowed the position of the block in the biosynthetic pathway to be identified. Without exception the genotypes were those expected from mitotic crossing over in the interval between gene and centromere.

Mitotic crossing over in human blood cells in culture was inferred by German (1964) to account for quadriradial configurations which he observed at metaphase (Fig. 62). The configurations were symmetrical and composed of homologous chromosomes, as expected if an exchange had occurred between homologous segments. The traditional interpretation of a quadriradial chromosome association at mitosis is a chromatid interchange. Superimposing the symmetry, the implication is mitotic crossing over between a chromatid from each chromosome (Fig. 62(c)). It is evident that mitotic crossing over is of widespread and perhaps general occurrence and that it resembles meiotic crossing over in occurring after chromosome duplication. Further aspects of mitotic crossing over are discussed in Sections 5 and 18.

A system for investigating the mechanism of mitotic crossing over in cultured mammalian cells has been developed by Wake and Wilson (1979). It involves the use of simian virus 40 (SV40) cultured in kidney cells of *Cercopithecus aethiops* (African green monkey). The viral genome is a circular duplex DNA molecule. DNA from differing SV40 mutants was cleaved *in vitro* into linear molecules with restriction endonucleases and then linked by T4 DNA ligase to form oligomers. Infection of the cultured *Cercopithecus* cells with these molecules gave rise to recombinants. Pairwise crosses with three temperature-sensitive mutants gave non-parental progeny with a frequency roughly proportional to the physical distance between the mutant sites.

3. NEGATIVE INTERFERENCE

Giles (1952) crossed two of Beadle and Tatum's ultraviolet-induced allelic inositol-requiring (*inl*) mutants of *Neurospora crassa* in the presence of flanking markers. The markers were an isoleucine-requiring mutant (*ilv-1*) mapping 17 units to the left and a *p*-aminobenzoic acid-requiring mutant (*pab-1*) mapping 3 units to the right of *inl*. Over 150 000 progeny ascospores were tested and 27 inositol-independent cultures were obtained. When these were scored for flanking marker genotype a surprising result was obtained: all four possible genotypes were present! Details are given in Table 11. Subsequently, Giles (1956) confirmed this result with more extensive data for the same pair of mutants. A similar result was obtained by Mitchell (1955a) with pyridoxin-requiring (*pdx-1*) mutants (Table 11) and, again, more extensive counts were published later (Mitchell, 1956). The data of St Lawrence (1956) for nicotinamide-requiring (*nic-1*) mutants, some of which are given in Table 11, were even more remarkable. She found that all four classes of flanking marker genotypes in the nicotinamide-independent recombinants were occurring with approximately equal frequency!

A detailed investigation was made by Pritchard (1955) into recombination of adenine-requiring mutants in *Aspergillus nidulans* that mapped at the *adE* locus. The flanking markers were a mutant (*yA*) giving yellow conidia, mapping 0.2 units to the left, and a biotin-requiring (*biA*) mutant mapping 6 units to the right. Different *adE* mutants were crossed and large numbers of progeny ascospores plated on minimal medium. In this way adenine-independent colonies were selected. These were scored for conidial colour and tested for biotin requirement. The initial plating was in fact made on minimal medium supplemented with biotin, so there was no selection for particular flanking marker genotypes. Altogether, seven pairs of *adE* mutants were tested. All gave results similar to those obtained with *Neurospora*, though with a higher proportion of recombinant flanking markers. The data for two of the crosses are given in Table 11.

If an allelic cross is denoted as

$$\frac{M\ 1 + N}{m + 2\ n},$$

where 1 and 2 are the alleles and M/m and N/n are the flanking markers, it is customary to call the four flanking marker genotypes for the wild-type recombinants P1, P2, R1 and R2, as follows:

$$
\begin{aligned}
M + + N &\quad \text{P1,} \\
m + + n &\quad \text{P2,} \\
m + + N &\quad \text{R1,} \\
M + + n &\quad \text{R2.}
\end{aligned}
$$

In other words, P1 and P2 have the flanking marker genotypes of the proximal and distal alleles, respectively; R1 has the genotype that would result from a simple exchange between the sites of the alleles; and R2 has the other crossover flanking marker genotype, requiring a triple exchange.

Pritchard (1955) pointed out that the occurrence of all four flanking marker genotypes in allelic recombinants implied negative interference. Selection had occurred for an exchange between the sites of the alleles. If no other exchanges occurred, all the wild types would be of R1 genotype. The occurrence of appreciable numbers of P1 and P2 genotypes, and even R2, indicated that the occurrence of recombination in the selected interval actually favoured recombination in neighbouring intervals.

The question of whether this negative interference was confined to alleles, or whether it extended to closely linked mutants in different genes, was investigated by Calef (1957) with the closely linked *adF* and *pabaA* genes of *Aspergillus nidulans*. The flanking markers were a proline-requiring (*proA*) mutant 10 units to the left and the yellow conidial mutant (*yA*) 16 units to the right. The technique was to plate the progeny ascospores on minimal medium supplemented only with proline, so that there was selection both for adenine independence and *p*-aminobenzoic acid independence, but not for flanking marker characters. The results are given in Table 11. They are similar to the results with allelic crosses

Table 11. Flanking marker genotypes of wild-type recombinants from crosses of alleles or other closely linked mutants.

Species	Gene, mutants and reference	Class of wild-type recombinant	Parental genotype, genotypes of wild-type recombinants, and numbers observed			Parental genotype, genotypes of wild-type recombinants, and numbers observed	
			$ilv\text{-}1$ 3 + $pab\text{-}1$ / + 6 +	(a)	(b)	$ilv\text{-}1$ 3 + + / + + 6 $pab\text{-}1$	(b)
Neurospora crassa	*inositol* (*inl*)						
	37401 (= 3)	P1	$ilv\text{-}1$ + + $pab\text{-}1$	3	6	$ilv\text{-}1$ + + +	2
	64001 (= 6)	P2	+ + + +	5	6	+ + + $pab\text{-}1$	6
	(a) Giles (1952)	R1	+ + + $pab\text{-}1$	14	21	+ + + $pab\text{-}1$	25
	(b) Giles (1956)	R2	$ilv\text{-}1$ + + +	5	5	$ilv\text{-}1$ + + $pab\text{-}1$	4
				27	38		37
			+ 6 + $col\text{-}4$ / $pyr\text{-}1$ + 3 +	(a)	(b)	$pyr\text{-}1$ 6 + + / + + 3 $col\text{-}4$	(b)
	pyridoxin-1 (*pdx-1*)						
	37803 (= 3)	P1	+ + + $col\text{-}4$	5	43	$pyr\text{-}1$ + + +	32
	39106 (= 6)	P2	$pyr\text{-}1$ + + +	13	28	+ + + $col\text{-}4$	29
	(a) Mitchell (1955a)	R1	$pyr\text{-}1$ + + $col\text{-}4$	7	31	+ + + $col\text{-}4$	34
	(b) Mitchell (1956)	R2	+ + + +	7	9	$pyr\text{-}1$ + + $col\text{-}4$	8
				32	111		103
			+ 1 + $os\text{-}1$ / $lys\text{-}3$ + 2 +	(a)		$lys\text{-}3$ 1 + + / + + 2 $os\text{-}1$	(b)
	nicotinic acid-1 (*nic-1*), 1, 2						
	St Lawrence (1956)	P1	+ + + $os\text{-}1$	25		$lys\text{-}3$ + + +	13
		P2	$lys\text{-}3$ + + +	14		+ + + $os\text{-}1$	8
		R1	$lys\text{-}3$ + + $os\text{-}1$	18		+ + + $os\text{-}1$	17
		R2	+ + + +	15		$lys\text{-}3$ + + $os\text{-}1$	7
				72			45

methionine-7 (met-7), 73
methionine-9 (met-9), 43t
Murray (1970)

	thi-3	73	+	+			+	73	+	wc	
	+		43t	wc			thi-3		43t	+	
P1	thi-3	+	+	+	107		+	+	+	wc	162
P2	+	+	+	wc	56		thi-3	+	+	+	92
R1	+	+	+	+	35		thi-3	+	+	wc	93
R2	thi-3	+	+	wc	6		+	+	+	+	18
					204						365

Aspergillus nidulans

adenine E (adE), 8, 10, 11
Pritchard (1955)

	+	11	+	biA			yA	8	+	+	
	yA		8	+			+		10	biA	
P1	+	+	+	biA	10		yA	+	+	+	3
P2	yA	+	+	+	43		+	+	+	biA	3
R1	yA	+	+	biA	136		+	+	+	+	14
R2	+	+	+	+	0		yA	+	+	biA	8
					189						28

adenine F (adF), 15, 17
p-aminobenzoic acid A (pabaA), 1
Calef (1957)

	+	15	+	+			+	17	+	yA	
	proA		1	yA			proA		1	+	
P1	proA	+	+	+	22		proA	+	+	yA	23
P2	+	+	+	yA	22		+	+	+	+	49
R1	proA	+	+	+	56		proA	+	+	yA	155
R2	+	+	+	yA	5		+	+	+	+	6
					105						233

and indicated that the negative interference was not an intragenic phenomenon as such, but could extend to a neighbouring gene.

Similar evidence for negative interference extending from one gene to another was obtained by Murray (1970) with the *met-7* and *met-9* loci in *Neurospora crassa*. These appear to be separate genes since they code for different enzymes in the methionine pathway and no mutants are known blocked in both activities. Such mutants would be expected if the *met-7 met-9* region coded for a single bifunctional polypeptide. In crosses between *met-7* and *met-9* mutants in the presence of flanking markers, all four genotypes were found among the methionine-independent recombinants (Table 11).

Further results obtained by selecting wild-type recombinants from crosses of closely linked mutants are discussed in Section 12.

4. ABERRANT SEGREGATION

Evidence that a single-gene character difference might not always segregate 2 : 2 among the four products of meiosis was obtained by Zickler (1934). He worked with the ascomycete fungus *Bombardia lunata* in which he obtained mutants affecting spore colour. This fungus resembles *Neurospora crassa* in having a linear sequence of eight spores in the ascus. Zickler found that in asci heterozygous for a pale mutant, the majority of the asci had four dark (wild-type) spores and four pale ones, as expected with Mendelian segregation, but occasionally asci with six dark and two pale spores, or two dark and six pale ones, were found. For this phenomenon Zickler used the term *conversion*, which had been introduced by Winkler (1930) on purely theoretical grounds.

The occurrence of 3 : 1 ratios within tetrads for single Mendelian character differences was rediscovered by Lindegren (1953) working with yeast, *Saccharomyces cerevisiae*. He also used Winkler's term for this phenomenon.

As Emerson (1969) has pointed out, Zickler's evidence for conversion was ignored and Lindegren's was disbelieved, because this idea contradicted the basic Mendelian concept that alleles did not interact with one another. So long as other possible explanations of 3 : 1 ratios in a tetrad had not been ruled out, for example, abnormal chromosome duplication, or polyploidy, these were preferred.

Definitive evidence for conversion was obtained by Mitchell (1955a,b), working with pyridoxin-requiring (*pdx-1*) mutants of *Neurospora crassa*. The difference between her work and that of Lindegren or Zickler was that she had the use of linked markers, whereas linked mutants were not available in yeast or *Bombardia*. The markers used by Mitchell were a pyrimidine-requiring mutant (*pyr-1*) mapping 1.6 units to the left of *pdx-1* and a colonial growth mutant (*col-4*) mapping 4.4 units to the right of *pdx-1*. Out of 246 asci analysed from a cross between the *pdx-1* mutant and wild type, two showed six wild-type and two mutant spores (Mitchell, 1955b). But the linked mutants behaved normallly in these asci, with both + :*pyr-1* and + :*col-4* segregating as 4 : 4. This showed that the *aberrant segregation*, as it was called, was confined to a limited region in the vicinity of the *pdx-1* locus, and could not have been caused by abnormal behaviour of the whole chromosome or by polyploidy.

Mitchell (1955*a*) analysed 585 asci from a cross between two *pdx-1* mutants and selected those with wild-type recombinants by germinating the spores on medium lacking pyridoxin. After 17 h at 25 °C any pyridoxin-independent spore-lings could be recognized. Four of the asci has such wild-type spores. Their genotypes are given in Table 12. The genotypes of the *pdx-1* spores in these asci were obtained by backcrossing to each parent and looking for further recombinants by plating random spores on minimal medium. As can be seen from the table, no double mutants were found and it was evident that the recombinants were arising by conversion and not by a reciprocal exchange. The double mutant would have been recognized, if present, by a failure to give pyridoxin-independent progeny when crossed with either parent.

The occurrence of conversion in mitotic recombination was demonstrated by Roman (1957) using adenine-requiring mutants of *Saccharomyces cerevisiae* at loci *ade3* and *ade6*. He found that wild-type segregants appeared in diploid strains heterozygous in *trans* for two allelic *ade* mutants. The wild-type recombinants were allowed to produce asci and the four spores from a number of such asci were isolated and scored for genotype by backcrossing to both haploid parents of the original diploid. It was found that the wild-type recombinants were heterozygous for one or other *ade* allele but never for the double mutant (Table 13(a)). As in Mitchell's work, this would have been recognized because it would fail to give recombinants with either single mutant. Roman concluded that the wild-type recombinants were arising during the mitotic cycle by conversion, that is, a non-reciprocal process, and not by crossing over (reciprocal exchange). Roman and Jacob (1957) found that the frequency of mitotic recombination of a pair of *ade3* mutants was increased more than 1000 times by irradiating the cells with ultraviolet light. Holliday (1964*a*) treated a diploid heterozygous in *trans* for two mutants at the *trp5* locus with the antibiotic mitomycin C and obtained a 10-fold increase in the frequency of tryptophan-independent colonies. He concluded that mitomycin C induced conversion in somatic yeast cells.

Mitotic recombination of alleles by conversion was also demonstrated in *Schizosaccharomyces pombe*. Leupold (1959) investigated recombination in a diploid heterozygous for mutants 257 and 273 at the *adenine7* (*ade7*) locus. The diploid gave rise with low frequency to wild-type mitotic recombinants, lacking the adenine requirement. Seventeen of these diploid prototrophs were allowed to sporulate and one ascus was analysed from each. The four haploid spores were isolated and germinated, and the colonies tested for genotype by backcrossing to both parental mutants and looking for prototroph formation. Mutants 257 and 273 differed in the temperature sensitivity of their adenine requirement, so tests of this on the colonies derived from the ascospores provided a check of their genotype. No double mutants were found (see Table 13(b)). The mitotic recombination of these alleles was evidently by conversion.

Conversion as a source of allelic mitotic recombinants in *Ustilago maydis* was

Table 12. Genotypes of recombinant asci from crosses heterozygous for a pair of alleles. The *pdx-1* mutants used by Mitchell (1955a) had isolation numbers 37803 and 39106 and are denoted by the last digit.

Reference:	Mitchell (1955a)	Roman (1958)		Case and Giles (1959)	
Species:	*Neurospora crassa*	*Saccharomyces cerevisiae*		*Neurospora crassa*	
Gene:	*pdx-1*	*ade3*		*pan-2*	
Parental genotype:	6 + / + 3	3 + / + 6	2 + / + 9	B5 + / + B3	B5 B3 / + +

Event	Ascus genotype and number observed				
(a) Conversion to + at proximal site	6 + + + + 3 + 3 2	3 + + + + 6 + 6 1	2 + + + + 9 + 9 1	B5 + + + + B3 + B3 4	B5 B3 + B3 + + + + 2
(b) Conversion to + at distal site	6 + 6 + + + + 3 1	3 + 3 + + + + 6 1		B5 + B5 + + + + B3 2	
(c) Conversion to + at both sites	6 + + + + + + 3 1				
(d) Conversion to mutant at distal site	—	2 + 2 9 + 9 + 9 1			B5 B3 B5 B3 + B3 + + 3
(e) Crossing over		2 + + + 2 9 + 9 1		B5 + + + B5 B3 + B3 2	
(f) Co-conversion	—			B5 + + B3 + B3 + B3 3	B5 B3 + + + + + + 2
(g) Conversion to + at proximal site in two chromatids					+ B3 + B3 + + + + 1
Non-recombinant asci	581	97	99	856	846

Table 13. Data on mitotic allelic recombination to give wild type from a *trans* cross. The homologous chromosome to that of wild-type genotype is expected not to have been involved in recombination in 50% of the recombinants (cf. Fig. 63). The results in (c) relate to ultraviolet-induced conversion with the exception of four of those showing conversion of mutant 13 and one of those showing conversion of mutant 6 in the cross between these mutants. These five conversions were of spontaneous origin.

Species and reference	Gene	Genotype of parental diploid	Genotypes of wild-type recombinants and number observed			
(a) *Saccharomyces cerevisiae* (Roman, 1957)	*ade3*	$\frac{1\ +}{+\ 4}$	$\frac{+\ +}{+\ 4}$	4	$\frac{1\ +}{+\ +}$	6
		$\frac{2\ +}{+\ 9}$	$\frac{+\ +}{+\ 9}$	4	$\frac{2\ +}{+\ +}$	2
	ade6	$\frac{2\ +}{+\ 3}$	$\frac{+\ +}{+\ 3}$	4	$\frac{2\ +}{+\ +}$	3
		$\frac{2\ +}{+\ 7}$	$\frac{+\ +}{+\ 7}$	10	$\frac{2\ +}{+\ +}$	1
		$\frac{2\ +}{+\ 12}$	$\frac{+\ +}{+\ 12}$	15	$\frac{2\ +}{+\ +}$	7
(b) *Schizosaccharomyces pombe* (Leupold, 1959)	*ade7*	$\frac{273\ +}{+\ 257}$	$\frac{+\ +}{+\ 257}$	5	$\frac{273\ +}{273\ +}$	12
(c) *Ustilago maydis* (Holliday, 1966)	*nar-1*	$\frac{13\ +}{+\ 6}$	$\frac{+\ +}{+\ 6}$	10	$\frac{13\ +}{+\ +}$	4
		$\frac{13\ +}{+\ 11}$	$\frac{+\ +}{+\ 11}$	4	$\frac{13\ +}{+\ +}$	3
		$\frac{6\ +}{+\ 11}$	$\frac{+\ +}{+\ 11}$	10	$\frac{6\ +}{+\ +}$	3

demonstrated in a similar way by Holliday (1966). He obtained mutants unable to grow on nitrate and believed to be defective in the structural gene for nitrate reductase. Five such *nar-1* mutants were used in recombination studies, and diploids heterozygous for pairs of them were synthesized. Low doses of ultraviolet light increased the mitotic recombination frequency 100-fold compared with the spontaneous rate. From the frequency of recombinants, able to grow on nitrate, which he obtained from each of the 10 heterozygotes, a linear map of the five sites was constructed. To find out what kind of event was giving rise to the recombinants, half-tetrad analysis was undertaken, that is, the *nar-1* genotype of the homologous chromosome in the wild-type recombinants was determined. This involved inoculating maize (*Zea mays*) seedlings with the wild types to obtain teliospores (in which nuclear fusion takes place); germinating them (meiosis); selecting a haploid of *nar-1* genotype; inoculating it, paired with each parental *nar-1* mutant in turn, into maize; and treating each resulting diploid with ultraviolet light, before plating on minimal medium to test for recombinants. No double mutants were found. They would have been recognized, if present, because they would have failed to give recombinants with either parental mutant. All the homologous chromosomes carried single *nar-1* mutants (see Table 13(c)). The wild-type recombinants had evidently arisen by conversion, not by crossing over.

Roman (1958), following his demonstration of conversion in mitotic recombination in *Saccharomyces cerevisiae*, went on to investigate recombination at meiosis. From a diploid heterozygous for alleles 3 and 6 at the *ade3* locus, 99 asci were analysed and from a diploid heterozygous for alleles 2 and 9 at the same locus, 102 asci were analysed. All except five of the asci showed normal 2:2 segregation. The genotypes of the others are given in Table 12. In addition to conversion to wild-type at either mutant site, such as Mitchell had found at the *pdx-1* locus in *Neurospora*, Roman found an example of conversion to mutant (Table 12(d)) and also of crossing over between the sites of the mutants (Table 12(e)).

Similar diversity of recombination events was obtained by Case and Giles (1959) from a comparable investigation using *Neurospora crassa* heterozygous for a pair of alleles at the *pantothenic acid-2* (*pan-2*) locus. A total of 1721 asci were analysed, comprising about equal numbers from a *trans* and from a *cis* cross, and 19 were found to be recombinant. Their genotypes are given in Table 12. In addition to recombination events matching most of those found by Mitchell, and all of those found by Roman, two new classes of event were found (Table 12(f),(g)). These will be discussed later (see Sections 9 and 18).

Conversion was demonstrated in *Aspergillus nidulans* by Strickland (1958a), who found three examples of asci with a 6:2 segregation for wild type and biotin requirement. He isolated large numbers of asci, germinated the spores, and established their genotype. Other linked markers segregated 4:4.

Evidence for another type of aberrant segregation was discovered by Olive (1956). He obtained three unlinked spore-colour mutants in the homothallic ascomycete *Sordaria fimicola* by treatment with ultraviolet light. The wild-type spores are black and the mutant spores were grey, yellow and hyaline, respectively. He found that in asci heterozygous for any of these mutants, five black spores and three mutant ones were occasionally present instead of four of each colour. He investigated these aberrant segregations by ascus analysis, using crosses in which at least one other character difference was also segregating (Olive, 1959). The second character difference behaved normally, with a 4:4 ratio in each ascus (Table 14). An ascus with three black and five mutant spores was also found, and here too the other character difference present in the cross segregated normally. Various mutants were used for the second character difference but all were unlinked or only loosely linked to the spore-colour mutant showing the aberrant segregation. Examination of progeny from self-fertilization or backcrosses confirmed that without exception the genotype for spore colour corresponded to its colour. No confirmation was possible, however, for the hyaline spores, which failed to germinate. The crosses studied by Olive also produced asci with 6:2 and 2:6 ratios of wild-type:mutant spores, and analysis of these asci revealed that the second character difference segregated normally. It was of particular interest that the *grey* mutant gave aberrant asci of all four kinds, that is, 6:2, 2:6, 5:3 and 3:5.

A fifth kind of aberrant segregation was discovered by Kitani, Olive and El-Ani (1961, 1962). In crosses between wild type and the *grey* spore-colour mutant of *S. fimicola*, asci were observed with a normal 4:4 colour ratio but with the spores in an abnormal order. These 'aberrant 4:4' asci were isolated and dissected, and the spores germinated and scored for a slow-growing mutant, *mat*, for which the asci were also heterozygous. The *mat* mutant was closely linked to the *grey* mutant, mapping 0.4 units to the left. The centromere is also to the left, but over 50 units away. The ascus genotypes are shown in Table 15. It is evident that wild type versus *mat* growth segregated normally with four spores of each genotype

Table 14. Data of Olive (1959) for eight asci of *Sordaria fimicola* showing a 5:3 or a 3:5 ratio of wild type:mutant spores. The mutants were *hyaline* (*h*) or *grey* (*g*). For each ascus the spore-colour sequence is given in the left-hand column and a second character difference which was also segregating is indicated by plus and minus signs in each right-hand column. Several different mutants were used for the second character, giving either a dwarf culture, or sterility (when homozygous). The *h* spores failed to germinate and so could not be scored for the second character.

```
(h)        (h)        (h)        +⌉        +⌉+       g⌉        g⌉        +⌉
(h)        (h)        (h)        +⌋        g⌋        g⌋+       +⌋-       +⌋-
+⌉+        (h)        +⌉+        g⌉-       +⌉        +⌉        +⌉        g⌉
+⌋         +  +       +⌋         +⌋        +⌋-       g⌋        +⌋        +⌋
+⌉         +⌉         +⌉         g⌉        g⌉+       +⌉        g⌉        g⌉
+⌋-        +⌋-        +⌋-        g⌋+       +⌋        +⌋-       g⌋+       g⌋+
(h)        +⌉         (h)        +⌉        +⌉-       +⌉        +⌉        g⌉
+   -      +⌋         +   +      +⌋        +⌋        +⌋        +⌋        g⌋
```

Table 15. Data of Kitani, Olive and El-Ani (1962) for nine aberrant 4:4 asci of *Sordaria fimicola* from a cross between wild type and a *grey* (*g*) spore-colour mutant. A slow-growing mutant, *mat* (*m*), mapping 0.4 units to the left of *g*, was also segregating. For each ascus the segregation pattern for +/*m* is shown to the left of that for +/*g*.

```
      ⎧ +        ⎧ g      ⎧ +      ⎧ +               ⎧ g      ⎧ g      ⎧ +      ⎧ g      ⎧ g
  + ⎨  g    m ⎨  +    m ⎨  g    m ⎨  g         m ⎨  +    + ⎨  +    + ⎨  g    + ⎨  +    m ⎨  g
      ⎩ g      + ⎩ g      ⎧ g      ⎧ g      m ⎨  +      ⎩ +      ⎩ +      ⎧ +      ⎩ +    + ⎨  +
      ⎧ +              + ⎨  g    + ⎨  g      ⎩ +                    ⎧ +      ⎧ +      ⎧ g      ⎩ g
  m ⎨  +    m ⎨  +      ⎩ +      ⎩ g         ⎧ g      ⎧ +    m ⎨  g    m ⎨  g    m ⎨  +    m ⎨  +
      +    + ⎨  g    m ⎨  +    m ⎨  +    + ⎨  g    m ⎨  g      ⎩ g      ⎩ g      ⎩ g    + ⎨  g
      g      ⎩ +      ⎩ +      ⎩ +      ⎩ g      ⎩ g                                      ⎩ +
```

grouped in pairs. This showed that the aberrant segregation of wild type versus *grey* resulted from abnormal behaviour at the *grey* locus and not from events at the level of the whole chromosome or the whole nucleus. In *S. fimicola* the spindles at anaphase of the mitosis, which occurs in each of the four nuclei following meiosis, sometimes overlap, or the spores may slip past one another as the ascus matures. This can give rise to a colour sequence such as that in the right-hand ascus in Table 15. But scoring for *mat* would distinguish such an ascus because it would reveal that, unlike the aberrant 4:4 ascus shown in the table, the *mat* spores as well as the *grey* ones would not be in pairs. All the aberrant 4:4 asci in Table 15, other than the right-hand one, have the aberrant behaviour in non-adjacent spore-pairs. This bias was less evident in later studies and seems to result from overlooking some of the aberrant 4:4 asci involving adjacent spore-pairs, the spore sequence being attributed to spindle overlap or spore slippage. Even without allowing for aberrant 4:4 asci that were overlooked, Kitani *et al.* (1962) pointed out that the aberrant 4:4 ascus frequency was over three orders of magnitude higher than would be expected from the independent occurrence of a 5:3 event and a 3:5 event in the same ascus. Such simultaneous occurrence could generate a 'false' aberrant 4:4 ascus (see Section 14).

The 5:3, 3:5 and aberrant 4:4 categories of aberrant segregation are remarkable in showing segregation of a character difference at the mitosis after meiosis. This *postmeiotic segregation* distinguishes them from the 6:2 and 2:6 categories discovered earlier. There is difference of opinion as to the usage of the term *conversion* for these phenomena. Some authors regard all five kinds of aberrants as examples of conversion, distinguishing 6:2 and 2:6 as 'whole-chromatid conversion' and the others as 'half-chromatid conversion'. Even before the discovery of postmeiotic segregation there was controversy about the use of the word conversion and Curt Stern (in Roman, 1957) stated: 'It seems fully permissible for different workers to use the same terms in different ways so long as each individual defines clearly the meaning he attaches to the term'. In what follows, the term conversion is restricted to the 6:2 and 2:6 aberrant segregations, postmeiotic segregation being used for the other categories. Also, in referring to ratios such as 6:2, the first figure will always be used for wild type and the second for mutant.

Shortly after Olive's discovery of postmeiotic segregation in *Sordaria fimicola*, similar findings were reported in *Ascobolus immersus* by Rizet, Lissouba and Mousseau (1960) and Lissouba (1961). The spores in this heterothallic fungus are not arranged in linear order within the ascus. The eight spores from individual asci are projected at maturity in a cluster (octad) and are readily collected by placing an agar surface above the asci. The wild-type spores are dark brown ('black') when mature, but mutants with colourless ('white') or pink spores arose spontaneously. Rizet *et al.* (1960) found that when mutant 60 of spore-colour gene *19* was crossed with alleles, the majority of the asci had white spores, as expected. Small numbers of recombinant octads with two black spores were found and also some with one black spore. About 15 of these 1:7 octads were analysed for genotype, by isolating and germinating the spores, crossing the mutant isolates with each parental mutant, and looking for further recombinants. They found that in each case the black spore was wild type, three of the white spores had the genotype 60, and the other four white spores had the genotype of the other allele. There were no double-mutant spores. In other words, there was a 5:3 segregation at the 60 site and a normal 4:4 segregation at the allelic site. Lissouba (1961) obtained comparable results with mutant 277 of spore-colour gene *46*. His data for crosses with three alleles are given in Table 16. Fourteen of the 1:7 octads were investigated further. Genotype analysis revealed that all arose from a 5:3 segregation at the site of 277 and a normal 4:4 segregation at the allelic site.

Postmeiotic segregation was discovered in *Neurospora crassa* by Case and Giles (1964) in the course of analysis of 1457 unselected asci from a cross heterozygous for three *pan-2* mutants (see Section 9). In *Podospora anserina* Marcou (1969) reported conversion and postmeiotic segregation shown by mutants at spore-colour locus *14*.

It might have been supposed that postmeiotic segregation could not be detected

Table 16. Octad counts made by Lissouba (1961) for crosses between white-spored mutant 277 and alleles in spore-colour gene *46* of *Ascobolus immersus*.

Allele with which 277 crossed	Class of ascus (wild type : mutant spores)			
	0 : 8	1 : 7	2 : 6	4 : 4
63	11 395	38	177	1
46	4088	24	48	0
W	2234	6	27	0

in yeasts with four-spored asci. Gutz (1971), however, was able to detect post-meiotic segregation, using adenine-requiring (*ade6*) mutants of *Schizosaccharomyces pombe*, because these mutants, when grown on media with limited amounts of adenine, accumulate a red pigment. The wild type gives white colonies. In the course of an analysis of thousands of unselected asci heterozygous for one, two or three *ade6* mutants, three examples of postmeiotic segregation were obtained. They were revealed by sectoring of the colonies derived from the individual spores. For example, in a cross between an *ade6* mutant (M375) and wild type, two of the spores of one ascus gave red colonies, one spore gave a white colony, and the fourth gave a colony with a white sector and a red sector. Backcrosses confirmed that this was a 3:5 ascus, using the octad nomenclature.

Similar results were obtained by Esposito (1971) with *adenine 8* (*ade8*) mutants of *Saccharomyces cerevisiae*. Mutants of the *adenine 2* (*ade2*) locus accumulate a red pigment, but fail to do so in the presence of *ade8* mutants. Postmeiotic segregation could therefore be detected by crossing *ade8* mutants with wild type in strains homozygous for an *ade2* mutant, analysing unselected asci, and looking for sectored colonies with one half red and the other half white. Mutant 18 at the *ade8* locus was investigated in this way. Out of 251 asci analysed, seven gave rise to two red colonies, one sectored one and one white one, equivalent to a 5:3 octad, and six the converse situation, that is, one red, one sectored and two white colonies (a 3:5 ascus). There were also three 3:1 asci and three 1:3 asci, the remainder having a normal 4:4 segregation. Another *ade8* mutant, in a similar analysis, gave only one ascus showing postmeiotic segregation – a 5:3 ascus. Hurst, Fogel and Mortimer (1972) detected postmeiotic segregation at an ochre suppressor locus (*Sup6*) in *S. cerevisiae* by finding sectored colonies, one part suppressing and the other not suppressing ochre mutants in the *lysine 1* (*lys1*) and *ade2* genes.

The equivalent of postmeiotic segregation for mitotic recombination was found by Wildenberg (1970) with X-ray-induced recombination in *S. cerevisiae*. Her technique was to synchronize cell division by isolating a uniform-sized cell fraction. This was done with diploid cells heterozygous for two *histidine 1* (*his1*) mutants and for flanking markers. Following X-ray treatment, histidine prototrophs were selected and the first few division products of each diploid cell were isolated to ensure the recovery of all genetic strands present at the time of recombination.

Postmeiotic segregation was detected in *Ustilago maydis* by Holliday and Dickson (1977) by finding mosaic colonies. Mutants 1 and 6 at the nitrate reductase (*nar-1*) locus were crossed and the diploid teliospores germinated on agar to give rise, following meiosis, to four haploid clones. Precautions were taken, however, to prevent the survival of more than a single meiotic product from each teliospore, by using crosses heterozygous for four or five unlinked auxotrophic mutants. The chances that more than one product of an individual meiosis will be able to grow on minimal medium are then very low. Wild-type recombinants lacking the nitrate reductase deficiency were selected by replica plating on nitrate minimal medium. Ten out of 69 *nar*[+] recombinants were found to be *nar⁺/nar⁻* mosaics. These had evidently arisen through postmeiotic segregation, because they had identical genotypes for four unselected markers for which the crosses were heterozygous. These markers were not linked to *nar-1* nor to one another.

The diversity of recombination events that may occur within a gene, first demonstrated with *Neurospora crassa* and *Saccharomyces cerevisiae* (see Table 12), was revealed in more detail by the work of Lissouba *et al.* (1962) with *Ascobolus immersus*. They studied numerous crosses between allelic spore-colour mutants at loci *19*, *46* and *75*. From the background of octads with all eight spores white, they isolated recombinant octads consisting of two wild-type (black) and six mutant (white) spores. From backcrosses to the parental strains the spore genotypes were established. It was found that conversion to wild type at either mutant site was responsible for many of the recombinant asci, but others contained a pair of double-mutant spores and had evidently arisen through crossing over (reciprocal exchange) between the mutant sites. For many crosses all these three alternatives were occurring, but their relative frequencies differed widely from one pair of mutants to another. Closely linked alleles, however, rarely showed crossing over between their sites: for example, none was found within three mutant clusters in gene *19*. This influence of site separation applied even to the same region of a gene: for example, in gene *75* four mutants mapped in the following sequence:

$$\text{—— 322 —— 278 —— 147 —— 1987 ——}$$

From the analysis of a total of about 70 recombinant octads from pairwise crosses of these mutants, the frequencies of crossing over, out of total recombination events, for the three marked intervals were, from left to right, 19%, 16% and 6%, but for a cross of the extreme mutants (322 x 1987) the proportion of crossovers was 76%. The relationship between crossover frequency and the distance between the mutant sites was revealed by further studies with gene *75* reported by Rizet and Rossignol (1966). Analysis of 2:6 asci from 24 of the 28 possible pairwise crosses of eight mutants gave percentages of crossing over, out of total recombinant asci (crossing over, or conversion to wild type at either mutant site), of 6, 6, 12, 27, 44, 59 and 96, with increasing distance between the mutant sites. The measure of distance was the number of mutants, ranging from zero to six, that mapped between the sites of those crossed.

The relative frequency of conversion and crossing over also depend on properties of the mutants other than their distance from one another. This was exemplified by a comparison, made by Lissouba *et al.* (1962), of mutants 137 and 277 in gene *46*. These mutants were closely linked, with similar recombination frequencies in crosses with alleles, including those called 63 and W. Mutant 277, however, when crossed with 63 or W, gave crossovers in about one-quarter of the recombinants, while 137 crossed with 63 or W showed no crossovers in 50 recombinant octads. Another peculiarity of mutant 277, namely, the occurrence of postmeiotic segregation, has already been described. From these studies by

Lissouba *et al.* it was evident that intragenic recombination was a complex process involving both mutant-specific and site-specific factors.

Another example of mutant specificity in the occurrence of intragenic crossing over was reported by Kitani and Olive (1969), working with the *grey* spore-colour gene in *Sordaria fimicola* (see also Kitani and Whitehouse, 1974; Whitehouse, 1974). In crosses heterozygous either in *trans* or in *cis* for alleles 1 and 4 at the *grey* locus, about 10% of the recombinant asci resulted from crossing over between the sites of the mutants. On the other hand, a derivative of mutant 4 called 4b showed no crossing over when crossed with mutant 1. Mutant 4b was of spontaneous origin and was derived from the odd mutant spore in a 3:5 ascus from a cross between mutant 4 and wild type. Mutant 4 had hyaline spores and 4b faintly yellowish spores.

5. ASSOCIATION BETWEEN ABERRANT SEGREGATION AND CROSSING OVER

Kitani, Olive and El-Ani (1961, 1962), working with *Sordaria fimicola*, obtained a mutant that they called *corona* which showed linkage to the *grey* spore-colour mutant and mapped on the opposite side of *grey* to the *mat* mutant, already described (Section 4). The *corona* mutant was so called because a distinctive ring of pigment developed around the site of inoculation. The recombination frequencies gave the following map:

$$\text{———}\ mat\ \underset{}{\overset{0.4}{\text{———}}}\ grey\ \underset{}{\overset{3.4}{\text{—————}}}\ corona\ \text{———}$$

With such flanking markers for a spore-colour locus it became possible for the first time to investigate the relationship between aberrant segregation and crossing over. The *grey* mutant was crossed with wild type in the presence of the markers, aberrant asci were isolated and dissected, and the spores germinated and scored for the flanking marker characters. The frequency of aberrant asci was about 0.12%. The results obtained are given in Table 17. They showed two features of outstanding interest.

The first discovery concerns the frequency of crossing over. The *mat* and *corona* markers span an interval of 3.8 units. Allowing for the fact that crossing over occurs at the four-chromatid stage (see Section 1), crossovers are expected in the *mat–corona* interval in 7.6% of meioses. Kitani and associates analysed a total of 122 aberrant asci, covering all five classes, and no less than 49 of them, or 40%, showed crossing over (Table 17). This was a highly significant difference from the frequency expected in a random sample of asci.

The second discovery concerns the chromatids taking part in crossing over. From the ascus analyses, the crossovers can be subdivided into those associated with the aberrant segregation as regards the chromatids involved, and those incidental to the aberrant segregation. With aberrant 4:4 asci both chromatids involved in aberrant segregation can be identified as they both give rise to postmeiotic segregation. With the other kinds of aberrant asci, either chromatid from the majority parent might have contributed and there is no means of knowing which. By majority parent is meant that of wild-type genotype for the

Table 17. Data of Kitani, Olive and El-Ani (1962) relating crossing over to aberrant segregation at the *grey* (*g*) spore-colour locus in *Sordaria fimicola*.

Aberrant segregation (wild type : mutant)	Chromatid genotype at *grey* locus		(a) Non-crossover	(b) Associated crossover	(c) Incidental crossover
Aberrant 4 : 4	1 2 3 4	+ + / g + / g g	2	2	0
5 : 3	1 2 3 4	+ + + / g g	16	16	3
	1 2 3 4	+ + / g + g	15	13	
3 : 5	1 2 3 4	+ + / g g g	6	0	1
	1 2 3 4	+ g + / g g	1	3	
6 : 2	1 2 3 4	+ + + g	30	8	2
2 : 6	1 2 3 4	+ g g g	3	1	0

(a) Chromatids 1 and 2 with flanking marker genotype of + parent, 3 and 4 with flanking marker genotype of g parent. (b) Chromatids 1 and 4 of parental genotype; chromatid 2 with proximal (left-hand) marker of + parent and distal (right-hand) marker of g parent; chromatid 3 the converse. (c) Chromatid 1 involved in a crossover in aberrant 4 : 4, 3 : 5 or 2 : 6 asci; chromatid 4 involved in a crossover in aberrant 4 : 4, 5 : 3 or 6 : 2 asci.

spore-colour gene in 5 : 3 and 6 : 2 asci, and that of *grey* genotype in 3 : 5 and 2 : 6 asci. By the expression 'associated crossovers' in Table 17 is meant those which appear to involve the same chromatids as the aberrant segregation, but with the reservation indicated above that there is always ambiguity about the chromatids of the majority parent. The expression 'incidental crossovers' refers to crossovers

Table 18. Total data of Kitani and associates for frequency of crossing over associated with aberrant segregation at the *grey* spore-colour locus in *Sordaria fimicola.* Incidental crossovers are omitted.

Aberrant segregation (wild type : mutant)	Total number of asci	Number of asci with crossing over associated with the aberrant segregation	Percentage crossover ± standard error (%)
4 : 4	577	283	49.0 ± 2.1
5 : 3	871	393	45.1 ± 1.7
3 : 5	475	205	43.2 ± 2.3
6 : 2	617	248	40.2 ± 2.0
2 : 6	101	39	38.6 ± 4.8
Total	2641	1168	44.2

that involve the chromatid of the minority parent not involved in the aberrant segregation. The minority parent is that of *grey* genotype in 5 : 3 and 6 : 2 asci, and of wild-type spore colour genotype in 3 : 5 and 2 : 6 asci. With aberrant 4 : 4 asci, incidental crossovers may involve one or both of the chromatids not contributing to postmeiotic segregation. The terms 'associated' and 'incidental' are defined more precisely in the footnote to Table 17. If it is assumed that incidental crossovers involve chromatids chosen at random irrespective of those involved in the aberrant segregation, it follows from the arguments above that half the incidental crossovers will be classified as associated ones, except with aberrant 4 : 4 asci where only one-quarter will be so classified. A correction can therefore be applied to allow for incidental crossovers that cannot be distinguished from associated ones. Subsequently, small numbers of aberrant asci were found with both associated and incidental crossovers. An additional correction can therefore be made (see below). In the data for the *grey* locus these two corrections effectively cancel out. Later studies by Kitani and associates (Kitani and Olive, 1967; Kitani and Whitehouse, 1974), using the *grey* mutant and several alleles crossed individually with wild type, confirmed the earlier results (Table 18).

The conclusion reached from the work of Kitani *et al.* (1962) was that about 40% of the aberrant asci were associated with crossing over, both as regards its position in the chromosome and as regards the chromatids involved in both events. Nevertheless, a majority of the aberrant asci were not associated with crossing over. Any hypothesis to account for the relationship between aberrant segregation and crossing over had also to account for aberrant segregation in its absence.

Comparable results were obtained by Stadler, Towe and Rossignol (1970) for the *white17* (*w17*) spore-colour locus in the Pasadena strain of *Ascobolus immersus.* The flanking markers were a mutant showing colonial growth at 37 °C, mapping 10 units to the left of *w17*, and a mutant giving resistance to *p*-fluorophenylalanine, mapping 3 units to the right of *w17*. Several different *w17* mutants were crossed individually with wild type in the presence of the flanking markers and totals of 109 octads with a 6:2 segregation and 100 with a 2:6 segregation were scored for flanking marker behaviour. The results are given in Table 19(i). Small numbers of octads were found with two crossovers in the marked interval, one associated with the aberrant segregation and one incidental to it. Such double crossovers are detected only if each involves a different pair of chromatids. If they share one or both chromatids, one or both of the crossovers will not be detected. Assuming the chromatids involved in the two crossovers are chosen at random (one being from each parent), the number of undetected associated crossovers can be estimated.

Let a be the observed frequency of non-crossover aberrant asci, b the frequency of aberrant asci with an associated crossover, c the frequency of aberrant asci with an incidental crossover, and d the frequency of aberrant asci with both an associated and an incidental crossover. Let p be the frequency of associated crossovers, and $1 - p$ be the frequency of non-crossovers. Let x be the frequency of non-crossovers with an incidental crossover, and $1 - x$ be the frequency of non-crossovers without an incidental crossover. Let y be the frequency of associated crossovers with an incidental crossover, and $1 - y$ the frequency of associated crossovers without an incidental crossover. As already indicated, assuming there is no chromatid interference, incidental crossovers in conjunction with non-crossover aberrant segregation will contribute equally to b and c, while incidental crossovers in conjunction with crossover aberrant segregation will contribute equally to a, b, c and d. Expressing this algebraically:

$$a = (1 - x)(1 - p) + \frac{y}{4}p,$$

$$b = (1 - y)p + \frac{x}{2}(1 - p) + \frac{y}{4}p,$$

$$c = \frac{x}{2}(1 - p) + \frac{y}{4}p,$$

$$d = \frac{y}{4}p.$$

It follows that $p = b - c + 4d$. Using this expression, the number and frequency of crossovers associated with conversion at the $w17$ locus have been estimated (Table 19(i)).†

Sang and Whitehouse (1979) studied segregation at the $buff$ spore-colour locus in the heterothallic species of $Sordaria$, $S.\ brevicollis$, using as flanking markers a

---

†The expression $p = b - c + (8d/5)$ has been used as the correction factor in Table 38 (see Section 16), which is concerned with data where a second allele is present. With this extra marker, an incidental crossover in conjunction with an associated one is more readily detected, y being distributed between a, b, c and d in the ratio $1:1:1:5$ instead of equally.

Table 19. Data relating crossing over to aberrant segregation obtained (i) by Stadler, Towe and Rossignol (1970) for the *white-17* (*w17*) spore-colour gene in the Pasadena strain of *Ascobolus immersus*, and (ii) by Sang and Whitehouse (1979) for the *buff* spore-colour gene in *Sordaria brevicollis*. The expressions 'associated crossover' and 'incidental crossover' have the same meanings as in Table 17.

Aberrant segregation (wild type: mutant)	Non-crossover frequency, a	Associated crossover frequency, b	Incidental crossover frequency, c	Associated and incidental crossover frequency, d	Total asci	Corrected number ($b - c + 4d$) and frequency of associated crossovers
(i) 6:2	47	40	18	4	109 }	79 (38%)
2:6	50	38	9	3	100 }	
(ii) 5:3	80	20	6	2	108 }	28 (16%)
3:5	44	13	7	0	64 }	
6:2 } 2:6	282	147	43	6	478	128 (27%)

methionine auxotroph mapping 5 units to the left and a nicotinamide auxotroph mapping 2 units to the right. The results are given in Table 19(ii) and reveal a lower frequency of associated crossovers than was found in the other investigations. Surprisingly, postmeiotic segregation was associated with a significantly lower frequency of crossing over than conversion.

In both *Ascobolus immersus* and *Sordaria brevicollis* aberrant 4:4 asci cannot be detected in a straightforward way and so were not included in the investigation of the frequency of associated crossovers. In *Ascobolus*, as already mentioned, the spores are not arranged in an orderly way in the ascus, and in *S. brevicollis* spindle overlap at the third nuclear division in the ascus is frequent. There are, however, indirect methods of detecting aberrant 4:4 asci in these fungi (see Section 14).

Rizet and Rossignol (1966) made an important discovery concerning the relationship between conversion and crossing over. They made crosses involving three spore-colour mutants of gene *75* of *Ascobolus immersus* simultaneously. Flanking markers were not available, but a pair of gene-*75* mutants was chosen, no. 1186 and either no. 889 or no. 1472, that showed little conversion to wild type. When these mutants were crossed, asci with wild-type spores arose almost entirely through crossing over between their sites. Three mutants (nos 278, 2029 and 147) showing frequent conversion in either direction were available that mapped between these sites, and one of these was used in each three-point cross. Denoting mutant 1186 as *a*, 889 or 1472 (whichever was used in a particular cross) as *c*, and the intervening allele (again, whichever one was used) as *b*, the crosses were of the form:

$$\text{(i)} \quad \frac{a++}{+bc} \quad \text{or} \quad \text{(ii)} \quad \frac{ab+}{++c} \quad \text{or} \quad \text{(iii)} \quad \frac{abc}{+++}.$$

In the two *trans* crosses, most of the octads will have all eight spores white, while in the *cis* cross most of them will have four black and four white spores. In all the crosses, octads with two black and six white spores will therefore be readily recognized. Such octads were isolated and the spores germinated and scored by crossing with the individual mutants. The results for those asci showing crossing over in the marked interval are given in Table 20. For the *trans* crosses there were few additional sources of 2:6 asci, since conversion to wild type at *a* or *c*, such as would give rise to a 2:6 ascus in crosses of types (i) and (ii) respectively, was rare. For the *cis* cross, however, there were numerous 2:6 asci, not shown in the table, resulting from conversion to mutant at one or more of the mutant sites. A striking feature of the 256 crossover asci shown in Table 20 is that 60 (23%) of them were associated with conversion at the intermediate site. Rizet and Rossignol (1966) concluded that crossing over and conversion should be regarded, not as two distinct events, but as representing two different aspects of a single event.

The association between aberrant segregation and crossing over, and the occurrence of aberrant segregation in the absence of crossing over, provide an explanation for the data on negative interference described in Section 3. The results reported there concerned individual products of meiosis and so provided

Table 20. Results obtained by Rizet and Rossignol (1966) for the genotypes of 2 : 6 asci showing crossing over in three-point crosses of gene-75 mutants of *Ascobolus immersus*.

Parental genotypes	Genotype of the two recombinant spore pairs	Number of octads	Description of event
(i) $\dfrac{a++}{+bc}$	*a b c* *+ + +*	60	Crossover in *a−b* interval
	a + c *+ + +*	21	Crossover in *a−c* interval and conversion of *b* to wild type
(ii) $\dfrac{a\,b+}{++c}$	*a b c* *+ + +*	94	Crossover in *b−c* interval
	a + c *+ + +*	23	Crossover in *a−c* interval and conversion of *b* to wild type
(iii) $\dfrac{a\,b\,c}{+++}$	*a + +* *+ b c*	16	Crossover in *a−b* interval
	a b + *+ + c*	26	Crossover in *b−c* interval
	a + + *+ + c*	6	Crossover in *a−c* interval and conversion of *b* to wild type
	a b + *+ b c*	10	Crossover in *a−c* interval and conversion to mutant at *b*

little information about how the wild-type recombinants originated. Conversion to wild type in the absence of crossing over would generate P1 and P2 flanking marker genotypes, such as those illustrated in Table 11. Either conversion to wild type associated with crossing over, or the occurrence of crossing over in the interval between the sites of the alleles, will give rise to R1 wild types. The occurrence of R2 wild types, also found to be frequent on occasion (Table 11), is less easily explained and will be discussed later (Section 16).

Information about the relationship between aberrant segregation and crossing over in mitotic recombination has been obtained by selecting for wild-type recombinants in diploids heterozygous for pairs of alleles and for flanking markers. Such investigations were made with *Aspergillus nidulans* by Pritchard (1955, 1960) using adenine-requiring (*adE*) mutants; by Putrament (1964) using *p*-aminobenzoic acid-requiring (*pabaA*) mutants of *Aspergillus*; and by Kakar (1963) and Hurst and Fogel (1964) using isoleucine-requiring (*ilv1*) and histidine-requiring (*his1*) mutants of *Saccharomyces cerevisiae*, respectively. The results are given in Table 21. The genotypes of the wild-type recombinants were established by analysis of the progeny from further crosses. The half-tetrad analysis that is possible by this means will not always reveal the genotype of the second chromatid that has been involved in the recombination event. Indeed, in 50% of the recombinant diploids the homologous chromosome to the wild-type one is expected to be of parental origin. This possibility is shown in Table 21, where all the expected genotypes for the recombinant diploids are given on the assumption that there may be conversion to wild type at either mutant site, with or without associated crossing over, or there may be crossing over between the sites of the alleles. The genotypes homozygous for the proximal (left-hand) mutant (classes Ib, IIb and IVb) would result from an incidental crossover in the chromosome segment between the site of the mutant and the centromere. The observed numbers of the various genotypes, including some (class V) not pre-dicted by the above assumptions, are given in the table. Diploids of class I evidently resulted from conversion at the proximal site, those of class II from conversion at the distal site, those of class III from crossing over (reciprocal exchange) between the sites of the mutants, and those of class IV could have resulted from any of these possibilities.

The data obtained by Pritchard (1955) for *adE* alleles 16 and 8 (Table 21) show that the proportion of the wild types resulting from crossing over occurring in the interval between the sites of the mutants was 40–50%. This frequency can be estimated either by doubling the number of class III, or by adding together the numbers in classes III and IVa, after subtracting the numbers in Ic and IIc (since conversion and associated crossing over will contribute to IVa as often as to Ic and IIc). Only a minority of the conversion events, however, were associated with crossing over. His data (Pritchard, 1960) for alleles 20 and 8 were similar to those for 16 and 8. The extensive results obtained by Putrament (1964) for *pabaA* were made possible by the use of *p*-fluorophenylalanine to induce haploidization. This facilitates genotype analysis. Her results indicate a low frequency of crossing over between the sites of the closely linked mutants 5 and 2. With the more distant mutants 5 and 18 reciprocal recombinants accounted for about 15% of the events. Again, most of the conversion events were not associated with crossing over. The data of Kakar (1963) and Hurst and Fogel (1964) for *ilv1* and *his1* mutants, respectively, of *Saccharomyces cerevisiae* are similar to the *Aspergillus* results, particularly those for *pabaA* mutants 5 and 2 with their low frequency of crossing over between the mutant sites (class III in Table 21). The occurrence of appreci-able numbers of wild-type recombinants with the R2 flanking marker genotype (class Vd) is not expected with the sources of recombinants discussed above, which generate genotypes in classes I–IV only. R2 genotypes were found, however, when prototrophs arising from meiotic recombination were selected in crosses between closely linked auxotrophic mutants (see Section 3). As already mentioned, the origin of R2 recombinants will be discussed later (see Section 16).

There is considerable similarity between the mitotic recombination data described above and the meiotic data, such as those of Lissouba et al. (1962) for *Ascobolus* described in Section 4. The mitotic data show the same variability from one pair of alleles to another in the proportion of wild-type recombinants arising

Table 21. Data on mitotic recombination obtained by Pritchard (1955, 1960) and Putrament (1964) for the adE and pabaA loci, respectively, in Aspergillus nidulans, and by Kakar (1963) and Hurst and Fogel (1964) for the ilv1 and his1 loci, respectively, in Saccharomyces cerevisiae.

Class	Chromatid number	Conversion at proximal site — Without crossing over	Conversion at proximal site — With crossing over	Conversion at distal site — Without crossing over	Conversion at distal site — With crossing over	Crossing over between sites of alleles	adE Site sequence 20-16-8 $16+/+8$	adE $20+/+8$	pabaA Site sequence 5-2-15-18-19 $5+/+2$	pabaA $5+/+15$	pabaA $5+/+18$	pabaA $2+/+19$	ilv1 $2+/+1$	his1 $7+/+1$
I a	1 / 2 / 3 or 4	$M1+N$ / $M++N$ / $m+2n$ / $m+2n$					4	2	48	50	26	11	14	57
I b	1 / 2	$M1+N$ / $M++N$					1	0	0	0	2	0	0	3
I c	2 / 3		$M+2n$ / $m++N$				0	0	4	2	1	1	2	16
II a	1 or 2 / 3			$M1+N$ / $m++n$ / $m++n$ / $m+2n$			5	1	24	28	21	26	11	60
II b	4				$M1+n$ / $m++N$		0	0	1	0	1	0	0	0
II c	2 / 3						2	1	3	1	1	5	2	9
III	2 / 3					$M12n$ / $m++N$	8	2	1	4	8	1	0	1
IV a	1 / 3		$M1+N$ / $m++N$ / $m++N$ / $m+2n$		$M1+N$ / $m++N$ / $m++N$ / $m+2n$	$M1+N$ / $m++N$ / $m++N$ / $m+2n$	16	3	12	10	8	12	16	22
IV b	3 / 4						0	0	1	1	3	5	0	20
V a	P1 } and incidental crossovers						1	0	5	0	2	2	6	4
V b	P2 }						0	0	3	3	3	0	5	9
V c	R1						0	0	0	0	0	0	0	13
V d	R2						1	1	25	6	14	1	11	12
V e	Both chromosomes wild-type						1	0	1	1	4	1	1	14
V f	Complex events						2	1	2	0	0	0	1	12
	Total						41	11	128	106	94	65	69	252

Other genotypes:

from crossing over between the sites of the mutants, and the same tendency for this proportion to increase with increasing site separation.

Detailed information about the association between aberrant segregation and crossing over can be obtained from the analysis of recombinant asci from pairwise crosses of alleles in the presence of flanking markers. Such investigations have been made in *Saccharomyces cerevisiae*, *Podospora anserina*, *Sordaria fimicola* and *S. brevicollis* and will be discussed later (see Section 16).

6. ABERRANT SEGREGATION IN EUKARYOTES OTHER THAN FUNGI

Evidence for aberrant segregation in *Drosophila* was obtained by Demerec (1926). He found a recessive sex-linked body-colour mutant in *D. virilis* which he called *reddish* (*re*). It resembled the previously known recessive body-colour mutant *yellow* (*y*) in many respects, and mapped in the same region of the X chromosome. When *re* and *y* were crossed, the F_1 females were yellow, as expected if these were alleles, but in F_2, in addition to 112 reddish and 129 yellow flies, there were 30 wild type. When flanking markers were used (*sepia* (*se*) mapping 2 units to the left of *y*, and *scute* (*sc*) mapping 0.7 units to the right), it was found that the majority of the wild types arose without associated crossing over. The results are given in Table 22. It appeared as if the *re* mutant was at the *y* locus and showed conversion (to use current terminology) to wild type with high frequency. Only about 14% of the conversion events were associated with crossing over (Table 22). Their genotype points to *re* being to the right of the *y* mutant. Later studies (Demerec, 1928) supported these conclusions: the wild types were formed only in females heterozygous for *re* and either another *y* mutant or wild type; they were not found in somatic cells, in females homozygous for *re*, or in males. Thus, all the observations pointed to a recombination process, though its

Table 22. Data of Demerec (1926) on the genotype of male progeny of females of *Drosophila virilis* of the genotype given at the top of the table, crossed with males of *re* or of *se y sc* genotype. *re*, reddish; *sc*, scute; *se*, sepia; *y*, yellow.

se y + sc	Number	Event
+ + re +	observed	
se y sc	4190	Parental
+ re +	4685	
+ y sc	74	Crossover in *se−y* interval
se re +	80	
se y +	36	Crossover in *y−sc* interval
+ re sc	25	
+ + + +	146	Conversion of *re* to + and no crossing over
+ + + sc	23	Conversion of *re* to + and associated crossover
Total	9259	

frequency was variable: 11% in the initial experiment above, 2% in the data in Table 22. Demerec (1928) found that conversion at *re* was associated with a 24-fold increase in crossing over in the *y–sc* interval.

Aberrant segregation at the *garnet* (*g*) eye-pigment locus, which is about 22 units from the centromere in the X-chromosome of *Drosophila melanogaster*, was reported by Chovnick (1958). Females heterozygous for pairs of *garnet* mutants were assayed by crossing to males carrying a *garnet* mutant. A total of 762 429 progeny were scored. Four *garnet* mutants were used and all six pairwise crosses were made, though one gave no recombinants. Flanking markers were also present in the crosses. Altogether, 19 non-garnet individuals were recovered. These were subjected to further breeding to verify their genotype. Two did not reproduce. Denoting the genotype of the heterozygous females as

$$\frac{M\ 1 + N}{m + 2\ n},$$

where 1 and 2 represent the alleles crossed and M/m and N/n are the flanking markers, there were 11 non-garnet individuals of genotype mN, that is, R1 according to the classification given in Section 3. The occurrence of these established the sequence of the mutant sites. The other non-garnets comprised MN (P1) 3, mn (P2) 2 and Mn (R2) 1.

Similar results were obtained by Hexter (1958), also using four *garnet* mutants, though two of the mutants differed from those used by Chovnick. Hexter made use of a *Drosophila* strain in which the two X chromosomes were joined at their centromeres (to the right of N/n). He recovered 12 wild-type females in 224 013 tested. Further tests showed the genotypes of the wild-type chromosomes to be R1 10; P1 2, in agreement with Chovnick's results. The use of the attached-X strain enabled a half-tetrad analysis to be carried out. One of the P1 wild types had a second X chromosome with the genotype of one parent ($M\ 1 + N$), and the other P1 wild type had a second X corresponding to the other parent ($m + 2\ n$). Five of the R1 wild types also had a second X chromosome of this genotype ($m + 2\ n$). In all these seven instances, therefore, the X chromosome attached to the wild-type one might have been derived from a meiotic chromatid not involved in the recombination event. The other five R1 wild-type females were of particular interest because the attached X chromosome had the reciprocal crossover genotype for the flanking markers. Extensive texts showed, however, that in each case it did not have the reciprocal genotype at the *garnet* locus. The genotype was

$$\left.\begin{array}{c} m + + N \\ M + 2\ n \end{array}\right\} \text{ and not } \left.\begin{array}{c} m + + N \\ M\ 1\ 2\ n \end{array}\right\}.$$

This distinction was possible because the left-hand *garnet* alleles in all Hexter's crosses gave orange-coloured eyes, where the other *garnet* mutants gave brown eyes. Progeny tests involving the scoring of over 350 000 males for eye colour failed to produce any with orange eyes. It was evident that much, and perhaps all, the recombination between *garnet* mutants was arising by conversion.

Chovnick (1961) reported an investigation similar to Hexter's, studying recombination at the *garnet* locus in attached-X females. In 157 669 female progeny there were 11 that were non-garnet. The wild-type chromosomes comprised two of flanking marker genotype P2 and nine of genotype R1. The attached X in four of the latter had the *mn* genotype. In the other five it was the reciprocal recombinant for the flanking markers (*Mn*). One of these individuals was tested for the presence of the *garnet* double mutant, that is,

$$\frac{m + + N}{M\ 1\ 2\ n} .$$

Following detachment of this X chromosome from its wild-type partner (achieved through a crossover near the centromere), crosses were made with each *garnet* mutant. No wild-type recombinants were obtained in a total of 273 946 progeny. This showed that the double mutant was present. Thus, contrary to Hexter's conclusion, Chovnick (1961) inferred that crossing over could occur between the sites of *garnet* mutants.

Aberrant segregation at the *white* (*w*) eye-colour locus, which is near the distal end of the X chromosome in *D. melanogaster*, was reported by Green (1960), but here, unlike *garnet*, much of the recombination appeared to be by crossing over between the sites of the alleles. Allele *a2* was crossed with each of three other *w* mutants. In 598 000 progeny there were 20 wild-type individuals comprising 14 with flanking marker genotype R1, two of genotype P1 and four of genotype P2. There were also 18 individuals of double mutant genotype, recognized by their white eyes, the parental mutants having eye colours intermediate between white and the wild-type red. The double mutant genotype of some of the white-eyed flies was confirmed by progeny tests. The flanking marker genotypes of the double mutants were 17 *M* 1 2 *n* and one *m* 1 2 *n*. The group of 17 were of complementary genotypes to the R1 wild types. Thus, of the 38 recombinants, all except seven appeared to result from reciprocal recombination.

Recombination at the *maroon-like* (*ma-l*) locus, which is situated about 1 unit from the centromere in the X chromosome of *D. melanogaster*, was investigated by Finnerty, Duck and Chovnick (1970). Mutants of this gene have brownish eyes and are deficient for several enzyme activities, including xanthine dehydrogenase (XDH). Females heterozygous for pairs of *ma-l* mutants were used and aqueous purine solution was added to the developing cultures. This kills *ma-l* larvae but allows wild-type recombinants, with their high level of XDH activity, to survive. Thus, recombinants could be selected. From 21 pairwise crosses about 6×10^7 zygotes were produced, of which 61 were wild type. Four of the crosses failed to give recombinants. Altogether, 25 of the wild types had a crossover genotype for the flanking markers and of the others 15 were P1 and 21 were P2. Thus 41% of the wild types showed flanking marker crossing over, although the markers spanned only 4.2 units on the linkage map.

250

Smith, Finnerty and Chovnick (1970) applied their selective techniques for *ma-l* recombinants to an attached-X strain. Two complementing *ma-l* mutants were used, giving normal eye colour but a low level of XDH activity. In an estimated 5×10^6 female progeny, 29 wild types were recovered following the purine treatment. Three genotypes for the attached-X recombinants are possible (Fig. 63). The numbers of each observed, on the basis of crosses to tester strains, are given (Fig. 63(a)). The diagram shows the likely alternatives for the origin of the recombinants, namely, crossing over between the sites of the alleles, or conversion of either of them to wild type, with or without associated crossing over. With crossing over there are two possibilities to be considered: either an exchange between attached chromatids, or between unattached chromatids. These alternatives are expected with equal frequency. From Fig. 63 it is evident that conversion to wild type at the site of the proximal mutant with respect to the centromere was responsible for at least half the recombinants, and conversion to wild type at the site of the distal mutant for the remainder. How many of the conversion events were associated with crossing over cannot be determined. None of the recombinants had arisen from crossing over between the sites of the alleles, to judge from the absence of double mutant chromosomes.

Chovnick *et al.* (1970) applied their selective technique for recombinants to the *rosy* eye-colour locus in *D. melanogaster*. This is the structural gene for XDH, so wild-type recombinants, with their high level of this enzyme, survive the purine treatment. The *rosy* gene is about 5 units from the centromere in the right arm of chromosome 3. Half-tetrad analysis was possible through the use of compound-3 strains having two right arms of chromosome 3 joined to one centromere and two left arms to another centromere. Compound-3 females heterozygous for *rosy* mutants 5 and 41 and for flanking markers were crossed with males homozygous for another *rosy* mutant. From 520 000 progeny 26 wild-types were obtained. Three failed to give progeny; the genotypes of the remainder are given in the right-hand column of Fig. 64. There were six wild types with R1 genotype for the flanking markers in the wild-type type chromosome and the reciprocally recombinant flanking marker genotype in the attached chromosome (last three rows of Fig. 64). One of these six was subjected to tests to identify the *rosy* mutants present. This involved detaching the chromosome arm (through crossing over near the centromere), followed by further search for wild types in females heterozygous for the detached *rosy* arm and for each parental mutant in turn. No wild types were found in 2 025 000 progeny. It was concluded that the original wild type carried the double mutant in the attached chromosome (last row of Fig. 64). The likely sources of the various recombinant genotypes are shown in Fig. 64. It was evident that at least 12 of the 23 recombinants arose through conversion and at least one through crossing over between the sites of the alleles.

Evidence was obtained by Chovnick, Ballantyne and Holm (1971) that, with increasing site separation, the events at the *rosy* locus that gave wild-type recombinants were more often associated with flanking marker crossing over. Seven mutants were used and 14 different pairs investigated. Three flanking marker genotypes (P1, P2, R1) were found for the wild-type recombinants. For pairs of

251

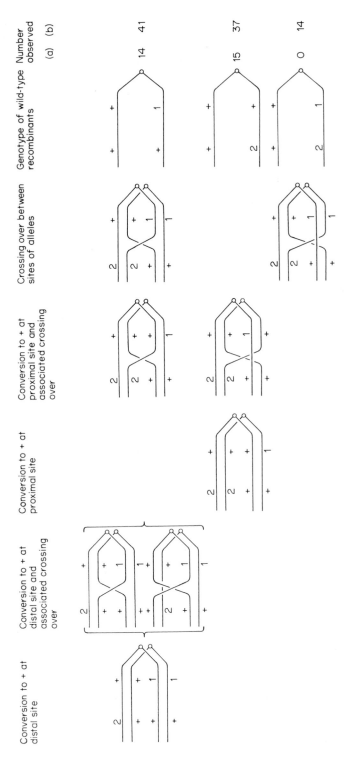

Fig. 63. Possible sources of the three classes of wild-type recombinants from attached-X *Drosophila melanogaster* heterozygous for a pair of alleles. Observed numbers, found (a) by Smith, Finnerty and Chovnick (1970) for two mutants at the *maroon-like* locus, and (b) by Carlson (1971) for six pairs of mutants at the *rudimentary* locus, are given.

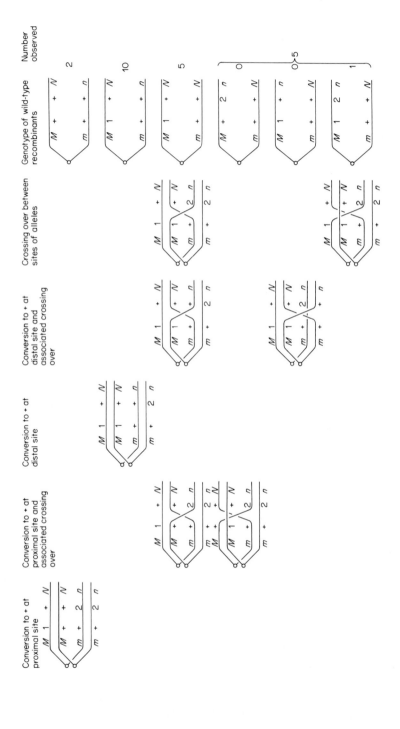

Fig. 64. Data of Chovnick *et al.* (1970) for 23 wild-type recombinants at the *rosy* eye-colour locus in the right arm of chromosome 3 of *Drosophila melanogaster*. Females heterozygous for alleles 5 (denoted by 1) and 41 (denoted by 2) were used in a strain having the pair of right arms of chromosome 3 attached at the centromere. The possible sources of the various genotypes are shown. *M/m* and *N/n* are flanking markers.

mutants with no intervening mutant site among the seven alleles used, there was one R1 in 32 recombinants (3%); with one intervening site, there were eight R1 genotypes in 39 recombinants (21%); and with two to five intervening sites there were 171 R1 genotypes in 386 (44%).

Two examples were reported by Chovnick *et al.* (1971) of wild-type recombinants that reproduced as *rosy* mutant homozygotes. These mosaic individuals were attributed to postmeiotic segregation.

Observations comparable to those of Chovnick *et al.* with the *rosy* locus were made by Carlson (1971) with *rudimentary* (*r*) wing mutants. This gene is in the X chromosome of *D. melanogaster* about 11 units from the centromere. Recombinants could be selected because only 1% of *r* mutant homozygotes survive, apart from a maternal influence of a wild allele. Altogether 45 *r* mutants were mapped. Flanking markers were available: *tiny chaetae* (*tc*), with short fine bristles, 3.5 units distal (that is, to the left on the map); and *forked* (*f*) bristles, 1.5 units proximal. Females heterozygous for a pair of *r* mutants were crossed with males carrying one of these *r* mutants, and the wild-type progeny were scored for the flanking markers. Several hundred pairs of *r* mutants were tested in this way. About 40% of the wild-type recombinants had a parental flanking marker genotype, except with very closely linked alleles (less than 5×10^{-5} wild recombinant progeny), where a higher proportion of parental genotypes was found. Attached-X females heterozygous for a pair of *r* mutants were used for a half-tetrad analysis of recombination. Altogether six pairs were tested in this way. All gave similar results, summarized in Fig. 63(b). About 30% of the recombinants evidently arose from reciprocal recombination between the sites of the alleles and the remainder from conversion to wild type at one or other site, with or without associated crossing over. In the progeny of a female heterozygous for two *r* mutants, Carlson found one example of a mosaic individual with one wing of mutant appearance and the other wild type. Its progeny were wild type. He concluded that it might have resulted from postmeiotic segregation.

The process of recombination in *Drosophila*, from the experimental results described above, appears closely to resemble the fungal mechanism. This is shown, for example, by the similarity between the *Drosophila* results in Fig. 64 and the fungal mitotic recombination data in Table 21. (The six classes in Fig. 64, reading down, correspond to classes Ia, IIa, IVa, Ic, IIc and III, respectively, in Table 21.) The phenomena that *Drosophila* and fungi share include the following: conversion of either mutant, with or without associated crossing over; postmeiotic segregation; crossing over between the sites of the alleles; and the occurrence of all these events within a group of alleles or even with a single pair of alleles. The *Drosophila* data, like the fungal results, also revealed an increased proportion of crossing over between the sites of alleles, relative to conversion, with greater site separation. Other recombination phenomena shared between *Drosophila* and fungi are discussed in Sections 9 and 12.

Evidence for conversion at the *waxy* (*wx*) locus in *Zea mays* (Maize) was obtained by Nelson (1962). This gene is near the centromere in the short left arm of chromosome 9. Waxy mutants have a different form of starch in pollen and endosperm from normal. Plants heterozygous for two wx mutants called 90 and C were used to pollinate C homozygotes,

$$\frac{bz\ C + v}{bz\ C + v} \times \frac{Bz + 90\ V}{bz\ C + v},$$

where Bz gives purple and bz (when homozygous) gives bronze kernels; V gives normal and v (when homozygous) gives virescent seedlings, the yellow colour turning green early in development. The bz gene is in the left arm of the chromosome about 18 units distal to wx, and v is about 7 units from wx and in the right arm. In one experiment 399 ears were obtained, which gave 117 819 kernels, of which 116 were of Wx phenotype. Of these, 108 were viable. Tests in this and a later experiment (Nelson, 1968) gave the flanking marker genotypes shown in Table 23. Over 40% of the recombinants were associated with flanking marker genotypes other than the R1 configuration expected from crossing over between the sites of the mutants. It was evident that some (and possibly all) the Wx progeny were arising through conversion. In order to allow for possible errors from incidental crossovers in the bz–wx interval, Nelson (1975) used a structural change, Tp9, which when heterozygous reduced the recombination frequency between bz and wx to less than 2%. In Tp9 a large segment of chromosome 3 is inserted between bz and wx. Heterozygosity for this structural change was found to reduce the frequency of wild-type recombinants of wx alleles C and 90 from about 100 to about 15 per 10^5 progeny. The flanking marker genotypes of the Wx recombinants are given in Table 14. The relative frequency of P1 and R1 was changed in favour of P1 in the presence of heterozygous Tp9, as expected if incidental crossovers in the bz–wx interval in the absence of Tp9 lead to some P1 events appearing as R1.

Table 23. Nelson's data for flanking marker behaviour in wild-type (Wx) recombinants from heterozygotes for pairs of $waxy$ mutants in $Zea\ mays$. Figures in parentheses are raw data before correction for the weak competitive ability of pollen tubes carrying the structural change Tp9.

Genotype of heterozygote	Flanking marker genotype of wild-type recombinants				
	P1	P2	R1	R2	Reference
$\dfrac{bz\qquad\ \ C + v}{Bz\qquad\ \ + 90\ V}$	27 / 9	15 / 1	63 / 18	3 / 1	Nelson (1962) / Nelson (1968)
$\dfrac{Bz\ \mathrm{Tp9}\quad C + V}{bz\ +\qquad + 90\ v}$	10 (5)	2 (3)	10 (15)	2 (1)	Nelson (1975)
$\dfrac{Bz\quad \mathrm{H21} + \quad V}{bz\qquad + \ \ C\quad v}$	0	5	3	0	Nelson (1968)

7. HYBRID DNA

When conversion was given a firm basis as a genetic phenomenon, most authors favoured copy-choice as an explanation. In the replication of the genetic material it was assumed that, as replication proceeded along the chromosome, a switch of template to the homologous chromosome was liable to occur. With the discovery that chromosome replication was semi-conservative, that is, the chromatids were not parent and daughter, but each was half-parent and half-daughter (see Section 1), the copy-choice hypothesis was undermined. It also failed to explain postmeiotic segregation. Moreover, the close association that was evident between aberrant segregation and crossing over posed a dilemma, when a copy-choice explanation of crossing over had been dismissed 25 years earlier on what appeared to be good evidence (see Section 1).

A radically different approach to an explanation of aberrant segregation was proposed by Holliday (1962, 1964b), Meselson (1965) and Whitehouse (1963, 1965). They suggested that recombination was by breakage and joining, and that aberrant segregation was a manifestation of the joining process. This joining was postulated to occur by annealing of complementary nucleotide chains derived one from each parent to produce heteroduplex or *hybrid DNA*. An attractive feature of this hypothesis was that it would automatically give great precision to the joining process, since the pairing of complementary bases to form duplex DNA would not allow addition or deletion of nucleotides, to which end-to-end joining might be vulnerable.

The hybrid DNA hypothesis provides a simple explanation for aberrant $4:4$ segregation and for the association between aberrant $4:4$ segregation and crossing over. This can be seen in Fig. 65(b). It is assumed that the axis of a chromatid consists of one duplex DNA molecule. Breakage of two duplexes, one from each parent, the breaks being in staggered positions in the two polynucleotides, followed by cross-annealing, could give the configuration shown in Fig. 65(b). If the site of a spore-colour mutant was in the overlap region between the staggered breaks, an aberrant $4:4$ ascus would result, as shown. This is because meiosis distributes the four chromatids to four different nuclei, and the succeeding DNA replication and mitosis lead to the formation of eight spores each carrying a duplex that corresponds to one of the eight polynucleotides present at the time of recombination. The eight spores in the ascus thus directly reflect events at the level of the DNA. An aberrant $4:4$ ascus with associated crossing over would thus imply, on the hybrid DNA hypothesis, that the crossover occurred by breakage at sites spanning the mutant site.

The $5:3$, $3:5$, $6:2$ and $2:6$ ratios found in asci heterozygous for a mutation are accounted for on the hybrid DNA model by supposing that after the cross-annealing has occurred (Fig. 65(b)) there is some excision from one parental polynucleotide in the overlap region, followed by replacement of the excised nucleotides by synthesis along the complementary strand of opposite parentage. If a mutant strand is excised in one of the heteroduplex molecules only, a $5:3$ ascus

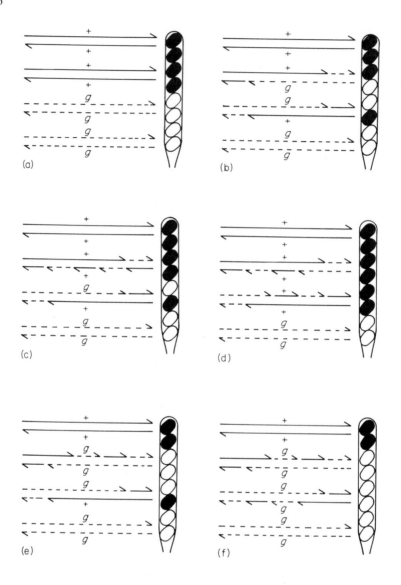

Fig. 65. Diagram to show how the formation of hybrid DNA will account for postmeiotic segregation, and how the subsequent correction of mispairing will explain conversion. The lines represent nucleotide chains at the pachytene stage of meiosis, broken lines distinguishing those of one parental genotype. Each chromatid is represented as a DNA duplex. The asci show the consequences for a *grey* (*g*) spore-colour mutant in *Sordari fimicola* of the event depicted at the DNA level. (a) No recombination; normal 4:4 segregation. (b) A reciprocal exchange between chromatids; aberrant 4:4 segregation. (c)–(f) A reciprocal exchange, followed by mismatch repair in favour of wild type in one (c) or both (d) hybrid DNA segments, giving 5:3 or 6:2 segregation, or in favour of mutant in one (e) or both (f), giving 3:5 or 2:6 segregation.

will result (Fig. 65(c)); if in both molecules, a 6:2 ascus will be produced (Fig. 65(d)). Similarly, if excision involves a wild-type strand in one molecule, the end result will be a 3:5 ascus (Fig. 65(e)) or if this happens in both molecules it will give rise to a 2:6 ascus (Fig. 65(f)). Another possibility, not shown in the diagram, is excision of a mutant strand in one molecule and of a wild-type one in the other molecule. This will restore a normal 4:4 segregation.

The excision and resynthesis postulated as the source of the aberrant segregations other than aberrant 4:4 might have come about in either one of two ways (cf. Chapter II, Section 3; Chapter V, Section 12). One possibility is excision triggered by, for example, a single-strand nick or gap in the molecule. Such excision will be detected genetically only if it includes a mutant site or its wild-type allele, but this would be an incidental consequence of the excision. The other possibility, which is that implied in Fig. 65, is that the excision is triggered by the mismatch of bases at the mutant site in the heteroduplex. This possibility as an explanation of conversion was proposed by Holliday (1962). He suggested that a repair mechanism may operate, and by adjusting the base sequences in order to restore normal pairing, could bring about gene conversion independent of the primary recombinational event.

Holliday (1964b) proposed that at certain points DNA molecules from opposite homologous chromatids unravel to form single strands, the strand separation beginning at a defined point but ending at any point. The strands broken would be of the same polarity in each molecule (Fig. 66(b)). Annealing of strands then occurs, followed by rejoining of the broken ends (Fig. 66(c)). Molecules with this appearance, with a strand from each exchanging partner, have been seen by Bell and Byers (1979) in electron micrographs of the DNA of a plasmid of *Saccharomyces cerevisiae* isolated during pachytene of meiosis. According to Holliday's hypothesis, at the point of partner exchange, breakage and reunion of strands of the same polarity takes place at corresponding positions. If the strands cut are the same as were broken initially, flanking markers would remain in a parental genotype, but the two cuts in each molecule would span a region of hybrid DNA (Fig. 66(d)). If the cuts at the node involve the strands of opposite polarity to those cut initially, flanking markers would be recombined and the hybrid DNA segments would mark the position of the crossover (Figs 65(b), 66(e)). This hypothesis will not only explain the occurrence of aberrant 4:4 asci associated with crossing over but also their occurrence in its absence. With the additional postulate of enzymes to recognize and excise mismatched bases, the other kinds of aberrant asci were accounted for, whether associated with crossing over (Fig. 65(c)–(f)) or not (Fig. 66(d)).

This hypothesis to account for aberrant segregation and crossing over is discussed further in Section 14.

Evidence for the formation of hybrid DNA was obtained by Rommelaere and Miller-Faurès (1975) and Moore and Holliday (1976). They labelled cells of *Cricetulus griseus* (Chinese Hamster) in culture with a density label by growing them for one round of replication in a medium containing 5-bromodeoxyuridine

258

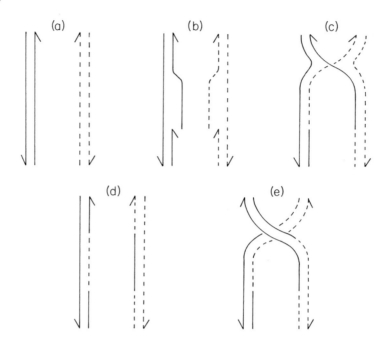

Fig. 66. Diagram to illustrate the hypothesis of Holliday (1964*b*) for the recombination mechanism. The broken lines distinguish the DNA strands of one of the recombining chromatids. (a) Chromatids before recombination. (b) Strands of the same polarity are broken in each molecule at an initiation site. (c) Hybrid DNA has formed in each molecule through cross-annealing of the broken strands, and the free ends have rejoined. (d), (e) Alternative outcomes depending on which strands are broken at the node in (c). (d) Result of breakage of the same two strands as were broken initially. (e) Result of breakage of the other two strands.

(BrdUrd). The DNA thus had one heavy (H) and one light (L) strand. Subsequent cell division in normal light medium would generate LL molecules as well as further HL molecules. On neutral caesium chloride gradients, however, a small fraction of DNA containing regions of HH DNA was detected. This could have arisen only by hybrid DNA formation from two HL molecules. Furthermore, Moore and Holliday (1976) found that this fraction was increased by treatment of the cells with mitomycin C, which also produced a similar increase in the frequency of sister chromatid exchanges. This correlation was evidence that HH DNA was hybrid DNA formed as an intermediate in a repair process involving recombination. The sister chromatid exchanges were detected by the staining method described in Section 1 involving two rounds of replication in the presence of BrdUrd, followed by staining with a fluorochrome. (See also Section 18.)

8.   FIDELITY OF CONVERSION

The identity or non-identity between parental and converted alleles was investigated by Roman (1957) with *ade3* mutants of *Saccharomyces cerevisiae*.

He analysed eight 1 : 3 asci by crossing each of the three adenine-requiring meiotic products to the parental adenine-requiring mutant. No recombinants were found. He concluded that the converted allele was not a new mutant but was identical with the parent mutant.

Case and Giles (1964), in their study of recombination at the *pan-2* locus in *Neurospora crassa*, also found that conversion did not generate new alleles. They analysed asci heterozygous for three *pan-2* mutants and tested the genotype of the progeny in recombinant asci by backcrossing to the individual mutants or to wild type. Interallelic crosses were scored for presence or absence of recombinants. The complementation pattern was observed when the ascus products with a requirement for pantothenic acid were paired with the three single mutants and the three double mutant combinations. Also, reverse mutation tests were carried out with ultraviolet light, the double mutants being stable and the single mutants showing reversion. There was no evidence for new mutants arising through conversion.

Zimmermann (1968) studied the fidelity of mitotic gene conversion in *S. cerevisiae*. A diploid strain heterozygous in *trans* for two alleles, *a* and *b*, at the isoleucine–valine *1* (*ilv1* locus) was treated with *N*-nitroso-*N*-methyl-urethane, which induces conversion. A total of 23 wild-type convertants had been subjected to genetic analysis (Zimmermann and Schwaier, 1967). Cells were sporulated, the asci formed digested with snail enzyme, and the spores isolated with a micromanipulator. The *ilv1* genotype of isoleucine-requiring spores was determined by mating with haploids carrying a known allele. It was found that conversion to wild type had occurred equally often at each site, with 11 examples of conversion of allele *a* only, 11 of *b* only, and one case of conversion to wild type at both sites. The *ilv1* gene codes for threonine dehydratase. Zimmermann (1968) examined the properties of this enzyme in all 23 convertants. He investigated the specific activity of the enzyme in crude extracts; its stability as shown by loss of activity during 6 h at ice-bath temperature; its degree of stabilization by the cofactor pyridoxal-5-phosphate; its sensitivity to inactivation by dilution; the extent to which degradation of the enzyme at low concentrations of L-threonine could be prevented by L-valine; the substrate affinity of the enzyme, found by determining the Michaelis constant; its feedback inhibition by L-isoleucine; the ratio of its serine dehydratase activity to its threonine dehydratase activity; and the pH dependence of the threonine dehydratase reaction over the pH range 7.0–8.5. All the convertants produced threonine dehydratase that was indistinguishable from wild-type enzyme as judged by all these criteria. Zimmermann concluded that conversion involved an accurate process of information transfer between homologous chromosomes.

Fogel and Mortimer (1970) investigated the fidelity of conversion to mutant shown by two ochre mutants, nos 4 and 17, at the *arg4* locus in *S. cerevisiae*. In a sample of 317 unselected asci from a diploid heterozygous for mutant no. 17 and wild type, 22 conversion asci were found, of which 11 segregated 1 : 3. The three auxotrophic segregants from each of these tetrads were tested for their response to ochre suppressors. All were found to be suppressible, as expected if the mutant nucleotide sequence had been transferred accurately to a chromatid of the wild-type parent during conversion. The assumptions underlying this conclusion are (1) that the ochre mutant has the nucleotide sequence that gives the stop triplet UAA in the messenger RNA, instead of a triplet that codes for an amino acid, with the result that the polypeptide coded by the gene terminates prematurely; and (2) that ochre suppressors have a mutant anticodon in a transfer RNA that pairs with UAA, such that it no longer functions as a stop signal when UAA is read by this transfer RNA, an amino acid being inserted again in the polypeptide at this position. If conversion had been imprecise, one of the three auxotrophs from the tetrad might be non-suppressible or suppressed by a different category of suppressors such as those affecting amber (UAG) mutants. No alterations in suppressibility were found, however, and Fogel and Mortimer (1970) concluded that the single base changes in the mutant codon, needed to affect suppressibility, had not taken place. Similar results were obtained with a *trans* cross of *arg4* alleles 4 and 17, where four 1:3 segregations for mutant 4 and 21 for mutant 17 were available. Measurements of reversion rates following ultraviolet irradiation, and response to complementation tests with alleles, also indicated identity of the converted mutant and the parent mutant.

Fogel and Mortimer (1970) also investigated the fidelity of conversion at an ochre-specific suppressor locus, *Sup6*. They used a diploid heterozygous at this locus and also for an amber mutant in the *trp1* gene. Altogether, 153 3:1 and 148 1:3 segregations for + : *Sup6* were found. They were scored on ochre mutants in the *lys1* and *ade2* genes, for which the parent diploid was homozygous, and they were tested for amber mutant suppression, using the *trp1* mutant. No amber suppression was found. The underlying suppositions in this work were that the wild-type suppressor locus codes for tyrosine transfer RNA and that the ochre-specific suppressor has undergone a particular base change in the DNA leading to recognition of messenger triplet UAA, but not UAG, by the anticodon of the transfer RNA. Lack of fidelity in conversion could generate a suppressor that recognized an amber mutant, such as that in the *trp1* gene. The failure to find amber suppression in the 301 *Sup6* convertants tested indicated fidelity of nucleotide sequence transfer in conversion.

Fidelity of conversion of deletion mutants was tested by Fink (1974), using three deletions at the *histidine4* (*his4*) locus in *S. cerevisiae*. From tetrad analysis 16 asci were obtained showing conversion to mutant (see Section 11). The end-points of the deletions in these conversion asci were tested by recombination with alleles and found to be identical to those of the parent deletion.

Conversion in *Schizosaccharomyces pombe* was subjected to tests for identity of converted alleles with parent alleles by Gutz (1971). The M26 mutant at the *ade6* locus is distinct from other mutants of this gene in giving high recombination frequencies in intragenic crosses. The 30 M26 strains from 10 asci with a 1:3 segregation at this site were crossed with alleles. All gave high and similar recombination frequencies. With M216, 17 asci with a 1:3 segregation at this site were investigated by crossing each M216 isolate with a set of complementing alleles. All gave similar and characteristic patterns of none, weak, and strong complementation. Comparable results were obtained with 15 asci that had a 1 : 3 segregation at the site of mutant L52. Thus, altogether, 62 examples of conversion to mutant were subjected to tests, either of recombination behaviour or of complementation behaviour, and no evidence was found that conversion occurred other than with complete fidelity.

In their investigation of recombination at the *rosy* eye-colour locus in *Drosophila melanogaster* using attached right arms of chromosome 3 (see Section 6), Chovnick *et al.* (1970) tested the fidelity of conversion to wild type. They measured the levels of activity of xanthine dehydrogenase coded by *rosy*, by recording the rate of change of fluorescence when enzyme extract is added to substrate. They did this assay for eight examples of non-crossover conversion of mutant 41 to wild type, five examples of non-crossover conversion of mutant 42 to wild type, and four examples of recombinant wild types of R1 flanking marker genotype (last three rows of Fig. 64). All 17 showed XDH levels that did not differ significantly from the controls, which were extracts of wild-type flies.

A more precise test of fidelity of conversion at the *rosy* locus was carried out by McCarron, Gelbart and Chovnick (1974). Females heterozygous for *rosy* mutants 5 and 41 were used for recombination test matings and the progeny reared on purine-enriched selective medium (see Section 6). A total of 62 wild-type recombinants were obtained in this way, comprising 11 P1, 35 P2 and 16 R1 flanking marker genotypes. (The relatively low frequency of recombinant flanking markers was no doubt because the females were also heterozygous for a large inversion spanning *rosy* and flanking markers; single crossovers within the inversion would give inviable gametes.) All of the 62 recombinants were found to have the same XDH, as revealed by acrylamide gel electrophoresis, as the wild-type enzyme. In a further experiment, wild-type strains from different sources were used that differed in their XDH electrophoretic pattern. Females heterozygous for two *rosy* mutants induced in different wild types were subjected to test matings, and wild-type recombinant progeny selected. From four different heterozygotes a total of 18 wild-type recombinants were obtained, consisting of four P1, five P2 and nine R1 flanking marker genotypes. Without exception the nine recombinants with a parental flanking marker genotype had the electrophoretic character for their XDH corresponding to that of the wild-type strain in which the mutant that had undergone conversion had been induced. Similarly, the R1 wild types gave a consistent result with the electrophoretic pattern corresponding to the genotype for the distal flanking marker, as expected if the site of the electrophoretic difference was distal to the crossover position. McCarron *et al.* (1974) concluded that conversion to wild type did not generate new, electrophoretically distinct wild types, but gave rise to a gene that was identical to that of the original strain in which the mutation that had undergone conversion had arisen. Later studies (see Section 9) showed that conversion of a *rosy* mutant to wild type was sometimes associated with conversion from one electrophoretic type to another. This observation, however, does not undermine the evidence for fidelity in conversion described above.

Evidence for new mutants arising in aberrant asci was given by Kitani and

Olive (1967) for the *grey* spore-colour locus in *Sordaria fimicola* and has been referred to in Section 4. They reported that mutants which they called 2a and 4b arose in 3 : 5 asci from crosses of mutants 2 and 4, respectively, with wild type. These secondary mutants were detected because the spores differed slightly in colour from the primary mutants, 2a having a bluish tint and 4b being faintly yellowish. Moreover, mutant 4b showed marked differences (see Section 4) from its parent mutant in recombination behaviour (Kitani and Olive, 1967, 1969; Kitani and Whitehouse, 1974). It is of particular interest that both of the secondary mutants arose as the odd mutant spore in the 3 : 5 asci, that is, they were products of postmeiotic segregation.

From the experimental results reviewed in this Section there is a substantial body of evidence that conversion takes place with complete fidelity. On the other hand, there is no comparable evidence for accurate transfer of nucleotide sequence in postmeiotic segregation and indeed some contrary evidence. Clearly, more investigations are needed on the fidelity of information transfer in postmeiotic segregation.

9. LINKED ABERRANT SEGREGATION

In their analysis of large numbers of unselected asci of *Neurospora crassa* heterozygous either in *trans* or in *cis* for two allelic *pan-2* mutants, Case and Giles (1959) found five examples of coincident conversion at both sites (Table 12(f)). In each case three of the products of meiosis had the genotype of one parent and the fourth that of the other parent. In the *trans* cross the implied conversion to wild type at one site and to mutant at the other, while in the *cis* cross there was conversion to wild type at both sites. Many other examples of this linked conversion, or *co-conversion*, have since been discovered. Case and Giles (1964) went on to investigate recombination at the *pan-2* locus in crosses heterozygous at three sites within the gene. The parental genotypes were

$$\frac{ylo\ ad\text{-}1\ 23\ +\ 36\ \ +}{+\ \ \ +\ \ \ +\ 72\ +\ trp\text{-}2}\ ,$$

where 23, 36 and 72 are *pan-2* mutants and *ad-1*, *trp-2* and *ylo* are loci that give, when mutant, an adenine requirement, a tryptophan requirement, and yellow (instead of wild-type orange) conidia, respectively. Altogether, 1457 asci from this cross were dissected. Ascus no. 78 showed co-conversion of two *pan-2* mutants and nos 565 and 581 of all three, the four products of meiosis having the genotypes

$$\begin{array}{c}
23\ +\ 36,\\
+\ 72\ +\ ,\\
+\ 72\ +\ ,\\
+\ 72\ +
\end{array}$$

at the *pan-2* locus. Asci nos 78 and 565 had a crossover associated with the conversion, revealed by the behaviour of the flanking markers *ad-1* and *trp-2*.

Two of the asci analysed by Case and Giles (1964) showed a new phenomenon: linked postmeiotic segregation. For example, the four products of meiosis in ascus no. 529 had the genotypes

$$23 + 36,$$
$$+ + 36,$$
$$+ \dfrac{+\ 36}{72\ +},$$
$$+\ 72\ +$$

at the *pan-2* locus, that is, a 6:2 segregation at the site of mutant 23 and linked 5:3 and 3:5 segregations at the other two sites. Their other example of linked postmeiotic segregation (ascus no. 43) showed a 3:5 ratio for mutant 72 and an aberrant 4:4 segregation for mutant 36.

Rizet and Rossignol (1963) analysed large numbers of 0:8 unselected octads from crosses between mutants in spore-colour gene *46* of *Ascobolus immersus*, and found small numbers of examples of co-conversion in each direction. For example, from a cross between mutants 63 and 137, the following tetrad genotypes were found:

$$
\begin{array}{cc}
63\ + & 63\ + \\
63\ + & +\ 137 \\
63\ + \quad \text{and} & +\ 137 \\
+\ 137 & +\ 137
\end{array}
$$

They also crossed the double mutant of alleles 1604 and 137 with wild type. The majority of the octads showed 4:4 segregation, but there were small numbers with 6:2 or 2:6 ratios. Some of the latter were analysed and four examples found of co-conversion to mutant:

$$
\begin{array}{cc}
1604 & 137, \\
1604 & 137, \\
1604 & 137, \\
+ & + \ .
\end{array}
$$

A comparison of observed and expected frequencies of co-conversion was made by Mousseau (1966) in *Ascobolus*. He took advantage of the fact that in a two-point *cis* cross of spore-colour mutants, that is, double mutant x wild type, co-conversion is readily detected, as Rizet and Rossignol (1963) had found. Simultaneous conversion of both mutants to wild type becomes visible as a 6:2 octad among the normal 4:4s, and 2:6 octads can be analysed to find how many arise from new mutations, how many from crossing over between the mutant sites, how many from conversion to mutant at one or other site, and how many from simultaneous conversion at both sites. Mousseau made such an investigation for five different two-point *cis* crosses, using mutants of spore-colour gene *19*. The results are given in Fig. 67. The mutants used were mapped by means of their recombination frequencies and their recombination, or lack of it, with a number of overlapping deletions. The expected frequencies of simultaneous conversion of

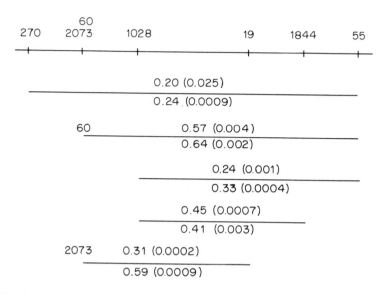

Fig. 67. Data of Mousseau (1966) on the frequency per 1000 octads of simultaneous conversion of pairs of gene *19* mutants in *Ascobolus immersus*. The genetic map of the mutants is given at the top. The pairs of mutants crossed (in *cis* configuration) with wild type are indicated by the horizontal lines. For each cross the frequency of co-conversion to wild type is given above the line and to mutant below it. The expected frequencies, on the assumption that conversion at one mutant site is independent of that at the other site, are given in parentheses.

each pair to wild type and to mutant, which are given in parentheses in the figure, were calculated by multiplying the observed conversion frequencies in one-point crosses of each mutant with wild type. These expectations are thus based on the assumption that conversion at one mutant site is independent of that at another. As is evident from Fig. 67, the observed frequencies were all much higher than the expected ones, often by several orders of magnitude. Mousseau concluded that simultaneous conversion at both sites resulted from a single event, not two coincident ones, and that it involved the whole segment between the two sites. Paszewski (1967) made a similar investigation, using two closely linked spore-colour mutants in *Ascobolus*, and found that the frequency of 6:2 asci in a *cis* cross (double mutant x wild-type), implying co-conversion to wild type, was seven times greater than that expected from independent conversion at each site.

An important discovery about co-conversion was made by Fogel and Mortimer (1969) with the *arginine 4* (*arg4*) gene in *Saccharomyces cerevisiae*. Four mutants at this locus, identification numbers 1, 2, 4 and 17, were mapped in the sequence

$$—— 4 ——— 1 ——— 2 ——— 17 ——$$

on the basis of (1) the frequency of X-ray-induced recombinants and (2) flanking marker behaviour in pairwise crosses. The former method indicated that interval

2–17 was shorter than interval 1–2. Unselected asci from diploids heterozygous for pairs of *arg4* mutants with various site separations were analysed. The results are given in Table 24(a). It is evident that the frequency of co-conversion, out of total conversion asci, declines rapidly with increasing distance between the mutant sites, ranging from 75% for mutants 2 and 17 to 6% for 4 and 17.

Comparable results were obtained by Gutz (1971) for the *adenine6* (*ade6*) gene of *Schizosaccharomyces pombe*. The three *ade6* mutants used in two-point (and three-point) ascus analyses mapped in the sequence

$$\text{—— M216 —— M26 ——— L52 —— ,}$$

with prototroph frequencies per 10^6 ascospores of 68 for M216 x M26 and 3620 for M26 x L52. The numbers of linked and unlinked conversions found in various crosses involving these mutants are given in Table 24(b). As can be seen, the results from a three-point cross agree well with those from the two-point crosses. The two closely linked mutants, M216 and M26, were remarkable in showing 100% co-conversion, no single-site conversion having been found in 53 conversion asci. With the more widely spaced mutants M26 and L52, however, co-conversion accounted for only about half the events, the remainder resulting from conversion of one or other mutant alone, or from independent (not linked) conversion of both, such as Mitchell (1955a) had found with *pdx-1* mutants of *Neurospora crassa* (Table 12(c)).

Gutz (1971) discovered one example of linked postmeiotic segregation, also from the M26 x L52 cross. It could be recognized because on medium low in adenine M26 gives red colonies and L52 pink colonies. The four spores in the ascus in question gave one red colony, two pink ones, and one with a red sector and a pink sector. Backcrosses confirmed the ascus genotype as

$$
\begin{array}{ll}
\text{M26} & + \quad , \\
\left\{ \begin{array}{ll} \text{M26} & + \\ \hline + & \text{L52} \end{array} \right. & , \\
+ & \text{L52} , \\
+ & \text{L52} ,
\end{array}
$$

that is, linked postmeiotic segregation, with (using the octad notation) a 5:3 ratio at the site of M26 and a 3:5 ratio at the site of L52.

In later studies with the *arg4* gene of *Saccharomyces cerevisiae*, Fogel and associates (Fogel, Hurst and Mortimer, 1971; Hurst, Fogel and Mortimer, 1972) made three- and four-point crosses, using closely linked alleles, some of which had not been used in the earlier work. From the analysis of thousands of unselected asci, numerous examples were obtained of co-conversion involving all three or all four alleles.

Since co-conversion in a *trans* cross does not give rise to any wild-type progeny (see, for example, Table 12(f)), its detection normally requires the analysis of unselected asci. In consequence, much effort is needed, as indicated above, to obtain significant results. Kitani and Olive (1969), however, were able to take

Table 24. Relative frequencies of linked and unlinked conversion in crosses of two or more alleles. Results obtained (a) by Fogel and Mortimer (1969) for *arg4* mutants in *Saccharomyces cerevisiae* and (b) by Gutz (1971) for *ade6* mutants in *Schizosaccharomyces pombe*. With the three-point cross in the latter investigation, (i) and (ii) refer to the same asci; in (i) M216 is the proximal mutant and M26 the distal, and in (ii) M26 is the proximal mutant and L52 the distal.

Species and gene	Alleles crossed (proximal to left, distal to right)	Total asci analysed	Crossing over between alleles	Unlinked conversion – Proximal mutant only 3:1	Proximal mutant only 1:3	Distal mutant only 3:1	Distal mutant only 1:3	Both mutants	Total	Co-conversion – Proximal mutant† 3:1	Proximal mutant† 1:3	Total No.	Total %‡
(a) *Saccharomyces cerevisiae*, *arg4*	2/+ , +/17	544	0	1	3	3	2	0	9	14	13	27	75
	1/+ , +/2	502	5	3	3	10	11	0	27	13	10	23	46
	4/+ , +/17	697	9	3	5	18	20	0	46	2	1	3	6
(b) *Schizosaccharomyces pombe*, *ade6*	M216/+ , +/M26	1032	0	0	0	0	0	0	⎫0	3	29	⎫53	⎫100
	M216/+ , +/M26 L52 (i)	719 { (i) 0		0	0	0	0	0	⎭	0	21	⎭	⎭
	M26/+ , +/L52 (ii)	(ii) 0		8	0	0	0	0	⎫26	13	0	⎫35	⎫57
	M26/+ , +/L52	1060	0	13	1	1	0	1§	⎬	19	1	⎬	⎬
	M26 L52/+ , +/+	199	0	2	0	0	0	0	⎭	2	0	⎭	⎭

† In *cis* crosses the distal mutant will show the same wild type : mutant ratio as the proximal mutant. In *trans* crosses the distal mutant will show the converse ratio to the proximal one, e.g. proximal 3 : 1, distal 1 : 3.

‡ Percentage of total number of asci showing conversion.

§ Both mutants showed 3 : 1 segregation.

advantage of diversity of spore colour in mutants at the *grey* spore-colour locus in *Sordaria fimicola* to observe co-conversion asci directly. For example, mutant no. 1 has grey spores and no. 2 has almost colourless spores. In a *trans* cross of these alleles, co-conversion asci will have six grey and two colourless, or two grey and six colourless spores, depending on the direction of co-conversion. Furthermore, exploitation of the allelic colour difference allowed them also directly to observe linked postmeiotic segregation. This was revealed as 5:3, 3:5, or aberrant 4:4 segregation for the two spore colours. In ascus analyses reported by Kitani and Olive (1969) and Kitani and Whitehouse (1974), a total of 10 crosses were investigated, all involving pairs of mutants differing in colour: grey (nos 1 and 6), light grey (no. 5), slightly tinted (no. 2), and colourless (nos 3, 4 and 4b). These crosses involved recombination frequencies ranging from 0.8 to 2.2 asci with wild-type spores per 1000 asci, and enabled the frequencies of linked postmeiotic segregation and co-conversion, out of total aberrant asci, to be plotted against site separation (Whitehouse, 1974). The results are shown in Fig. 68. There is clear evidence of a decline, both in linked postmeiotic segregation and in co-conversion, with increasing site separation.

Co-conversion was discovered in *Drosophila melanogaster* by Gelbart *et al.* (1974) in the course of an investigation of the sites of genetic variation at the *rosy* eye-colour locus found in different wild-type strains. This gene codes for xanthine dehydrogenase (XDH) and, as already pointed out (Section 8), *rosy* mutants had been induced in wild types that differed in their XDH electrophoretic pattern. In order to map the sites of wild-type difference, females heterozygous for pairs of *rosy* mutants induced in different wild types were subjected to test matings, and

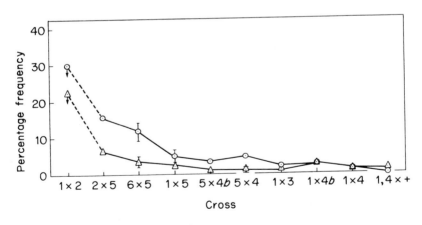

Fig. 68. Results obtained by Kitani and associates from crosses of *grey* spore-colour mutants in *Sordaria fimicola*. The total frequencies of asci showing linked postmeiotic segregation (⊙) and of those showing co-conversion (△) are plotted as percentages of all aberrant asci for each cross. The crosses are in order of increasing separation of the mutant sites, as far as this is known. Standard errors are given where possible. The points plotted in the 1 × 2 cross are the maximum values and are probably in excess of the true ones. (Reproduced by permission of Cambridge University Press from H. L. K. Whitehouse, *Genet. Res.* **24**, 257 (1974).)

wild-type recombinant progeny were selected by use of a purine-enriched medium lethal to XDH-deficient larvae (see Section 6). The wild-type recombinants had P1, P2 or R1 flanking marker genotypes. The R1 class allowed the mutants to be mapped, and the electrophoretic character of the XDH produced by the wild-type recombinants provided information about the position in the gene of the sites of difference between the wild types. In particular, some of the P1 and P2 progeny had the XDH electrophoretic pattern, not of the wild-type strain in which the mutant that had undergone conversion had been induced, but of the wild type in which the other mutant had been induced. This result implied co-conversion of mutant and electrophoretic site. For example, females of genotype

$$\frac{M\ 102\ +\ 1\ N}{m\ +\ 41\ 0\ n}\ ,$$

where 41 and 102 are *rosy* mutants induced in the original wild type (0) and wild type 1, respectively, and M/m and N/n are the flanking markers, gave rise to 12 wild-type recombinants of P2 flanking marker genotype, that is, $m\ n$. Seven of these had the XDH electrophoretic pattern corresponding to the parentage implied by the flanking markers $(m\ +\ +\ 0\ n)$, but the other five showed co-conversion at the mutant 41 and electrophoretic sites $(m\ +\ +\ 1\ n)$. The data on co-conversion obtained by Gelbart *et al.* in this way are given in Fig. 69. The frequencies range from 0% to 100% of the conversions at the mutant site.

Gelbart *et al.* (1974) had inferred that the *rosy* gene was large, about 3000 nucleotides in length, because the XDH polypeptide had been found to have a molecular weight of 130 000–150 000 daltons, implying 1000 amino acids. It was

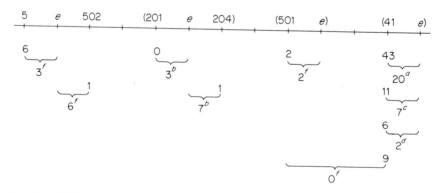

Fig. 69. Linkage map of the *rosy* eye-colour locus in *Drosophila melanogaster*, and numbers of progeny showing conversion or co-conversion, from the data of Gelbart *et al.* (1974). Figures above the line are the isolation numbers of rosy mutants; *e* indicates a site of difference between wild-type strains, detected by electrophoresis of the gene product, xanthine dehydrogenase; parentheses indicate uncertainty of sequence. Pairs of numbers below the line show the number of single-site conversions to wild type (upper figure) and of co-conversions (below the bracket); superscripts *a–d, f* indicate that a mutant induced in the original wild-type strain was crossed with a mutant induced in wild types 1–5, respectively.

evident from their results that co-conversion was taking place over distances measured in hundreds rather than in thousands of nucleotides. A similar conclusion can be drawn from the other examples of co-conversion described above, notably the *arg4* data in *Saccharomyces cerevisiae* (Fogel and Mortimer, 1969), the *ade6* data in *Schizosaccharomyces pombe* (Gutz, 1971), and the *grey* data in *Sordaria fimicola* (Kitani and Olive, 1969; Kitani and Whitehouse, 1974). The results for *grey* also indicated that linked postmeiotic segregation likewise usually extended only for a fraction of the length of the gene, although its extent outside

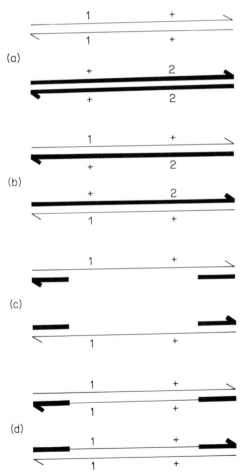

Fig. 70. Diagram to explain linked aberrant segregation through the formation of hybrid DNA. The lines represent polynucleotide chains, their thickness indicating their parentage. The numbers 1 and 2 show mutant sites. (a) Parental chromatids. (b) Reciprocal hybrid DNA formation. (c) Excision of strands of one parentage over both mutant sites. (d) Resynthesis using the intact strands as template. If the hybrid DNA were confined to one chromatid, as discussed in Section 13, co-conversion would merely require excision to extend to both mutant sites in the appropriate strand of that chromatid.

the gene is unknown. In contrast to these conclusions, Lawrence *et al.* (1975) found examples of co-conversion in *Saccharomyces cerevisiae* extending from one gene to another, for example, involving mutations in the *cyc1* and *rad7* genes, which are closely linked and possibly adjacent. DiCaprio and Hastings (1976) likewise found long conversion lengths to be frequent, extending from the *Sup6* gene (an ochre suppressor) into adjacent marked loci.

Linked postmeiotic segregation is readily accounted for if recombination involves the formation of hybrid DNA (Section 7). If the heteroduplex resulting from the annealing of nucleotide chains of each parentage extends over more than one mutant site and there is no mismatch correction, linked postmeiotic segregation will result (Fig. 70(b)). Co-conversion is accounted for if, following heteroduplex formation extending to more than one mutant site, mismatch correction involves extensive degradation of one strand of the duplex (Fig. 70(c)), followed by resynthesis using the other strand as template (Fig. 70(d)). If the excision reaches a second mutant site, co-conversion will result.

Evidence for co-conversion in mammalian cells was obtained by Miller, Cooke and Fried (1976). They constructed heteroduplex DNA from variants of polyoma virus differing at four mutant sites. One of these was a temperature-sensitive mutant that governs the ability to produce infectious virus at 38.5 °C. The other sites affected cleavage by restriction enzymes or the electrophoretic mobility of the resulting fragments. One of the mutants was believed to involve the addition of about 18 nucleotides. Embryo cells of *Mus musculus* were infected with individual heteroduplexes and the progeny virus scored for the parental characters. Markers separated by more than 600 nucleotides segregated independently, whereas two markers about 90 nucleotides apart did not separate. Miller *et al.* (1976) concluded that mouse cells have an efficient mismatch correction mechanism and that correction at each mutant site, including the multisite mutant, occurs equally often in each direction. The excision length appeared to be in the range 90–600 nucleotides, giving co-conversion of the closely linked markers.

Co-conversion is discussed further in Sections 10 and 13.

10. MAP EXPANSION

Holliday (1964*b*) drew attention to a peculiarity of recombination frequencies obtained from crosses of alleles, namely, that mutants mapping far apart in a gene often gave higher recombination frequencies than the sum of the frequencies obtained from crossing pairs of mutants mapping at intervening positions. Rizet, Lissouba and Mousseau (1960) had found that mutants of spore-colour gene *19* of *Ascobolus immersus* mapped in three clusters, A, B and C. Crosses between mutants in A and B, or in B and C, gave 2:6 asci with a frequency of about 1.6 per thousand, whereas the corresponding frequency for A and C was about 19 per thousand, six times the expected value. Wild-type recombinants obtained by crossing allelic auxotrophic mutants and plating the progeny ascospores on minimal medium gave similar results, for example, with the *adenine7* (*ade7*) gene of *Schizosaccharomyces pombe* mapped by Leupold (1961) and the *pabaA* gene of *Aspergillus nidulans* (Siddiqi, 1962). Holliday called this phenomenon *map* *expansion* and pointed out that it implied that the mutant sites were interfering with the process of recombination. This was because in a three-point cross of alleles, for example,

$$\frac{1+3}{+2+},$$

the recombination frequency for mutants 1 and 3 cannot exceed the sum of that for the two shorter intervals. The map expansion evident with two-point crosses implied that mutants 1 and 3 were showing less recombination in the three-point cross than in the two-point (or, less likely, the converse differences with the shorter intervals).

The way in which mutant sites interfere with the recombination process was likely to be through the mismatched bases that occur when a mutant site is in hybrid DNA. Co-conversion is believed to result from strand degradation from the site of excision of a mismatched base, followed by resynthesis using the intact strand as template. If the degradation extended to the second mutant site, co-conversion would result (see Section 9). Holliday (1968) pointed out that this provided an explanation for map expansion. Co-conversion does not give rise to allelic recombinants and so depresses the recombination frequency of mutants whose mismatches are excised together, compared with more distant ones. This explanation of map expansion was analysed algebraically by Fincham and Holliday (1970). They found it necessary to assume that the degradation of one strand of the duplex in mismatch correction extended in both directions along the DNA, in order to account for the fact that closely linked alleles show additivity of recombination frequencies. This additivity seemed to be general, for it was evident in data obtained by various authors for genes in *Ascobolus, Aspergillus, Neurospora, Saccharomyces* and *Schizosaccharomyces*. Map expansion was shown by more widely spaced alleles (see Fig. 71). Fincham and Holliday (1970) argued that if degradation of single DNA strands could proceed in one chemical direction only, for example, 3' to 5', from a mismatched base pair, two mutant sites in hybrid DNA could undergo independent correction when degradation was divergent no matter how close together they were, with

$$\frac{1+}{+2} \quad \text{leading to} \quad \frac{+}{+}$$

and, following resynthesis,

$$\frac{+\ +}{+\ +}.$$

Divergent excision would be expected in 50% of cases. The contribution to wild-type recombinants arising from independent correction would then be substantial however close together the mutants were situated. In other words, recombination frequencies would be almost independent of distance at very short intervals, extrapolating to a positive frequency at zero distance. Experimental data, however, showed a linear relationship (for a tenth to a half of the length of the gene) that extrapolated to zero (Fig. 71).

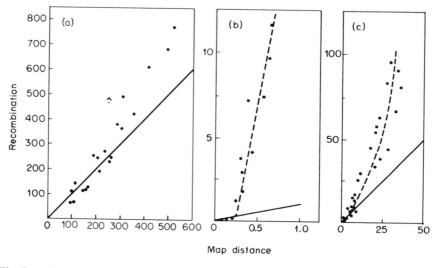

Map distance

Fig. 71. Map expansion in (a) the *ade7* gene of *Schizosaccharomyces pombe*, (b) spore-colour gene *19* of *Ascobolus immersus* and (c) the *trp-3* gene of *Neurospora crassa*. The diagrams are from (a) Holliday (1964*b*) and ((b), (c)) Fincham and Holliday (1970), where the sources of the data are given. The map distance, or expected recombination frequency, is the sum of the map intervals between mutants two or more intervals apart. Recombination is the actual frequency of recombinants observed when such mutants are crossed. The solid lines show the relationship between the map distance and recombination which would be observed if the map was strictly additive. The frequencies of wild-type recombinants are (a) per 10^6 ascospores, (b) per 10^3 tetrads, and (c) per 10^5 ascospores. (Part (a) is reproduced by permission of Cambridge University Press from R. Holliday, *Genet. Res.* **5**, 291 (1964); parts (b) and (c) are reproduced by permission of Springer Verlag from J. R. S. Fincham and R. Holliday, *Molec. gen. Genet.* **109**, 316 (1970).)

Support for the idea that excision proceeds in both directions along the DNA from a site of mismatch is given by the results obtained by Gutz (1971) with the *ade6* gene of *Schizosaccharomyces pombe*. He analysed asci from a three-point cross heterozygous for mutant M26 at this locus and for two alleles that spanned the site of M26 (Table 24). The M26 mutant, through co-conversion, has a dominant effect on the frequency and direction of conversion of alleles (see Section 15). The presumption, therefore, is that mismatch correction is initiated predominantly at this site. Gutz found 13 examples of a co-conversion involving all three mutants (Table 24). This implied that excision was extending in both directions along the DNA from the site of M26 (cf. Chapter II, Section 3).

For their algebraic analysis of mismatch correction as an explanation of map expansion, Fincham and Holliday (1970) made several assumptions which they believed were not strictly true, but which they thought would not invalidate the analysis. These assumptions were the following:

(1) Hybrid DNA segments are distributed at random along the DNA, with the result that each mutant site in a gene has the same chance of being included in hybrid DNA; why this is believed not always to be true is discussed in Section 12.

(2) The probability of excision of a mismatched base is the same for all mutants, and results in conversion to wild-type or to mutant with equal frequency; evidence that this is not always true is given in Section 11.

(3) If two mutant sites are both in hybrid DNA and are sufficiently close to one another for degradation of one strand of the DNA, initiated at one site, to extend to the other, then a second independent excision at that site does not occur; in other words, exonucleolytic breakdown is rapid once a mismatch has been recognized.

Fincham and Holliday's analysis, based on these assumptions, is set out in Table 25. From a cross between two closely linked mutants there are five sources of wild-type recombinants (column 1). These depend on the distribution of hybrid DNA in relation to the mutant sites, and on whether or not mismatch correction

Table 25. Relative contributions of various sources to wild-type recombinants (from Fincham and Holliday, 1970). Key to symbols: c, proportion of hybrid regions which become sites of crossing over; d, distance between mutant sites; h, length of hybrid DNA; p, probability of excision of mismatched bases.

Sources of wild-type recombinants	Sites closer than excision length	Sites further apart than excision length but within hybrid DNA length	Sites further apart than hybrid DNA length
(i) Hybrid DNA reaching one site only; no mismatch correction	$d(1-p)$†	$d(1-p)$†	$h(1-p)$†
(ii) Hybrid DNA reaching one site only; mismatch correction occurring	dp	dp	hp
(iii) Both sites in hybrid DNA; independent mismatch correction at each site	—	$\frac{1}{2}p^2(h-d)$	—
(iv) Both sites in hybrid DNA; mismatch correction at one site only	—	$p(1-p)(h-d)$†	—
(v) Hybrid DNA falling entirely between the sites	—	—	$c(d-h)$‡

†Sources of wild-type recombinants involving postmeiotic segregation will be doubled in frequency if wild/mutant mosaics arising from this cause are scored as wild type, as is likely in yeasts.

‡The relative contribution to total wild-type recombinants of crossing over between the mutant sites will be doubled if hybrid DNA is non-reciprocal (see Section 13).

has occurred. The occurrence of both sites in hybrid DNA, either without mismatch correction or with excision extending from one site to the other to give co-conversion, are possibilities that are not included in the table since they will not give rise to recombination of the mutants. Three alternatives for the distance between the mutant sites, relative to the length of nucleotide chain excised in mismatch correction and to the length of the hybrid DNA segment, were treated initially as if sharply separated from one another (columns 2–4). For sites closer together than the excision length in mismatch correction, recombinants will arise only from hybrid DNA ending between the mutant sites, and the recombination frequency will be directly proportional to the distance between the sites (column 2, sources (i) and (ii)). This would correspond to the region of additivity shown by closely linked alleles (Fig. 71). As site separation increases, an abrupt increase in recombination frequency is predicted when events involving hybrid DNA at both sites start to contribute (sources (iii) and (iv) of Table 25). This would happen when the distance between the mutant sites begins to exceed the excision length, preventing co-conversion and allowing the sites to be corrected independently. This burst of recombinants would account for map expansion and for its abrupt onset (Fig. 71). Holliday (personal communication) has pointed out that map expansion will be evident only if the excision length is constant, or approximately so; if the repair enzyme has a fixed probability of falling off the DNA, map expansion will not occur. The data point to an excision length of about 100 nucleotides: see discussion by Holliday (1968).

Fincham and Holliday's algebraic analysis predicts that map expansion will not continue with further site separation, but a decline in map expansion with widely spaced alleles has not been observed experimentally (Fig. 71). A possible explanation is the occurrence of crossing over between the mutant sites (source (v) in Table 25), which seems often to contribute substantially to wild-type recombinants at site separations that also allow recombinants to arise from the other sources given in the table: see, for example, the discussion in Section 4 of results obtained by Rizet and associates with spore-colour gene 75 of Ascobolus. Crossing over between mutant sites gains in importance in this respect (see second footnote to Table 25) from a feature of eukaryote recombination discussed in Section 13.

Map expansion was found at the *rudimentary* (*r*) wing locus in *Drosophila melanogaster* by Carlson (1971). Interestingly, the degree of expansion was rather slight, though uniform and well marked, and the region of additivity shown by closely linked fungal alleles (Fig. 71) was evident only with *r* mutants showing 1.5 or less wild-type recombinants per 10^5 progeny. This limit represents only 5% of the length of the gene and suggests that the excision length in mismatch correction may be shorter in *Drosophila* than in fungi.

11. MUTAGEN SPECIFICITY IN ABERRANT SEGREGATION PATTERN

In studies of the frequencies of 6:2 and 2:6 segregations in crosses of spore-colour mutants of *Ascobolus immersus* with wild type, striking variations were found both in relative frequency and total frequency. Results obtained by Kruszewska and Gajewski (1967) and Rossignol (1969) for genes Y and 75, respectively, are given in Table 26. Examples were found in each gene of mutants mapping at the same position but showing different conversion patterns and frequencies. Rossignol (1969) pointed out that if the gene-75 mutants were classified by conversion pattern, the conversion frequency increased progressively from left to right on the gene map for each pattern (Table 26). There are indications that the gene-Y mutants might be behaving similarly. Progressive changes in conversion frequency with map position are well known and are discussed in Section 12. Rossignol concluded that the nature of the mispairing in hybrid DNA might be responsible for the conversion pattern. It could also affect the conversion frequency since mismatch correction could restore a normal 4:4 segregation, which would not be detected (Section 7). The strain of *Ascobolus* used in these investigations shows a high frequency of spontaneous mutation to give spore-colour mutants. All the mutants used arose in this way. To investigate a possible relationship between conversion pattern and the nature of the mispairing in hybrid DNA, it is necessary to use mutants obtained with particular mutagens.

Leblon (1972a) used a different strain of *Ascobolus immersus* that gave a relatively low frequency of spontaneous mutation to colourless ascospores. This

Table 26. Data of Kruszewska and Gajewski (1967) and Rossignol (1969) for conversion frequencies of mutants at spore-colour loci Y and 75, respectively, in *Ascobolus immersus*. Braces indicate mutants mapping at the same site.

Spore colour gene	Mutants in sequence from left to right on map	Frequency of aberrant asci per 1000 asci		
		$\dfrac{6:2}{2:6}=\dfrac{1}{14}$	$\dfrac{6:2}{2:6}=\dfrac{4}{3}$	$\dfrac{6:2}{2:6}=1$
Y	Y	0.42		
	183	0.63		
	73	0.70		
	77		4.3	
	775		1.9	
75	1186	7		
	302			32
	1493			43
	278		23	
	2029		32	
	1245	13		
	147			52
	1573		42	
	889	16		
	1983	23		
	1848		67	
	1472	25		

meant that mutagens could be used to obtain such mutants. He used *N*-methyl-*N'*-nitro-*N*-nitrosoguanidine (NG), the acridine mustard ICR 170, and ethyl methane sulphonate (EMS). Investigation was restricted to mutants at two spore-colour loci, *b1* and *b2*. The mutants were crossed with wild type and the numbers of aberrant asci recorded. In addition to 6:2, 2:6, 5:3 and 3:5 asci, small numbers of 7:1, 1:7, 8:0 and 0:8 segregations were recorded. These could reasonably be regarded as simultaneous occurrences of two aberrant segregation events involving different chromatids, that is 7:1 as a 6:2 and a 5:3 combined; 8:0 as two 6:2 events; and similarly with the others. Such double occurrences were not unexpected because the frequencies of aberrant asci were high: 5–64 per 1000 asci with *b1* mutants and 105–284 per 1000 asci with *b2* mutants except for three spontaneous *b2* mutants which gave lower frequencies. In spite of the difference between the two genes in conversion frequencies, the pattern of conversion was similar, with each mutagen giving characteristic results. Details are given in Fig. 72. Three kinds of mutants could be recognized: class A with predominantly 6:2 asci; class B with predominantly 2:6 asci; and class C with 6:2, 2:6, 5:3 and 3:5 asci all about equally frequent. With A and B postmeiotic segregation was rare or absent. As can be seen from Fig. 72, NG gave class C mutants only; ICR 170 gave B mutants only, but including some that gave nearly as much conversion to wild type as to mutant; and EMS gave chiefly A and C mutants. Several spontaneous mutants were obtained, and these were diverse, including A, B and C.

In order to establish the molecular nature of the A, B and C classes of mutants, Leblon (1972*b*) investigated their reversion behaviour, using the same mutagens as had been used to induce the mutants. One A, two B and two C mutants were investigated. One of the C mutants (194) reverted through mutations at unlinked suppressor loci. Two of these suppressors were tested, and found to suppress some of the C mutants but none of the A or B mutants at each of three spore-colour loci (*b1*, *b2* and *b3*). In other words, the suppressors were allele-specific but not gene-specific. Leblon concluded that the C mutants that were suppressed were base substitutions that gave rise to a stop triplet in the messenger RNA, the suppressors being mutations in genes for particular transfer RNAs such that the stop triplet was misread and an amino acid inserted again in the polypeptide at this point. Further evidence that the C mutants were base substitutions was the known ability of EMS and particularly of NG to induce this type of alteration in DNA.

Revertants of the A mutant (47E) and of one of the B mutants (A38) had pale brown spores instead of the dark brown of wild type. The original white-spored mutants could be recovered from these strains and also a second white-spored mutant mapping in the same gene (*b2*). Re-association of the two white mutants restored the pale brown phenotype. It was evident that reversion had occurred by a second mutation acting as an intragenic suppressor. Four such suppressors were obtained for the B mutant, and crosses with one another established that all four were at different sites. When the suppressor mutations were crossed with wild type, an important discovery was made: the A-mutant suppressor was a B mutant, and the four B-mutant suppressors comprised three A mutants and one C

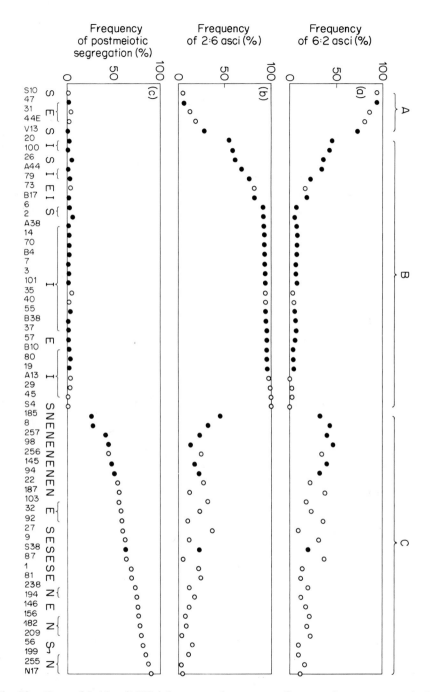

Fig. 72. Data of Leblon (1972a) for conversion pattern of spore colour mutants at loci b1 (○) and b2 (●) in *Ascobolus immersus* induced with various mutagens: E, ethyl methane sulphonate; I, ICR170; N, nitrosoguanidine; S, spontaneous. The isolation numbers of the mutants are shown below the diagram. The frequencies of (a) 6:2 asci, (b) 2:6 asci and (c) 5:3 and 3:5 asci are shown as percentages of total aberrant asci. The mutants in classes A and B are in order of increasing frequency of 2:6 asci, and those in class C in order of increasing frequency of postmeiotic segregation.

mutant. Leblon (1972b) concluded that A and B mutants were frameshift mutants of complementary type, such that each could be suppressed by a second mutation close to it and in the opposite direction. The messenger RNA would then be misread only between the sites of the nucleotide deletion and the compensating addition, and the phenotype would be wild type or near wild type. This conclusion was supported by the known ability of EMS and of the acridine mustard ICR170 to produce frameshift mutants. These mutagens evidently produce largely complementary types of frameshifts, as EMS and ICR170 gave rise to A and B mutants respectively (Fig. 72). In agreement with this, the highest frequency of reverse mutation of the A mutant 47E was obtained with ICR170 and of the B mutant A38 with EMS. From the known behaviour of alkylating agents such as EMS in deleting nucleotides, and of acridines in adding them, Leblon (1972b) concluded that A mutants were deletions of one (or more) nucleotides and B mutants the corresponding additions.

Leblon's general conclusions from his investigation of aberrant segregation shown by *Ascobolus* spore-colour mutants induced with chemical mutagens were as follows: (1) that mismatches in hybrid DNA arising from base substitutions often remained uncorrected, resulting in the frequent postmeiotic segregation found with C mutants; (2) that such correction, if it occurred with substitution mismatches, was equally likely to favour wild type or mutant, accounting for the equality of 6:2 and 2:6 with C mutants; and (3) that mismatches resulting from frameshift mutants in hybrid DNA were readily recognized and were corrected by excision of the shorter strand, leading to the wild type being favoured (6:2 asci) with deletions (A mutants), and the mutant being favoured (2:6 asci) with additions (B mutants).

Leblon and Rossignol (1973) took advantage of the ability to recognize some double mutants in gene *b2* of *Ascobolus* by their spore colour to investigate the interactions between mutants of different kinds in two-point crosses. One of the double mutants investigated had pale brown spores and was one of the intragenic suppressors of B mutant A38 already referred to. This suppressor, when separated from the B mutant, behaved as an A mutant. In Table 27 are given the numbers of aberrant asci observed when the double mutant was crossed with wild type, and also when each single mutant was crossed with wild type, both in the absence of the other mutant and when homozygous for it. In the latter crosses the A mutant showed over 90% of the aberrant asci with conversion to wild type and the B mutant over 90% conversion to mutant. In the two-point *cis* cross there was much co-conversion in each direction, with the result that the bias in favour of wild type shown by the A mutant and that in favour of mutant shown by the B mutant largely cancelled out. It appeared as if mismatch correction could be initiated at either site and then often extended to the other one.

A second double mutant investigated by Leblon and Rossignol (1973) arose through the spontaneous occurrence of a B mutant in an EMS-induced C mutant strain. The double mutant was recognized by its pink spores, either mutant alone having white spores. The B mutant appeared to be identical in map position and conversion behaviour to the ICR170-induced B mutant A38 used in the other double mutant investigation. The analysis of the interaction between the B and C

Table 27. Numbers of aberrant asci found by Leblon and Rossignol (1973) in one- and two-point crosses involving an A and a B mutant of gene *b2* of *Ascobolus immersus*. The figures are estimates based on genotype analysis of a sample of each phenotype, since phenotype alone can be misleading.

Cross:	$\dfrac{A\ B}{+\ +}$				$\dfrac{A\ B}{+\ B}$		$\dfrac{A\ +}{+\ +}$	
	Segregation: + : B							
Segregation: + : A	Normal 4:4	6:2	2:6	Percentage of total aberrants	Number	Percentage	Number	Percentage
Normal 4:4	—	1	3	—	—	—	—	—
6:2	6	481	2	56	114	91	119	92
2:6	3	0	377	44	11	9	10	8
Percentage of total aberrants	—	56	44	—	—	—	—	—

Cross:								
$\dfrac{A\ B}{A\ +}$ Number		22	267					
Percentage		8	92					
$\dfrac{+\ B}{+\ +}$ Number		84	1059					
Percentage		7	93					

Table 28. Numbers of aberrant asci found by Leblon and Rossignol (1973) in one- and two-point crosses involving a B and a C mutant of gene $b2$ of *Ascobolus immersus*. The figures are estimates based on genotype analysis of a sample of each phenotype, since phenotype alone can be misleading.

Cross:		$\dfrac{B\ C}{+\ +}$				$\dfrac{B\ C}{B\ +}$		$\dfrac{+\ C}{+\ +}$	
		Segregation: + : B			Percentage of total aberrants	Number	Percentage	Number	Percentage
	Segregation: + : C	Normal 4:4	6:2	2:6					
$\dfrac{B\ C}{+\ +}$	Normal 4:4	—	1	22	—	—	—	—	—
	Aberrant 4:4	19	0	8	1	—	—	—	—
	5:3	70	11	0	3	88	40	100	42
	3:5	96	0	275	15	108	50	103	43
	6:2	7	133	0	6	4	2	15	6
	2:6	4	0	1896	75	18	8	21	9
	Percentage of total aberrants	—	6	94	—	—	—	—	—
$\dfrac{B\ C}{+\ C}$	Number	—	3	85	—				
	Percentage	—	3	97	—				
$\dfrac{B\ +}{+\ +}$	Number	—	17	133	—				
	Percentage	—	11	89	—				

mutants was made on similar lines to that between A and B, and the results are given in Table 28. Aberrant 4:4 segregation of the C mutant could be recognized in the two-point cross since the B mutant and mating type provided markers to identify sister spores (cf. Table 15). A striking feature of the results from the cross of the double mutant with wild type was that the proportion of aberrant segregation at C that involved postmeiotic segregation fell from over 80% in the one-point crosses to less than 20%. On the other hand, the behaviour of the B mutant was unchanged. This is the result expected if mismatch correction initiated at the site of the B mutant often extends to the site of the C mutant and leads to correction there that would not otherwise have taken place. Co-conversion was indeed frequent in the two-point cross and was responsible for the much stronger bias in favour of conversion to mutant shown by the C mutant in the two-point cross compared with the one-point crosses. This increased bias at C is also evident in the 5:3 and 3:5 segregations. That both these increases in bias are caused by the B mutant is also indicated by their absence when B shows normal 4:4 segregation (Table 28).

Results comparable to those of Leblon (1972a) for *Ascobolus* were obtained by Yu-Sun, Wickramaratne and Whitehouse (1977) with *Sordaria brevicollis*. Mutants at three spore-colour loci, *grey-3*, *grey-4* and *grey-5*, were investigated. The results obtained from counts of aberrant asci in crosses with wild type are shown in Fig. 73. Aberrant ascus frequencies ranged from 1–4% for *grey-4* to 4–10% for *grey-3* mutants. The same three classes of mutants as in *Ascobolus* could be recognized, with characteristic relative frequencies of the different kinds of aberrants, irrespective of absolute frequencies. The distinction between the A and B classes was clear cut, as no intermediates were found. The mutagen specificity was similar to that found with *Ascobolus*. ICR170 gave rise to two A mutants as well as numerous B mutants. Only C mutants were obtained with EMS. Ultraviolet light produced C mutants and one B mutant.

In *Aspergillus nidulans*, *Neurospora crassa* and *Saccharomyces cerevisiae* the A, B and C classes of mutants have not been recognized, and the regular intragenic gradients in frequency of conversion to wild type shown by these fungi (Section 12) indicate that there are no prominent marker effects. Indeed, in *S. cerevisiae* Fogel, Hurst and Mortimer (1971) found that for every mutant tested conversion to wild type and to mutant were equally frequent. This was true over the whole range of conversion frequencies observed, which varied from 0.5% for mating type to 20% for a stop-mutant suppressor, *Sup6*. Fink and Styles (1974) and Fink (1974) crossed deletion mutants 15, 26 and 29 of *his4* in *S. cerevisiae* with a point mutant, 290, mapping at the other end of the gene, and analysed asci from each cross. They made a similar analysis with a point mutant, 39, in place of the deletion mutants. The results are given in Table 29. The deletion mutants showed conversion to wild type and to mutant equally often. Deletion 15 was comparatively short, both ends lying within the proximal part of the gene, but deletions 26 and 29 were longer, both extending into *his4* from the proximal end (29 much further than 26) and for an unknown distance outside it. These two deletions, unlike 15, reduced the frequency of conversion of mutant 290. A deletion involving the entire *cyc1* locus (the structural gene for iso-1-cytochrome *c*) of *S.*

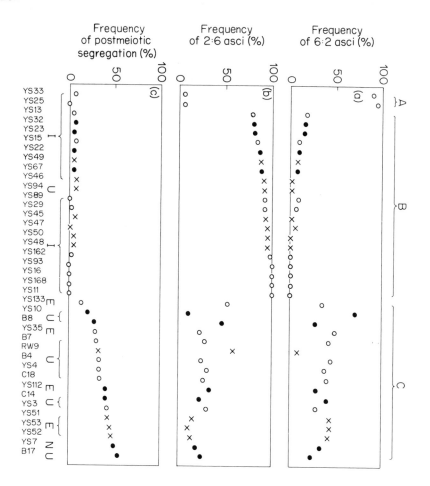

Fig. 73. Data of Yu-Sun, Wickramaratne and Whitehouse (1977) for conversion pattern of spore-colour mutants at *grey-3* (×), *grey-4* (●) and *grey-5* (○) loci in *Sordaria brevicollis* induced with various mutagens: E, ethyl methane sulphonate; I, ICR170; N, nitrosoguanidine; U, ultraviolet light. The isolation numbers of the mutants are shown below the diagram. The frequencies of (a) 6:2 asci, (b) 2:6 asci and (c) 5:3 and 3:5 asci are shown as percentages of total aberrant asci. The mutants in classes A and B are in order of increasing frequency of 2:6 asci, and those in class C in order of increasing frequency of postmeiotic segregation.

cerevisiae and at least part of a nearby gene, *rad7* (radiation sensitive), was found by Lawrence *et al.* (1975) to show conversion in each direction. Two 3:1 asci and one 1:3 ascus were found in 297 analysed.

The equality in the frequency of conversion in the two directions shown by these deletions is in contrast to the results obtained by Rossignol (1969) with spore-colour gene *75* of *Ascobolus*. He found that deletions W141 and 1303, which extend into the gene from the left and right ends, respectively, showed conversion to wild type about 25 times more often than to mutant. A similar bias in

Table 29. Results obtained by Fink and Styles (1974) and Fink (1974) from the analysis of random samples of asci of *Saccharomyces cerevisiae* heterozygous for pairs of *his4* mutants. Mutants 15, 26 and 29 are deletions.

Isolation numbers of his4 mutants crossed		Parental	Conversion of 1		Conversion of 2		Co-conversion		Crossing over	Total
1	2	1 + 1 + + 2 + 2	1 + + + + 2 + 2	1 + 1 + 1 2 + 2	1 + 1 + + + + 2	1 + 1 2 + 2 + 2	1 + + 2 + 2 + 2	1 + 1 + 1 + + 2	1 + 1 2 + + + 2	
15	290	366	11	2	7	8	3	4	0	401
26	290	119	4	4	0	3	0	0	1	131
29	290	417	3	5	2	4	0	1	1	433
39	290	355	16	10	3	11	2	5	2	404

favour of wild type was found by McKnight, Cardillo and Sherman (1981) with a deletion mutant, H3, of *cyc7*, the structural gene for iso-2-cytochrome *c* in *S. cerevisiae*. In this mutant about 5000 nucleotide pairs are deleted, encompassing part of the regulatory region of *cyc7* and two adjacent genes. In 294 tetrads, there were 10 with conversion of H3 to wild type and none with conversion to mutant. Seven of the conversion tetrads were from a cross of H3 with an allele, H2. The H2 mutation was caused by the insertion of about 5500 nucleotide pairs of DNA into the regulatory region of *cyc7* not more than 150 nucleotides from the site of H3. In all the seven tetrads the deletion and insertion mutants showed co-conversion. Surprisingly, the insertion did not show disparity in the direction of conversion despite its similar size and proximity to H3. From crosses of H2 with wild type or with H3 (omitting the co-conversion asci), there were six tetrads with conversion of H2 to wild type and four with conversion to mutant.

More precise information about mismatch correction and other marker effects can be obtained if the gene product has been sequenced in wild-type and mutant strains. The amino acid sequence of iso-1-cytochrome *c* of *S. cerevisiae* has been determined and the molecular basis of a number of mutants in *cyc1*, its structural gene, has been established. Moore and Sherman (1975) investigated the recombination frequencies of *cyc1* mutants. Some of their results are given in Table 30. It is evident that mutants 13 and 239 show an extraordinarily high frequency of mitotic recombination compared with other combinations of alleles. A similar result was obtained with X-ray- and ultraviolet-induced mitotic recombination, and to a lesser extent with meiotic recombination. Subsequently, the investigation was extended to include other mutants mapping in the same part of the gene (Moore and Sherman, 1977). The results were remarkable (Table 31). The deletion frameshift mutant 239 gave a relatively high frequency of recombination compared with transversions 179 and 76. This result would be accounted for if 239 behaved like deletion frameshifts in *Ascobolus* and *Sordaria* and showed conversion to wild type more often than to mutant. The probable transversion 163 also gave a relatively high frequency of recombination with alleles, in several instances much higher than was found with other base substitutions at the same site (mutants 13, 74 and 85). Moore and Sherman (1977) concluded that the nature of the mismatched bases influences the recombination process, but not in a way that can be simply interpreted. It seems as if the two mismatches may interact in complex ways to influence the outcome of the recombination event. Thus, the high frequency of mitotic recombination shown by mutants 13 and 239 cannot be predicted from the results of crosses of either of these mutants with other alleles (Table 30).

Marker effects apparently related to mismatch correction have also been demonstrated in *Schizosaccharomyces pombe*. Hofer *et al.* (1979) found that two opal suppressors, in the *sup3* and *sup9* genes, respectively, each arose by mutation at the anticodon site of a gene for a transfer RNA for serine. Each mutation was from TGA to TCA and both enhanced the local frequency of meiotic recombination five- to 10-fold. Tetrad analysis by Thuriaux *et al.* (1980) revealed that both mutants gave a high frequency of postmeiotic segregation in one-factor crosses

Table 30. Marker effects on recombination frequency revealed when site separations are precisely known. The yeast data refer to mitotic recombination. The molecular basis of mutant 14 at *am-1* in *Neurospora* was described by Fincham and Baron (1977).

Species; gene product and number of amino acid residues it contains; reference	Mutant Isolation number	Mutant Amino acid number	Triplet and amino acid Wild type	Triplet and amino acid Mutant	Numerator: number of prototrophs per 10^9 cells (*Saccharomyces*) or per 10^6 spores (*Neurospora*) Denominator: number of internucleotidic links between the mutant sites
Saccharomyces cerevisiae; *cyc1*; iso-1-cytochrome c; 108; Moore and Sherman (1975)	13	0	AUG (Start)	AUPy or AUA (Ile)	$\dfrac{690}{12}=57.5$; $\dfrac{33}{25}=1.3$
	239	4	AAG (Lys)	Deletion of G	$\dfrac{20}{13}=1.5$; $\dfrac{1100}{211}=5.2$
	179	9	AAG (Lys)	UAG (Stop)	$\dfrac{710}{199}=3.6$
	76	71	GAG (Glu)	UAG (Stop)	$\dfrac{470}{186}=2.5$
Neurospora crassa; *am-1*; NADP-specific glutamate dehydrogenase; 452; Fincham (1976)	14	20	CUPy (Leu)	CAPy (His)	$\dfrac{660}{363}=1.8$; $\dfrac{180}{367}=0.5$
	19	141	AAG (Lys)	AUG (Met)	$\dfrac{0.27}{4}=0.07$
	2	142	CAPy (His)	CAPu (Gln)	$\dfrac{240}{515}=0.5$
	17	313	GAA (Glu)	UAA (Stop)	$\dfrac{380}{511}=0.7$

Table 31. Results obtained by Moore and Sherman (1977) for the frequency of wild-type recombinants arising at meiosis from crosses between mutants in the structural gene (*cyc1*) for iso-1-cytochrome *c* in *Saccharomyces cerevisiae*. For each cross the frequencies are given per 10^5 asci and, in parentheses, per base pair separating the mutant sites.

Mutant isolation no.:		9	239	179	76
Amino acid no.:		2	4	9	71
Wild-type codon:		GAA	AAG	AAG	GAG
Mutant codon:		UAA	AA	UAG	UAG
Nucleotide no.:		4	12	25	211
Isolation number	131 GUG	1.3 (0.22)	30 (2.1)	19 (0.70)	488 (2.3)
and codon	51 CUG	1.8 (0.30)	53 (3.8)	22 (0.81)	522 (2.5)
of mutations	133 AGG	0.41 (0.08)	—	12 (0.46)	401 (1.9)
in the start	13 AUA?	0.12 (0.03)	77 (6.4)	21 (0.84)	142 (0.67)
codon (AUG)	74 AUA?	0.63 (0.16)	60 (5.0)	23 (0.92)	486 (2.3)
which precedes	85 AUC/U?	0.50 (0.13)	32 (2.7)	16 (0.64)	641 (3.0)
codon no. 1	163 AUU/C?	0.43 (0.11)	137 (11)	140 (5.6)	695 (3.3)

and, at least for the *sup3* mutant, also in a cross with an allele. This would account for the increased recombination, because 97% of the conversion events were found normally to be co-conversions that spanned the whole gene and hence failed to give recombination of alleles. Postmeiotic segregation was normally rare. Thuriaux *et al.* (1980) concluded that G · G or C · C mismatches (or both) at the anticodon site in each gene were poorly repaired. They had no information as to whether this poor repair applied to all such mismatches in *S. pombe* or whether the surrounding nucleotide sequence also played a part. Another *sup3* mutant, r10, increased allelic recombination up to 50-fold, apparently by shortening the excision length in mismatch correction (or shortening the hybrid DNA itself) with the result that single-site conversion was more frequent and co-conversion less frequent than normal.

A remarkable marker effect was discovered in *Neurospora crassa* by Fincham (1976) in the *amination-1* (*am-1*) gene, which codes for NADP-specific glutamate dehydrogenase. Mutant no. 19 at this locus showed consistently higher recombination frequencies than a closely linked mutant, no. 2, in crosses with alleles to their left, and lower than no. 2 with alleles to their right. From the amino acid sequence in each mutant polypeptide, however, the site of mutant 19 was found actually to be four nucleotides to the left of that of no. 2. One example of this anomaly is shown in Table 30. Fincham suggested that the unexpectedly low recombination frequency to the right of 19 and to the left of 2 indicated a favoured direction for co-conversion of each of these mutants and the mutant with which it was crossed, for co-conversion would give no recombinants. Mutant 19 arose by a transversion in the direction purine to pyrimidine and mutant 2 by the opposite change (Table 30). The contrary directions favoured for co-conversion could therefore be accounted for by assuming (1) that a pyrimidine–pyrimidine mismatch was recognized with a different frequency from a purine–purine mismatch, and (2) that excision of the mutant strand at the favoured mismatch occurred predominantly in one molecular direction (5' to 3' if pyrimidine mismatches are favoured for recognition; 3' to 5' if purine mismatches are favoured). It might ultimately be possible to test these ideas by ascus analyses, if transversions were available in a gene showing aberrant segregation with sufficiently high frequency.

12.   POLARITY

Lissouba and Rizet (1960) discovered a remarkable polarity in recombination in spore-colour gene *46* of *Ascobolus immersus*. Five mutants in this gene were found to map, on the basis of recombination frequencies, in the following sequence:

$$ \text{—— 188 —— 63 —— 46 —— W —— 138 ——} $$

All 10 pairwise crosses between these mutants were made. Most of the resulting asci had eight white spores, but small numbers of recombinant asci with two black spores were found in all the crosses. A total of 130 of these 2 : 6 octads were analysed (Lissouba, 1961). The cultures derived from the white spores in these octads were backcrossed to each parent. Their genotype was determined on the basis of whether or not further 2:6 asci were formed. No double-mutant spores were found, so it was clear that the recombinants were arising by conversion, and not by crossing over between the mutant sites. The remarkable feature of the results was that the conversion was always at the right-hand site of the pair (Fig. 74). Thus, in a cross between mutants W and 138, all the 2:6 octads tested had four spores of genotype W and two of genotype 138; in a cross of 46 and W, there were in each recombinant octad four of genotype 46 and two of genotype W; and similarly with all the other pairs. In other words, each recombinant octad showed normal 4:4 segregation at the left-hand mutant site and 6:2 segregation at the right-hand site.

Polarity in recombination was discovered independently by Murray (1960) from pairwise crosses of methionine-requiring mutants of *Neurospora crassa* that mapped at the *met-2* locus in the right arm of linkage group IV. Her results (Murray, 1960, 1963) showed an excess of P2 over P1 for 42 out of 44 pairs of *met-2* mutants that were crossed (Fig. 75). These results were based on the selection of methionine-independent progeny by plating the progeny ascospores from allelic crosses on minimal medium lacking methionine. The wild-type recombinants were isolated and scored for flanking markers. Those with one or other parental flanking marker genotype were attributed to conversion to wild type of the mutant with that parentage. On this basis, the slope of each line in Fig. 75 shows the relative frequency of conversion of the two mutants crossed. There

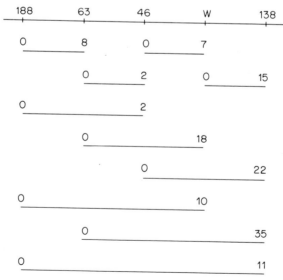

Fig. 74.   Data of Lissouba (1961) for 130 2:6 octads from crosses of spore-colour mutants at locus *46* in *Ascobolus immersus*. At the top is shown the sequence of the mutant sites on the basis of recombination frequencies. The lines below indicate the mutants crossed. The number of 2:6 octads analysed from each cross is shown, the position of the numbers indicating the mutant showing conversion to wild type.

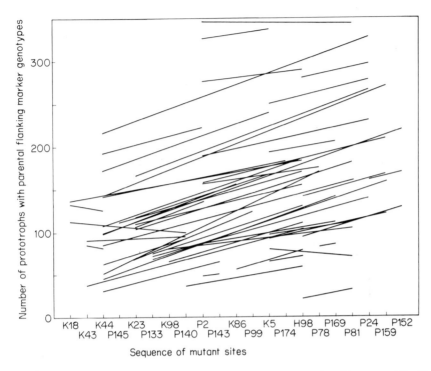

Fig. 75. Data of Murray (1960, 1963, 1969) for 6467 prototrophs with parental flanking marker genotypes from pairwise crosses of *met-2* mutants of *Neurospora crassa*. The *met-2* gene is in the right arm of linkage group IV about 20 units from the centromere. Cross K44 × P24 is plotted at three-quarters of the true number of progeny scored. Two lines are shown for K5 × P81 as the data from two crosses were heterogeneous.

The sequence of the mutant sites within the gene is plotted as the abscissa. The ends of each line show the mutants crossed. The site sequence is based primarily on the numbers of R1 and R2 prototrophs from the pairwise crosses. The number of convertants with parental flanking marker genotypes is plotted as the ordinate: the lowest point of each line refers to this axis. The slope of the lines indicates the relative frequency of conversion of the mutants crossed, as follows: a vertical rise of 1 unit at a distance of 10 units horizontally indicates 60% conversion at the mutant corresponding to the upper end of the line, 40% at the lower; a vertical rise of 2 in 10 indicates 70% conversion of the mutant at the upper end; and so on. Thus, 100% conversion of one mutant, 0% of the other, is indicated by a line rising 5 units in 10 measured horizontally, or at 26°34′ (tan 26°34′ = 0.5).

were also prototrophs with recombinant flanking markers, but they do not provide information about the relative frequency of conversion at the sites of the mutants crossed.

Polarity in recombination was found in the *pabaA* gene of *Aspergillus nidulans* by Siddiqi (1962). He crossed *pabaA* mutants with one another, plated the progeny ascospores on minimal medium, isolated wild-type recombinants selected in this way, and scored them for flanking marker characters. The results for the prototrophs with non-crossover flanking markers are given in Fig. 76. There is a

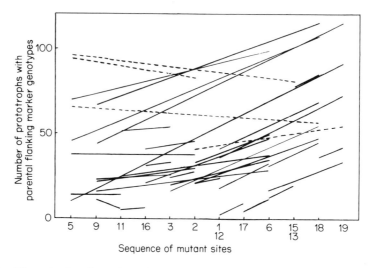

Fig. 76. The unbroken lines show the data of Siddiqi (1962) for 968 prototrophs with parental flanking marker genotypes from pairwise crosses of *pabaA* mutants of *Aspergillus nidulans*. The *pabaA* gene is in the right arm of linkage group I about 30 units from the centromere. The broken lines show the data of Putrament (1964) for 262 prototrophs with parental flanking marker genotypes resulting from mitotic recombination (see Table 21). For an explanation of the lines see the caption to Fig. 75.

well marked distal polarity, though it is less evident in crosses between mutants in the proximal half of the gene.

A feature of polarity in recombination, evident in Lissouba's, Murray's and Siddiqi's results, is that mutants show a lower frequency of conversion when they are the left-hand member of a pair than when they are the right-hand member. Indeed, in Lissouba's data left-hand mutants showed no conversion at all. The behaviour of a mutant seems to depend on its position and not to be an intrinsic property of the mutant itself. In other words, the gradient results, not from marker effects of individual mutants, but from a polarity in the recombination process.

The weaker polarity at the proximal end of *pabaA* in *Aspergillus*, compared with the distal end (Fig. 76) could indicate that, if more proximally located mutants were available, the direction of polarity had reversed. Whitehouse and Hastings (1965), in reviewing data on flanking marker genotype in prototrophs selected from crosses of allelic auxotrophs, found several examples of such polarity reversal, always with proximal polarity at the proximal end of the gene and distal polarity at the distal end. Mousseau (1966) mapped 11 mutants at the spore-colour locus *19* in *Ascobolus* using recombination frequencies and a series of overlapping deletions (as mentioned in Section 9). He found that conversion frequencies gave a U-shaped curve when plotted against the map position of the mutants. There were irregularities, however, due to marker effects (see Section 11). Murray (1969) found that two mutants at the proximal end of *met-2* in *Neurospora* showed polarity reversal when crossed with other alleles mapping near that end of the gene (Fig. 75). In two other genes, *met-6* and *met-7*, she also found evidence for polarity reversal, the sites of preferential conversion being located at each end of each gene. The best documented example of polarity reversal is the work of Pees (1967) with the *lysine F* (*lysF*) gene in the right arm of linkage group I of *Aspergillus nidulans* (Fig. 77). This gene is 4.5 units proximal to *pabaA*. Almost every combination of 12 mutants was investigated, and a symmetrical polarity pattern revealed. By contrast, the *methA* gene in the left arm of linkage group II, studied by Putrament, Rozbicka and Wojciechowska (1971), showed a steep and uniform distal polarity with no indication of reversal at the proximal end (Fig. 78).

A question of basic importance for an understanding of polarity in recombination is whether the polarity is determined locally, or is imposed from a distance, for example, by the centromere. Murray (1968) investigated this question in *Neurospora* by making use of a structural change in which a long segment of the right arm of linkage group I was inserted in inverted sequence into the right arm of linkage group V. The *met-6* gene and its flanking markers are in the part of

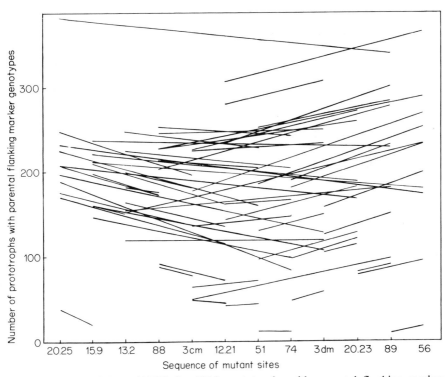

Fig. 77. Data of Pees (1968) for 9712 prototrophs with parental flanking marker genotypes from pairwise crosses of *lysF* mutants of *Aspergillus nidulans*. The *lysF* gene is in the right arm of linkage group I about 26 units from the centromere. For an explanation of the lines, see the caption to Fig. 75.

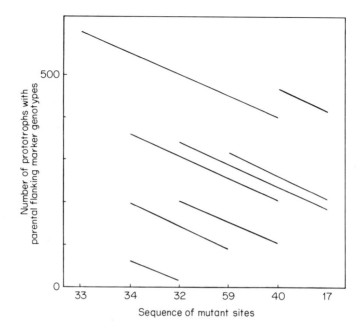

Fig. 78. Data of Putrament, Rozbicka and Wojciechowska (1971) for 2069 prototrophs with parental flanking marker genotypes from pairwise crosses of *methA* mutants of *Aspergillus nidulans*. The *methA* gene is in the left arm of chromosome II, so the centromere is to the right in the diagram. The *methA*–centromere interval is about 29 units. Cross 40 × 17 is plotted at half the true number of progeny scored. For an explanation of the lines, see the caption to Fig. 75.

linkage group I involved, so she was able to transfer these to the structural change. In this way, the orientation of *met-6* and its flanking markers was reversed with respect to the centromere. Recombination at *met-6* was studied in a strain homozygous for the inversion. She found that the polarity was unchanged with respect to the flanking markers, but reversed with respect to the centromere: proximal polarity when structurally normal became distal polarity in the inversion. She concluded that polarity is intrinsic to the local region of the chromosome and not imposed by the centromere or chromosome ends.

Another basic question about polarity in recombination is whether it is strictly intragenic, or whether it can extend from one gene into a neighbouring one. This question was investigated by Murray (1970) with methionine auxotrophs of *Neurospora* at the *met-7* and *met-9* loci. These are believed to be separate genes (see Section 3), though closely linked in the right arm of linkage group VII. The polarity pattern within the *met-7* gene was shown by Murray (1969) to be strongly distal in the distal part of the gene, reversing to be weakly proximal at the proximal end. The *met-9* gene maps distal to *met-7*, but the strong distal polarity at the distal end of *met-7* did not extend into *met-9*. This was evident from crosses made by Murray (1970) between *met-7* and *met-9* mutants in the presence of flanking markers: the *met-9* mutants showed a lower frequency of conversion than distal *met-7* mutants. On the basis of the relative frequencies of P1 and P2 recombinants, the conversion frequencies were found to be as follows:

―― centromere ―― intermediate ― low ― high ― intermediate ――
$\underbrace{}_{met\text{-}7}$ $\underbrace{}_{met\text{-}9}$

Holliday (1964*b*) and Hastings and Whitehouse (1964) suggested that polarity was a consequence of fixed initiation points for recombination with a variable extent of hybrid DNA from these points of primary breakage of polynucleotides. Conversion would then be of higher frequency near the initiation points, declining in frequency with distance. The polarity patterns observed indicated that the initiation points were outside the genes. Unipolar genes such as *methA* in *Aspergillus* would be affected by a single initiation point near the end of the gene with the higher conversion frequency, while bipolar genes such as *lysF* would have an initiation site at or beyond each end.

According to this hypothesis, Murray's finding that the distal polarity at the distal end of *met-7* is not continued into *met-9* would be accounted for if there was an initiation site for recombination between *met-7* and *met-9*. There is evidence, however, from two sources that hybrid DNA can extend from the one gene to the other. First, crosses involving mutants in each gene show negative interference (Section 3). Secondly, Murray (1970) used a temperature-sensitive *met-9* mutant, 43t, to study recombination in *met-9* when there was selection for wild type in crosses of *met-7* mutants. By carrying out the selection at low temperature, that is, permissive conditions, there was no selection against 43t. The genotype at this site could be scored by testing for methionine requirement under non-permissive conditions (high temperature). The results are given in Table 32. They show that 43t gave 1.7% co-conversion with mutant 331 which is situated in the middle of *met-7* and 5.2% co-conversion with mutant 271 which is nearer the distal end of *met-7*. Moreover, about 12% of the *met-7* prototrophs with recombinant flanking markers involved the *met-9* mutant in the recombination event. Murray concluded that the simplest explanation of the coincident events in *met-7* and *met-9* was to suppose that hybrid DNA could be initiated distal to *met-9* and could extend through *met-9* into *met-7*. Since it had already been inferred from the polarity data that there was an initiation site for recombination between *met-7* and *met-9*, it appeared that hybrid DNA initiated elsewhere could extend through this site. A further implication was that excision of one strand of the heteroduplex, and resynthesis using the other strand as template, if that was the explanation of co-conversion, could also extend across the initiation site.

Where data are available for meiotic and mitotic recombination in the same gene, it has been possible to compare them from the point of view of polarity. Surprisingly, in every case substantial differences have been found. As already pointed out, meiotic data for *pabaA* in *Aspergillus* indicated distal polarity, weakening towards the proximal end. In mitotic data, proximal polarity is much more in evidence (Fig. 76). Meiotic data show distal polarity for the *adE* gene in

Table 32. Data of Murray (1970) for *met-9* genotype in *Neurospora crassa* when wild-type recombinants are selected in crosses of *met-7* mutants.

Genotype of cross and of *met-7* prototrophs					Numbers of *met-7* prototrophs of each *met-9* genotype			
					met-9 and *wc* with a parental genotype		*met-9* and *wc* with a recombinant genotype	
	met-7		*met-9*					
thi-3	56	+	+	+				
+	+	331	43t	*wc*				
P1 thi-3	+	+		+	+	245	43t	0
P2 +	+	+		*wc*	43t	523	+	9†
R1 +	+	+		+	+	74	43t	6
R2 thi-3	+	+		*wc*	43t	15	+	6
thi-3	331	+	43t	+				
+	+	271	+	*wc*				
P1 thi-3	+	+		+	43t	78	+	0
P2 +	+	+		*wc*	+	291	43t	16‡
R1 +	+	+		+	43t	44	+	7
R2 thi-3	+	+		*wc*	+	12	43t	0

†1.7% co-conversion at sites of 331 and 43t.
‡5.2% co-conversion at sites of 271 and 43t.

Aspergillus and strong proximal polarity for the *ilv1* and *his1* genes of *Saccharomyces cerevisiae*, but mitotic data for these three genes show no polarity (Fig. 79). The inference from these results is that if polarity is a consequence of the initiation of recombination at fixed points outside the genes, initiation of meiotic and mitotic recombination occur at different points or, if from the same points, then with different frequencies.

The hypothesis of initiation sites for recombination at or beyond each end of a gene gained support from the discovery of repressors of recombination affecting specific localized regions. Jessop and Catcheside (1965) discovered genetic differences in stocks of *Neurospora* in the frequency of wild-type recombinants when *histidine-1* (*his-1*) mutants were crossed. The differences could be attributed to one gene, *recombination-1* (*rec-1*). The allele for low frequency, *rec-1*⁺ was dominant, only the homozygote for the other allele, *rec-1*, having the higher frequency of *his-1* recombination. The difference in frequency was more than 10-fold. Flanking markers were available for *his-1*, which is in the right arm of chromosome V. The markers used were *amination-1* (*am-1*) 5 units to the left and *inositol* (*inl*) 4 units to the right of *his-1*. The flanking marker genotypes were

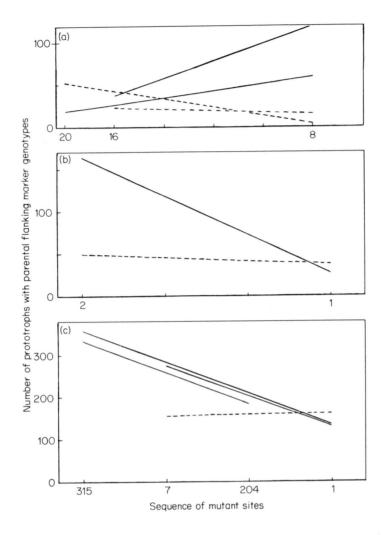

Fig. 79. Comparison of polarity in meiotic and mitotic recombination. The lines indicate the relative frequency of conversion to wild type of pairs of mutants, meiotic data being shown with unbroken lines and mitotic data with broken lines. For an explanation of the lines, see the caption to Fig. 75. (a) Data of Pritchard (1955, 1960) for mutants 8 and 16, and 8 and 20, at the *adE* locus in *Aspergillus nidulans*. This gene is about 46 units from the centromere in the right arm of linkage group I, and so is about 16 units distal to *pabaA*. (b) Data of Sherman and Roman (1963) for meiotic recombination and of Kakar (1963) for mitotic recombination of mutants 1 and 2 at the *isoleucine–valine 1* (*ilv1*) locus in *Saccharomyces cerevisiae*. This gene is about 52 units from the centromere in the right arm of linkage group V. (c) Data of Fogel and Hurst (1967) for meiotic recombination and of Hurst and Fogel (1964) for mitotic recombination at the *histidine 1* (*his1*) locus in *Saccharomyces cerevisiae*. The *his1* gene is about 18 units proximal to *ilv1*, that is, about 34 units from the centromere in the right arm of linkage group V. Cross 315 × 1 is plotted at half the true number of progeny scored, and cross 7 × 1 (meiotic data) at one-quarter the true number.

scored in histidine prototrophs obtained from pairwise crosses of *his-1* mutants, both in *rec-1*⁺ and homozygous *rec-1* backgrounds. The relative frequency of the two parental flanking marker genotypes was found to differ, depending on the *rec-1* genotype. The results are shown in Fig. 80. In the *rec-1* homozygote there was proximal polarity, and in *rec-1*⁺ it was distal. The simplest explanation of this result was that *rec-1*⁺ repressed recombination in *his-1* by blocking an initiation point for recombination situated proximal to *his-1* and responsible for the proximal polarity revealed in the absence of *rec-1*⁺ (Whitehouse, 1966). The distal

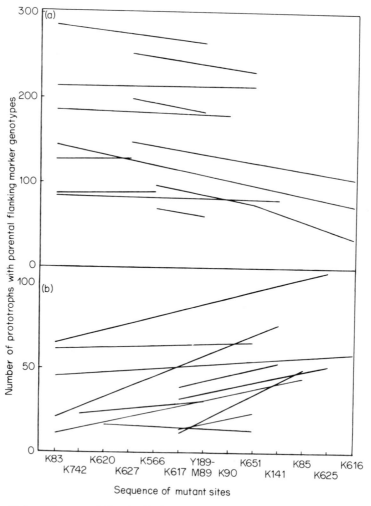

Fig. 80. Data of Jessop and Catcheside (1965) for 2269 prototrophs with parental flanking marker genotypes from pairwise crosses of *his-1* mutants of *Neurospora crassa*: (a) in *rec-1* homozygotes; (b) in *rec-1*⁺ genotypes. In (a), the results for the cross K83 × K651 are plotted at half the true number of progeny scored. For an explanation of the lines, see the caption to Fig. 75.

polarity in *his-1* in *rec-1*⁺ strains could be attributed to a second initiation point for recombination on the distal side of *his-1*, this initiation site being insensitive to *rec-1*⁺.

A second recombination repressor locus, *rec-2*, was discovered by Smith (1966) and a third, *rec-3*, by Catcheside (1966). The dominant allele of *rec-2* reduced the frequency of recombination between *pyr-3* and *his-5* in the right arm of linkage group IV and increased the strength of distal polarity in *his-5*. This effect would be expected if an initiation site proximal to *his-5* was closed by the repressor. The *amination-1* (*am-1*) gene in the right arm of linkage group V was affected by *rec-3*, recombination of *am-1* alleles being reduced by *rec-3*⁺ by a factor of 10–25. Smyth (1971, 1973) investigated the effect of *rec-3*⁺ on flanking marker genotypes of *am-1* prototrophs from pairwise crosses of *am-1* mutants, and found that proximal polarity shown by *am-1* in a *rec-3* homozygote was replaced by distal polarity when recombination was repressed by the dominant allele, *rec-3*⁺. The situation was thus exactly comparable to that of *his-1* affected by *rec-1*.

The three repressor loci were found to affect recombination at other sites in addition to those where their effect was discovered (see Table 33). Polarity was found to be affected in the neighbourhood of the sites of action of the repressors, whenever it was possible to test this. Thus, Angel, Austin and Catcheside (1970)

Table 33. Sites of action of recombination repressors in *Neurospora crassa*, from results obtained by Catcheside and associates (see Catcheside, 1977). I, IV, V, linkage groups; L, R, left, right arms; numerals after L and R are approximate centromere distances.

Repressor locus and its location	Sites of action of repressor and effect on polarity	Number of loci or regions tested and found not to be affected
rec-1, V R 50	(a) Proximal end of *his-1* (V R 30); proximal polarity becomes distal. (b) At or near *nit-2* (I L 32).	19
rec-2, V R 21	(a) Between *pyr-3* (IV R 12) and *his-5* (IV R 14); distal polarity in *his-5* strengthened. (b) Distal end of *his-3* (I R 3); distal polarity becomes proximal; interval between *his-3* and *ad-3A* (I R 5) also affected. (c) Between *arg-3* (I L 3) and *sn* (I L 1).	15
rec-3, I L 8	(a) Proximal end of *am-1* (V R 25); proximal polarity becomes distal. (b) At or near *his-2* (I R 2); interval between *sn* and *his-2* also affected.	12

found that *rec-2*[+] reduced the frequency of recombination of alleles at the *his-3* locus in the right arm of linkage group I, and reversed the polarity from distal to proximal. Moreover, they discovered a site, mapping on the distal side of *his-3*, through which the effects of *rec-2* on *his-3* appeared to be mediated. Their results are discussed in Section 15. The discovery that the *Neurospora* recombination repressors not only reduce the frequency of recombination within specific genes, but also modify the polarity of recombination within those genes, has strengthened the hypothesis that polarity results from the initiation of recombination at fixed points outside the genes. In particular, the locating of a site distal to *his-3* mediating the effect of *rec-2* was in agreement with the prediction from the direction of the polarity change that the initiation site affected by *rec-2* was distal to *his-3*. Also, the locations for initiation sites inferred from the effects of a repressor on polarity were in agreement with the locations inferred from the effects on intergenic recombination (see Table 33, *rec-2*, sites (a) and (b)). All the data are consistent with the hypothesis that the repressors block the initiation of recombination at specific sites. No example was found of different repressors affecting the same site. Many genes were tested with negative results (see Table 33) and it is clear that, on the evidence available, about one gene in eight is affected by any one repressor. The sites affected were unlinked, or only loosely linked, to the relevant repressor locus and to one another.

Polarity in recombination in *Drosophila melanogaster* was reported by Carlson (1971) with mutants at the locus for *rudimentary* (*r*) wings in the X-chromosome, and by Chovnick, Ballantyne and Holm (1971) with mutants at the *rosy* (*ry*) eye-colour locus in chromosome 3. At the *r* locus, Carlson (1971) found strong proximal polarity. This was shown by a marked excess of P1 over P2 flanking marker genotypes (average ratio 3.6 : 1) in wild-type recombinants from numerous pairwise crosses. Large numbers of such recombinants could be selected because most of the *r* homozygotes failed to survive (see Section 6). Although there was no indication of polarity reversal at the distal end of the gene from the two-point crosses, the results from some three-point crosses pointed to its probable occurrence. By contrast, polarity reversal was well marked at *ry*. Chovnick *et al.* (1971) found that mutants near each end of the gene gave higher conversion frequencies than those in mid-gene. The highest frequency was at the distal end. It is of interest that the *Drosophila* polarity data match so closely those for fungi, with diversity of pattern from one gene to another. In one respect, however, there seems no parallel. Carlson (1971) found that in crosses of *r* mutants, the P1 flanking marker genotype was consistently about 3.4 times as frequent as P2, except with closely linked alleles. For such mutants P2 wild-type recombinants were less frequent, declining progressively to zero for the shortest intervals. A steeper polarity gradient for short intervals than for longer ones does not seem to have been found in fungal genes.

13. NON-RECIPROCAL HYBRID DNA

Soon after the idea had been proposed that recombination in eukaryotes depended on the formation of hybrid DNA (Section 7), it became evident that the reciprocal formation of a heteroduplex in each of the recombining chromatids was not an adequate explanation of the data. It is not suggested that hybrid DNA never occurs in both chromatids at a mutant site: this provides a satisfactory explanation of aberrant 4:4 asci. But there is evidence of at least four kinds that hybrid DNA formation is often non-reciprocal.

First, the recombination of some mutants was found always to be non-reciprocal. Examples of this from the work of Rizet and associates with spore-colour gene *46* of *Ascobolus immersus* were referred to in Section 4 and 12. Mutants 63 and 138, for example, showed only non-reciprocal recombination when crossed, apparently through conversion each time at the site of mutant 138. If hybrid DNA formed reciprocally, however, it would be expected that mismatch correction would sometimes occur in opposite directions in the two chromatids, that is, to wild type in one and to mutant in the other, as pointed out in Section 7. This would restore a normal 4:4 segregation, but in 50% of cases it would be expected to give rise to reciprocal recombination with a nearby marker not included in hybrid DNA. For the cross between mutants 63 and 138 the hypothetical sequence of events would be

$$
\begin{array}{ll}
63 & + \\
63 & + \\
+ & 138 \\
+ & 138
\end{array}
$$

giving reciprocal heteroduplexes

$$
\begin{array}{ll}
63\ \ \ + & \quad 63\ \ + \\
63\ +/138 & \quad 63\ 138 \\
+\ \ +/138 & \quad +\ \ \ \ + \\
+\ \ \ \ \ 138 & \quad +\ \ \ 138
\end{array}
\quad \text{and then}
$$

through mismatch correction in opposite directions. The failure to find such reciprocal recombination implied that hybrid DNA was confined to one chromatid at the site of mutant 138. The sequence of events would then be

$$
\begin{array}{ll}
63 & + \\
63 & + \\
+ & 138 \\
+ & 138
\end{array}
$$

giving a non-reciprocal heteroduplex

$$
\begin{array}{ll}
63\ \ \ + & \quad 63\ \ + \\
63\ \ \ + & \quad 63\ \ + \\
+\ +/138 & \quad +\ \ + \\
+\ \ \ \ 138 & \quad +\ \ 138
\end{array}
\quad \text{and then}
$$

through mismatch correction. Since octads with wild-type spores were being selected, the occurrence of the other non-reciprocal heteroduplex,

$$
\begin{array}{ll}
63 & + \\
63 & +/138 \\
+ & 138 \\
+ & 138
\end{array}
$$

could not be detected.

The second argument for believing that hybrid DNA was often confined to one chromatid is based on the behaviour of flanking markers when alleles show reciprocal recombination. If this reciprocal recombination resulted from mismatch correction in opposite directions, as discussed above, flanking markers would be expected often to show a parental genotype with one another, as happens with conversion or postmeiotic segregation. It was found, however, that reciprocal recombination of alleles was usually associated with coincident recombination of flanking markers. Published data for several genes in *Aspergillus*, *Neurospora* and *Saccharomyces*, and including examples of mitotic as well as meiotic recombination, gave a total count of six P1, five P2, 147 R1 and five R2 flanking marker genotypes (Whitehouse, 1967). Thus, more than 90% of the reciprocal recombinations of alleles appeared to be the result of crossing over between their sites, the recombination of flanking markers being caused by that of the alleles. The rarity of events attributable to reciprocal mismatch correction pointed to hybrid DNA often being confined to one chromatid.

The third line of evidence for non-reciprocal formation of hybrid DNA is based on the genotype of 5:3 and 3:5 asci with non-crossover flanking markers. As indicated in Table 17, two genotypes are possible for such asci. One has two of the products of meiosis identical and so may be called *tritype*, while the other has all four products of meiosis different: *tetratype* (Table 34). If there is hybrid DNA in both recombining chromatids at the mutant site, these two alternatives are expected to be equally frequent because each would require mismatch correction in one heteroduplex and not in the other one: they differ merely in the parentage of the chromatid showing correction. Thus, for 5:3 asci, using the symbols of Table 34, the sequence of events would be

$$
\begin{array}{l}
M\,b\,N \\
M\,b\,N \\
m+n \\
m+n
\end{array}
$$

giving reciprocal heteroduplexes

$$
\begin{array}{lll}
M\ \ b\ \ N & M\ \ b\ \ N & M\ \ b\ \ N \\
M\ b/+N & M\ b/+N & M\ \ +\ \ N \\
m\ b/+n \quad \text{and then} & m\ \ +\ \ n \quad \text{or} & m\ b/+n \\
m\ \ +\ \ n & m\ \ +\ \ n & m\ \ +\ \ n
\end{array}
$$

through mismatch correction in one heteroduplex. On the other hand, if the hybrid DNA were confined to one chromatid, the other having been repaired to restore an intact chromatid, only the tritype genotype is possible and this would arise

Table 34. Numbers of non-crossover tritype and tetratype $5:3$ and $3:5$ asci from crosses of mutants and wild type. The parental genotype was

$$\frac{M \quad b \quad N}{m \ + \ n},$$

where b represents the mutant and M/m and N/n are the flanking markers.

Species and reference	Gene	+:b		Tritype	Tetratype	Percentage tetratype
		5:3	$\begin{cases} M \ b \ N \\ M \ b/+ \ N \\ m \ + \ n \\ m \ + \ n \end{cases}$		$\begin{array}{ccc} M & b & N \\ M & + & N \\ m & b/+ & n \\ m & + & n \end{array}$	
		3:5	$\begin{cases} M \ b \ N \\ M \ b \ N \\ m \ b/+ \ n \\ m \ + \ n \end{cases}$		$\begin{array}{ccc} M & b & N \\ M & b/+ & N \\ m & b & n \\ m & + & n \end{array}$	
Sordaria fimicola (Kitani and Olive, 1967; Kitani and Whitehouse, 1974)	*grey*	5:3	313		165	35
		3:5	204		66	24
Sordaria brevicollis (Sang and Whitehouse, 1979)	*buff*	5:3	83		3	3
		3:5	50		1	
Ascobolus immersus (Pasadena strain) (Stadler and Towe, 1971)	*w17*	5:3	21		0	2
		3:5	32		1	
Saccharomyces cerevisiae (DiCaprio and Hastings, 1976)	*Sup6*	5:3	4		0	0
		3:5	6		0	
	cdc14	5:3	1		0	

without mismatch correction. The sequence of events for $5:3$ asci would be

$$\begin{array}{ccc} M & b & N \\ M & b & N \\ m & + & n \\ m & + & n \end{array}$$

giving a non-reciprocal heteroduplex

$$\begin{array}{ccc} M & b & N \\ M & b/+ & N \\ m & + & n \\ m & + & n \end{array}$$

Experimental results have been obtained for several fungi and are given in Table 34. Without exception tritypes have been significantly more frequent than tetratypes, though with considerable variation in degree. In *Sordaria fimicola* tetratypes accounted for about one-third of the asci, implying that heteroduplex formation is reciprocal in two-thirds of the events. The results for *S. brevicollis*, the Pasadena strain of *Ascobolus immersus*, and *Saccharomyces cerevisiae* are similar to one another and indicate reciprocal heteroduplex formation in about 5% of the events (possibly none in yeast). The data, however, for all these fungi are so limited that the results might apply only to the particular genes or mutant sites that have been investigated. Indeed, there is evidence for variation in the relative frequency of reciprocal and non-reciprocal hybrid DNA within a gene. This is discussed in Section 14.

A fourth source of evidence that hybrid DNA formation is often non-reciprocal is provided by the frequency of aberrant 4:4 asci relative to 5:3 and 3:5. Where aberrant 4:4 frequency can be determined, it has sometimes been found to be much lower than would be expected from the frequencies of 5:3 and 3:5 asci. Examples of this have been found in *Sordaria brevicollis* (see Sang and Whitehouse, 1979) and *Saccharomyces cerevisiae* (see Section 14).

As pointed out in the caption to Fig. 70, co-conversion can be explained more simply if hybrid DNA is confined to one chromatid, since all that is required is that excision of the appropriate strand in mismatch correction should extend to the second site.

14. THE MESELSON–RADDING MODEL

In Holliday's model for the recombination process, described in Section 7, there is a reciprocal exchange of strands between two DNA molecules to give hybrid DNA in both molecules, and a node where strands exchange partners (Fig. 66(c)). Sigal and Alberts (1972) constructed space-filling molecular models of such a strand exchange between two duplex DNA molecules and showed that all the bases in the two double helices could remain paired. Furthermore, they showed that the positions of the two pairs of strands – those that cross and those that do not – can be interchanged by rotation, as implied also by Emerson (1969). Sigal and Alberts (1972) also concluded that the cross-connection could diffuse along the joined helices by a zipper-like action in which, if the movement is upwards, the two identical bases above the cross-connection exchange places. Meselson (1972) proposed that this process could be driven by rotary diffusion to give long tracts of hybrid DNA. He calculated that, under reasonable assumptions, such diffusion would be rapid enough to be relevant to genetic recombination. Diffusion of the nodal point might occur in either direction along the two double helices, so the hybrid DNA segments might be lengthened or shortened by this process.

In order to account for non-reciprocal heteroduplex formation inferred from the fungal recombination data (Section 13), Moore (1974) suggested that at the nodal point one of the strands broke (Fig. 81(b)). The broken ends could then become the sites of action of a DNA polymerase on one DNA molecule and of an

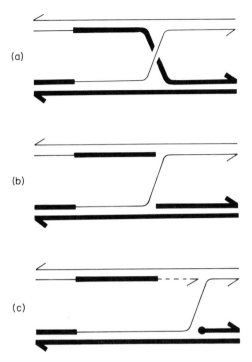

Fig. 81. Hypothesis of Moore (1974) to account for non-reciprocal heteroduplex formation. The broken line shows new synthesis and the dot the site of single-strand degradation by an exonuclease. Line thickness indicates parentage. Between (a) and (b) one strand has broken. The broken ends have become the sites of action of a DNA polymerase and an exonuclease, giving the result shown in (c).

exonuclease on the other. As the exchange point moved along, it could be associated with strand synthesis in one molecule and strand degradation in the other, with the result that hybrid DNA would be formed in one molecule only (Fig. 81(c)). Wagner and Radman (1975) proposed that recombination was initiated at inverted complementary repeats. These palindromic sequences would form loops and be associated with pairing, nicking and cross-annealing. The heteroduplex thus formed in each molecule would extend by branch migration. If a nick was encountered, the heteroduplex might continue beyond it in one molecule only.

A different idea to accommodate both reciprocal and non-reciprocal hybrid DNA formation was proposed by Meselson and Radding (1975). They suggested that a single-strand break in one DNA molecule becomes the site of strand displacement by a DNA polymerase (Fig. 82(a)). The displaced strand pairs with the complementary sequence in another molecule of DNA and induces a single-strand break in the strand of the same polarity as the invading strand. This induction of breakage in a homologous region of a DNA molecule by a strand with a free end is analogous to the behaviour shown in *Escherichia coli* in the presence of

304

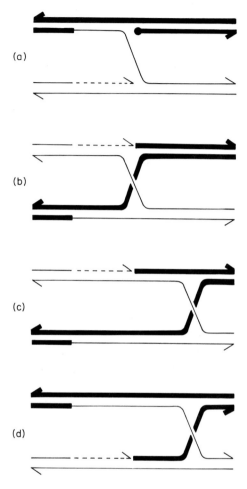

Fig. 82. The Meselson–Radding model, showing ((a), (b)) the asymmetric phase and ((c), (d)) the symmetric phase. The flanking arms have a parental configuration in (a) and (d) and a recombinant one in (b) and (c). Broken lines indicate newly synthesized strands and the dot the site of single-strand degradation by an exonuclease. Line thickness indicates parentage.

the *recA* gene product. Exonucleolytic action on the broken strand in the recipient molecule, in conjunction with the polymerase action in the donor molecule, allows continued displacement of the donor strand and its assimilation into the recipient molecule (Fig. 82(a)). Because this reaction produces a heteroduplex in only one of the interacting molecules, Meselson and Radding called it *asymmetric strand transfer*. They considered that the two enzymic activities, polymerase and exonuclease, might be in one molecule, and that the affinity of this enzyme for the substrate – the structure shown in Fig. 82(a) – determines the extent of asymmetric strand transfer. Dissociation of the enzyme allows the structure to rearrange, or isomerize (Fig. 82(b)). The arms that flank the site of strand exchange would now be in the recombinant configuration, as a result of rotation of a pair of them 180° about an axis between and parallel to them. Such rotation would require the interruption present in one of the four strands. The cross-connection in the isomerized molecule would be free to migrate in either direction as a consequence of rotary diffusion of the two molecules, both rotating in the same sense. This branch migration, if it occurs in the direction away from the site of initial breakage, will give rise to a heteroduplex in both chromatids (Fig. 82(c)). Meselson and Radding called this the *symmetric phase* of their model, to distinguish it from the earlier *asymmetric phase*. These two phases neatly accommodate the requirements of the genetic data that recombination should involve both reciprocal and non-reciprocal heteroduplex formation. A further isomerization would restore a parental configuration to the flanking arms (Fig. 82(d)). This isomerization is that discussed by Sigal and Alberts (1972) and would occur more slowly than the earlier one ((a) to (b) in Fig. 82) because, without a broken strand at the exchange point, extensive motions of the arms are required, as Sobell (1974) pointed out. Nevertheless, Meselson and Radding concluded from a hydro-dynamic calculation that this isomerization was unlikely to be so slow as to preclude it from the recombination process.

The sequence from (a) to (d) of Fig. 82 may be ended at any stage by cutting the crossed strands. Depending on when this happens, the genetic outcome may be non-reciprocal or reciprocal hybrid DNA at a mutant site, with or without flanking marker crossing over. It was assumed that at some unspecified time a polynucleotide ligase closed the interruption to one strand shown in Fig. 82. It was also assumed that mismatched base pairs in heteroduplex regions might be subject to enzymic repair, to account for conversion, and that there might be preferential initiation of recombination at fixed positions, to account for polarity.

Meselson and Radding pointed out that branch migration driven by rotary diffusion should be blocked by long regions of non-homology of nucleotide sequence. In consequence, long deletions or long substitutions would not be expected to enter the symmetric phase, and so should not show the genetic phenomena, such as aberrant 4:4 segregation, that require reciprocal hetero-duplex formation.

The occurrence of unequal numbers of tritype and tetratype non-crossover odd-ratio asci (Table 34), implying that in different cells both reciprocal and non-reciprocal heteroduplex formation may occur at a mutant site, is readily accounted for on the Meselson–Radding model by supposing that there is variability in the dissociation of the polymerase-nuclease from its substrate. Delayed dissociation might mean that the asymmetric strand transfer continued past the mutant site, giving a tritype ascus, while in another meiosis earlier dissociation, followed by migration of the node, could give reciprocal hybrid DNA at the mutant site, with the possibility of a tritype or a tetratype ascus (Fig. 83). The data point to early dissociation at the *grey* spore-colour locus in *Sordaria fimicola* and late dissociation in the other fungal genes for which data are available (Table 34). Indeed, in *Saccharomyces cerevisiae* there is no clear

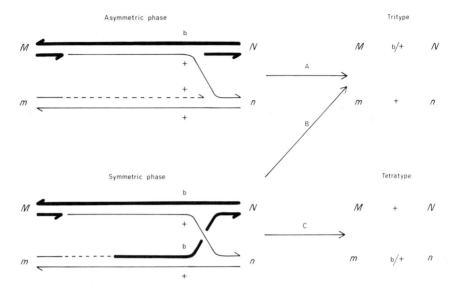

Fig. 83. The possible origin of tritype and tetratype 5:3 asci. Parents are

$$\frac{M \text{ b } N}{m + n},$$

where b is a spore-colour mutant and M/m and N/n are flanking markers. Line thickness indicates parentage. Broken lines indicate newly synthesized strands. A, no mismatch repair. B, mismatch repair to + in mn molecule. C, mismatch repair to + in MN molecule. (Reproduced by permission of Springer Verlag from H. Sang and H. L. K. Whitehouse, *Molec. gen. Genet.* **174**, 333 (1979).)

evidence for reciprocal heteroduplex formation. Fogel *et al.* (1979) reported 16 aberrant 4:4 asci in 9944 aberrant segregations, but this frequency is so low that Fogel *et al.* considered that some – and perhaps all – of them were false aberrant 4:4s caused by independent 5:3 and 3:5 events. Meselson and Radding (1975) suggested that in yeast the polymerase-nuclease has a particularly high affinity for its substrate, with the result that the symmetric phase makes only a small contribution to recombination.

A prediction of the Meselson–Radding model is that non-reciprocal hybrid DNA will predominate near the initiation site of recombination, with reciprocal hybrid DNA more frequent far from the intiation site. Other things being equal there should therefore be a gradient of aberrant 4:4 frequency through a gene in the opposite direction to the polarity gradient of total aberrant ascus frequency. Aberrant 4:4 asci for a spore-colour mutant can be detected even in an unordered ascus if sister spores can be identified through the segregation of another spore character, as indicated in Table 15 with a mycelial marker. Under the most favourable circumstances two-thirds of aberrant 4:4 asci will be revealed by a second spore marker; in the remaining third, both spore pairs showing postmeiotic segregation will have the same genotype for the other spore character. Kitani and Olive (1969) used a mutant giving a bluish tint to the spore wall to facilitate recognition of aberrant 4 :4 segregations at the *grey* locus in *Sordaria fimicola*, and Ghikas and Lamb (1977) used a mutant giving granular spore pigmentation to detect aberrant 4 :4 segregation of a white-spored mutant, w-78, in the Pasadena strain of *Ascobolus immersus*. Paquette (1978) used a round-spored mutant and a granular-spored mutant to detect aberrant 4 :4 segregation at the *b2* white-spore locus in the second French strain of *Ascobolus*. The *b2* gene is homologous with gene *75* of the first French strain, according to Rossignol and Paquette (1979). When both the round and granular mutants were segregating, Paquette was able to detect eight-ninths of the aberrant 4 :4 segregation at *b2*. The high frequency of aberrant 4 :4 segregation which he found at this locus made it a suitable gene to test the Meselson–Radding model.

Paquette and Rossignol (1978) carried out such a test using 15 C class *b2* mutants. The mutants were mapped, on the basis of recombination frequencies and with the help of two overlapping deletions, and were found to fall into seven groups. It was not possible to establish the sequence of the mutant sites within the groups. The numbers of aberrant asci of each kind were counted in several crosses of each mutant with wild type. A total of 295 061 octads were scored. A correction was applied for the fraction of aberrant 4 :4 asci that would not have been detected. The total frequency of aberrant asci of all kinds is plotted in Fig. 84(a) against the map position of the mutant. The mutants mapping towards the left gave about 300 aberrants per 1000 asci, while with the remaining mutants aberrants were about half as frequent. The frequency of aberrant 4 :4 asci out of total aberrants is plotted in Fig. 84(b). It shows an increase towards the right. The inverse relationship between aberrant ascus frequency and aberrant 4 :4 frequency evident from these graphs is in agreement with the predictions of the Meselson–Radding model.

To provide a better test of the model, Paquette and Rossignol (1978) used the frequencies of each kind of aberrant ascus to estimate the frequencies of mismatch correction, of correction to wild type, of asymmetric hybrid DNA formation, and of total hybrid DNA formation. The latter frequency would include an estimate of the number of asci with a normal 4 :4 segregation resulting from the restoration of a normal ratio as a result of mismatch correction. Their underlying assumptions were (1) that asymmetric hybrid DNA has the same probability of forming in each chromatid, (2) that the probability of mismatch correction taking place and (3) of its occurring to wild type were the same in each chromatid. These assumptions are probably not true in general, but appear to be justified, at least approximately, for the mutants they studied. With these assumptions, the contributions made to each kind of aberrant ascus by asymmetric and symmetric hybrid DNA can be expressed in terms of three basic quantities: asymmetry frequency, correction frequency, and correction direction (Table 35). From their counts, Paquette and Rossignol (1978) estimated these quantities for each mutant. The estimated frequencies of hybrid DNA per meiosis are plotted against map

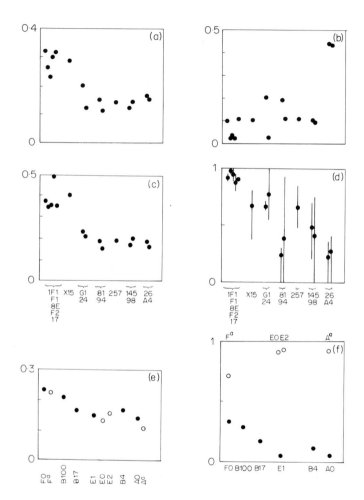

Fig. 84. Results obtained by Rossignol and associates with the *b2* spore-colour gene of *Ascobolus immersus*. (a)–(d) Data of Paquette and Rossignol (1978) with 15 C class mutants. The mutants mapped in seven groups and their site positions are plotted as abscissae in each graph. The sequence of the sites within each group is unknown. (a) Frequency of aberrant asci in crosses with wild type. (b) Frequency of aberrant 4:4 asci out of total aberrants. (c) Estimated frequency of hybrid DNA at each mutant site per meiosis. (d) Estimated frequency of asymmetric hybrid DNA out of total hybrid DNA; the lines show the range of variation between crosses involving the same mutant. (e), (f) Data of Rossignol, Paquette and Nicolas (1979) with 4 A (O) and 6 B (●) class mutants. The map positions of the mutants are plotted as abscissae. (e) Frequency of aberrant asci in crosses with wild type. (f) Frequency of 6:2 asci out of total aberrants.

Table 35. Algebraic analysis of recombination made by Paquette and Rossignol (1978). α, frequency of asymmetric hybrid DNA per recombination event; p, frequency of mismatch correction per mismatch; v, frequency of correction to wild type per mismatch correction.

Segregation (wild type : mutant)	Contribution from symmetric and asymmetric hybrid DNA	
	Asymmetric	Symmetric
Aberrant 4 : 4	–	$(1 - \alpha)(1 - p)^2$
5 : 3	$\dfrac{\alpha}{2}(1 - p)$	$2(1 - \alpha)p(1 - p)v$
3 : 5	$\dfrac{\alpha}{2}(1 - p)$	$2(1 - \alpha)p(1 - p)(1 - v)$
6 : 2	$\dfrac{\alpha}{2}pv$	$(1 - \alpha)p^2 v^2$
2 : 6	$\dfrac{\alpha}{2}p(1 - v)$	$(1 - \alpha)p^2(1 - v)^2$
Normal 4 : 4 restored	$\dfrac{\alpha}{2}p$	$2(1 - \alpha)p^2 v(1 - v)$
Total	α	$1 - \alpha$

position in Fig. 84(c). They differ somewhat from the frequencies of aberrant asci (Fig. 84(a)) due to the inclusion of the estimated numbers of restored normal 4 : 4 segregations. The hybrid DNA frequencies are about 39% for the two left-hand groups of mutants and about 19% for the other five. Paquette and Rossignol inferred that there was an initiation region for recombination to the left of the gene. The estimated frequencies of asymmetric hybrid DNA are plotted in Fig. 84(d). A striking feature is that the left-hand group of mutants showed about 90% asymmetric, 10% symmetric, while the other mutants gave a lower proportion of asymmetric hybrid DNA, with indications of a progressive decline with distance through the gene: the right-hand group of mutants showed about 30% asymmetric, 70% symmetric. These results provide strong support for the Meselson–Radding model, hybrid DNA being predominantly non-reciprocal near the initiation site and becoming reciprocal more often with increasing distance from it. The hypotheses of Moore (1974) and Wagner and Radman (1975), discussed above, predict reciprocal hybrid DNA at the recombination origin, non-reciprocal further away, and so are contradicted by these results.

Paquette and Rossignol drew attention to the uniformity in hybrid DNA asymmetry of different crosses involving the same mutant shown by the left-hand group of mutants, and the variability between crosses shown by the other mutants (Fig. 84(d)). They suggested that the occurrence of symmetrical hybrid DNA was very sensitive to variations of the genetic background.

Rossignol and Paquette (1979) extended the study of the *b2* gene to A and B mutants. A total of four A and six B mutants was mapped, each mutant being mapped in relation to the others and to the C mutants investigated earlier. Aberrant asci were counted in crosses of each mutant with wild type. The total frequency of aberrant asci, when plotted against the map position of the mutant, showed a decline in frequency from left to right over the left-hand part of the gene, and a constant relatively low value in the right-hand part (Fig. 84(e)). This result was similar to that found with the C mutants (Fig. 84(a)) and strengthened the conclusion that recombination is initiated to the left of the gene.

The numbers of 5:3 and 3:5 asci found in crosses of the A and B mutants with wild type were negligibly small and, unlike the C mutants, there were no aberrant 4:4 asci. A and B mutants that mapped in the left-hand part of the gene showed appreciable conversion in each direction, while those that mapped in the right-hand part showed conversion almost entirely in one direction (to wild type for A mutants and to mutant for B mutants). Moreover, there was a gradient in the left-hand part of the gene, the disparity in conversion in the two directions being least at the left-hand end (Fig. 84(f)). In other words, arranging the mutants in order of increasing conversion in one direction, as in Fig. 72, placed them in map sequence. Rossignol and Paquette (1979) explained this result as a consequence of the gradient of increasing symmetric hybrid DNA from left to right, already inferred from the C mutants. The relationship expected for A and B mutants between conversion pattern and hybrid DNA symmetry is shown in Table 36, which corresponds to part of Table 35. Rossignol, Paquette and Nicolas (1979), in discussing these results, assumed that when hybrid DNA is assymetric either chromatid may act as donor equally often. (As already mentioned, this assumption is probably not generally true.) They pointed out that a B mutant which showed mismatch correction to mutant three times as often as to wild type (that is, $v = 0.25$ in Table 36) would then show this ratio of 2:6 to 6:2 asci if the hybrid DNA was entirely asymmetric, but the square of this ratio (2:6 nine times as frequent as 6:2) if it was entirely symmetric. They inferred that over the left-hand part of the *b2* gene, asymmetric hybrid DNA was progressively replaced by symmetric as one moved from left to right. This result strengthened the earlier inference of such a progressive change (Fig. 84(d)) and fitted the Meselson–Radding model perfectly.

In a further investigation of the *b2* gene, Rossignol and Haedens (1980) crossed a double mutant, 17 A4, at this locus with wild type and analysed aberrant asci. Both 17 and A4 are C mutants, 17 being near the left-hand end of the gene and A4 near the right-hand end (Fig. 84). They found that among asci showing aberrant 4:4 segregation at either site, the most frequent aberrant segregations at the other site were 5:3 and 3:5. Furthermore, in a sample of about 40 such asci, the aberrant 4:4 segregation was at A4 and the 5:3 or 3:5 at 17 about 10 times more often than the converse arrangement. Rossignol and Haedens (1980) concluded that the asymmetric hybrid DNA favoured at the left-hand end of the gene and the symmetric hybrid DNA favoured at the right-hand end were physically

Table 36. The relationship expected for A and B mutants between conversion pattern and hybrid DNA symmetry. For each alternative as regards hybrid DNA distribution, the frequency of the alternative segregation patterns is given in terms of v, the frequency of correction to wild type per mismatch correction.

			Segregation (+ : b) at mutant site		
Symmetry of hybrid DNA	Genotype of donor chromatid	Genotype before mismatch correction	6 : 2	Normal 4 : 4 restored	2 : 6
			+ + + b	+ + + b or b + b + b b	+ b b b
Asymmetric	Wild type	+ + +/b b	Correction to wild type v	Correction to mutant $1 - v$	—
Asymmetric	Mutant	+ +/b b b	—	Correction to wild type v	Correction to mutant $1 - v$
Symmetric	—	+ +/b +/b b	Correction to wild type in both chromatids v^2	Correction to wild type in one chromatid and to mutant in the other $2v(1 - v)$	Correction to mutant in both chromatids $(1 - v)^2$

associated in the same recombination event, as predicted by the Meselson–Radding model. Their results contradicted the idea that asymmetric and symmetric hybrid DNA occurred in different events.

Evidence that hybrid DNA is confined to one chromatid in the vicinity of an initiation site for recombination has also been obtained with *Sordaria brevicollis*. MacDonald and Whitehouse (1979) investigated a *buff* spore-colour mutant, YS17, which shows conversion to wild type with a much higher frequency than alleles (Plate 5). There is reason to believe that the YS17 mutation acts as a recognition site for an enzyme that initiates recombination (see Section 15). In crosses between YS17 and alleles, aberrant segregation at the allelic sites is increased in frequency compared with crosses lacking YS17, but no examples were found of aberrant 4:4 segregation nor of tetratype non-crossover 3:5 segregation (see Section 13) at the allelic site. The non-crossover postmeiotic segregations at the allelic site were all tritype and confined to the chromatid which showed conversion to wild type at YS17. In other words,

$$
\begin{array}{llll}
M & + & YS17 & N \\
M & + & YS17 & N \\
m & b & + & n \\
m & b & + & n
\end{array}
$$

gave rise to

$$
\begin{array}{llll}
M & + & YS17 & N \\
M & +/b & + & N \\
m & b & + & n \\
m & b & + & n
\end{array}
$$

but not to

$$
\begin{array}{llll}
M & + & YS17 & N \\
M & b & + & N \\
m & +/b & + & n \\
m & b & + & n
\end{array}
\quad \text{or} \quad
\begin{array}{ll}
+ & YS17 \\
+/b & + \\
+/b & + \\
b & +
\end{array}
\quad \text{irrespective of flanking marker genotype,}
$$

where b is a C class *buff* mutant and M/m and N/n are flanking markers. This is the result expected on the Meselson–Radding model with asymmetric hybrid DNA initially.

There are other aspects of this model that are in conflict with experimental data. These are discussed in Sections 15 and 16.

15. THE INITIATION OF RECOMBINATION

As mentioned in Section 12, Angel, Austin and Catcheside (1970) found that the $rec-2^+$ recombination repressor in *Neurospora crassa* reduced the frequency of recombination of mutants at the *histidine-3* locus in the right arm of linkage group I and reversed the polarity of this gene from distal (Fig. 85(a), ignoring the dotted lines, which are discussed later) to proximal (Fig. 85(c)). Angel *et al.* (1970) also discovered a genetic factor, mapping 1 unit distal to *his-3*, that affected the sensitivity of *his-3* mutants to the repressor. They called it a *recognition (cog)* factor, and found that when recombination was derepressed (*rec-2* homozygote) the dominant variant cog^+ permitted a higher frequency of recombination of *his-3* mutants than the recessive variant *cog* by a factor of about six to eight. Moreover, the polarity was affected, *cog* allowing weak proximal polarity (Fig. 85(b)). It was evident that cog^+ was a dominant promoter of recombination. When recombination was repressed ($rec-2^+$) the *cog* alleles had no effect. Angel *et al.* (1970) concluded that *cog* was a recognition site for an enzyme that initiates recombination and that cog^+ was recognized more readily than the recessive variant. They postulated that the repressor, the product of $rec-2^+$, acted on a control site situated either (1) alongside the structural gene for a nuclease postulated to initiate recombination at *cog* or (2) near *his-3*. The repressor would then block either

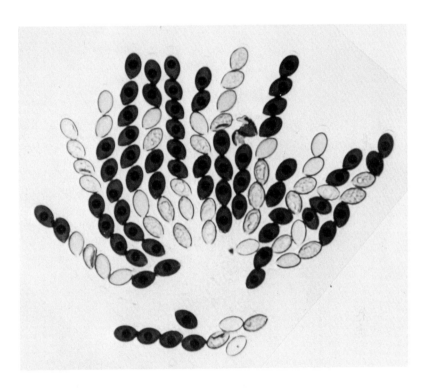

Plate 5. Cluster of asci of *Sordaria brevicollis* from a cross between *buff* spore-colour mutant YS17 and wild type. The asci with six black spores result from conversion of YS17 to wild type.

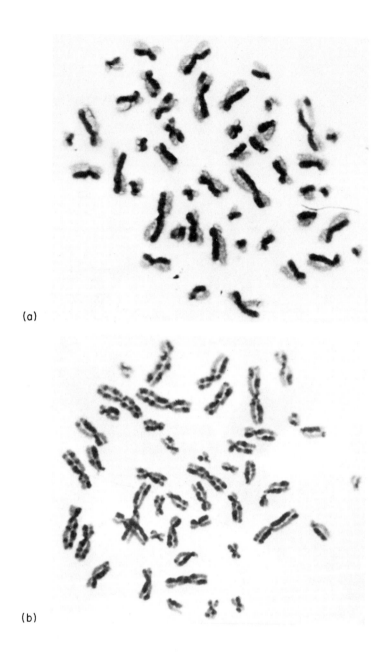

(a)

(b)

Plate 6.   Mitotic metaphase in human lymphocytes grown in culture and treated with Hoechst fluorochrome and Giemsa stain, following incorporation of 5-bromodeoxyuridine into both polynucleotides of one chromatid and one polynucleotide of the other. The latter chromatid stains more deeply. (a) Normal cell, with nine sister-chromatid exchanges. (b) Cell from a patient with Bloom's syndrome, showing about 100 sister-chromatid exchanges. (Courtesy of J. German and the National Academy of Sciences of the USA, from *Proceedings of the National Academy of Sciences*, **71**, 4511 (1974).)

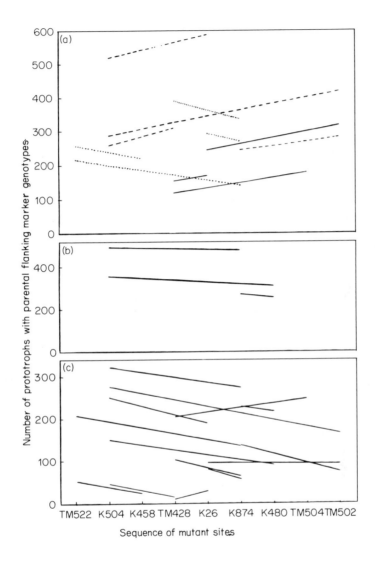

Fig. 85. Data of Angel, Austin and Catcheside (1970) for 5498 prototrophs with parental flanking marker genotypes from pairwise crosses of *his-3* mutants of *Neurospora crassa* with various genotypes for *rec-2* and *cog*: (a) *rec-2* homozygote, *cog*⁺; (b) *rec-2* homozygote, *cog*; (c) *rec-2*⁺. In (a), unbroken lines refer to crosses homozygous for *cog*⁺; dashed lines (– – –) to crosses in which the right-hand (distal) mutant was *cog*⁺ and the left-hand (proximal) was of *cog* genotype; and dotted lines (· · · · · ·) to crosses in which the proximal mutant was *cog*⁺ and the distal was *cog*. For an explanation of the lines, see the caption to Fig. 75.

(1) the synthesis or (2) the action of the nuclease, respectively. They did not favour the idea that the repressor acted directly on *cog* because of the lack of effect of the *cog* difference in a *rec-2*$^+$ background.

A distinction between these hypotheses became possible in a rather unexpected way. Catcheside (1975) discovered a third allele at the *rec-3* locus, which he called *rec-3*^{L} as it occurred in a strain isolated from the wild by Lindegren. As pointed out in Section 12 (see Table 33), *rec-3*$^+$ represses recombination in *am-1* and *his-2*. Catcheside found that *rec-3*^{L} affected these genes differently. The relative yields of prototrophs with *rec-3*$^+$, *rec-3*^{L} and *rec-3* were 1, 8 and 25, respectively, for *am-1* and 6, 6 and 40, respectively, for *his-2*. In other words, *rec-3*^{L} was similar to *rec-3*$^+$ in repressing recombination in *his-2*, but less efficient than *rec-3*$^+$ in *am-1*. He concluded that the repressor blocks the action rather than the synthesis of the nuclease, the control sites near *his-2* and *am-1*, to which the repressor was presumed to bind, differing slightly in nucleotide sequence. Evidently the recombination operators, to which the repressor binds, are in the chromosome regions where recombination is affected and not at the structural gene for a nuclease. Applied to *rec-2*, the implication of this result was that there was a recombination operator near *his-3* to which the *rec-2*$^+$ repressor bound, thereby blocking the action of the nuclease at *cog*. A further implication was that a single nuclease might be responsible for all the initiation of recombination in *Neurospora*.

In their investigation of the *cog* recombination promoter, Angel et al. (1970) discovered an interesting peculiarity of crosses heterozygous for *cog*$^+$ and *cog*. If the *his-3* mutant associated with *cog*$^+$ was the distal member of the pair, the polarity was distal (broken lines in Fig. 85(a)) and in those crosses in which the *his-3* mutant associated with *cog*$^+$ was proximal, the polarity was proximal (dotted lines in Fig. 85(a)). In other words, *cog*$^+$ promoted conversion in the *his-3* gene in the chromatid in which it occurred, irrespective of the position of the *his-3* mutant involved. If *cog*$^+$ is favoured over *cog* for recognition by a single-strand endonuclease, the *cog*$^+$ chromatid would be expected, on the Meselson–Radding model, to become the donor of a broken strand to the *cog* chromatid (cf. Fig. 82(a), where the molecule shown with thin lines would be *cog*$^+$ and that with thick lines would be *cog*). The model then predicts that, in the asymmetric phase, hybrid DNA and hence conversion would be confined to the *cog* chromatid. This is the opposite of what is observed. It seems as if the molecule recognized initially somehow becomes the recipient of a strand from the other parent rather than the donor of a strand to it.

Further discoveries relating to *cog* were made by Catcheside and Angel (1974) with a *his-3* mutant involving an interchange. In this mutant, TM429, the distal quarter of *his-3* together with the *cog* site are transferred to linkage group VII. The results of crosses between TM429 and alleles in various *cog* and *rec-2* backgrounds are shown in Table 37. When TM429 was crossed with distal alleles in a *cog*$^+$ homozygote (top row of table), a normal result was obtained, that is, *rec-2* derepression raised the level of *his-3* recombination 20–30-fold. If the structurally normal chromosome was *cog*, however, the response to derepression was less and the fraction of the prototrophs with recombinant flanking markers was increased

Table 37. Results obtained by Catcheside and Angel (1974) from crosses between the *his-3* structural mutant TM429 and other *his-3* mutants in *Neurospora crassa*. In TM429 the distal quarter of *his-3* and the *cog* site are transferred from linkage group I to VII.

Position of *his-3* mutant crossed with TM429	Genotype at *cog* site in structurally normal chromosome	Meiotic configuration prior to recombination	Frequency of histidine prototrophs in *rec-2* homozygote compared with *rec-2+*	Percentage frequency of various flanking marker genotypes in histidine prototrophs in *rec-2* homozygote			
				P1	P2	R1	R2
Distal to TM429	*cog+*		×20–30	5	45	36	14
	cog		×10–12	8	23	59	10
Proximal to TM429	*cog+*		×20–30	56	7	35	2
	cog		×1	34	1	64	1

(second row of table). To account for these results, Catcheside and Angel suggested that cog^+ is very strongly competitive over cog as the site for initiating recombination. A lower yield of prototrophs, and a higher proportion associated with crossing over, would then be expected when cog^+ was confined to the interchange chromosome compared with the cog^+ homozygote.

Surprising results were obtained when TM429 was crossed with proximal alleles: in the cog^+ homozygote, derepression raised the level of recombination to the normal extent, despite the site of the structural change lying between cog and the region where recombination was being promoted (third row of Table 37); and, contrariwise, when the structurally normal chromosome had the cog genotype, there was no promotion of recombination when derepression occurred (fourth row of table). If the event that occurred at cog^+ when recombination was initiated was the formation of a heteroduplex, this clearly could not pass the site of the interchange. To account for recombination events initiated at cog^+ not being blocked by the chromosomal discontinuity, Catcheside and Angel (1974) suggested that an endonuclease nicked one strand of the duplex at cog^+ and that nick migration then took place, that is, the site of the nick moved along the chromatid as a result of exonucleolytic breakdown ahead of it and DNA synthesis behind it, using the other strand of the duplex as template. If the initial single-strand break was at cog^+ in the structurally normal chromosome and if the nick migration was in the direction of his-3, once the nick had passed the site of the TM429 discontinuity in the other chromosome, heteroduplex formation could take place and lead to recombination of the his-3 mutants. The fact that this promotion of recombination did not occur when the structurally normal chromosome was of cog genotype confirmed that cog^+ was strongly competitive over cog as an initiation site, and showed that initiation was not taking place elsewhere in the his-3 region when rec-2 derepression occurred.

An alternative to nick migration as an explanation of the TM429 data is migration of the endonuclease without nicking (Markham and Whitehouse, 1982). This has the advantage that it will also accommodate the observation that cog^+, though favoured over cog for recognition, promotes conversion in its own chromatid, as though it was the recipient of a strand from the cog chromatid, not a donor to it. If the action of the endonuclease that recognized cog^+ was to nick the homologous molecule, that is, of cog genotype, the latter would become the donor of a strand to cog^+, as the data seem to require. If this action was delayed until the endonuclease had migrated from cog^+ into his-3 and had reached a point beyond the site of TM429, the failure of this mutant to block derepression of recombination with proximal alleles would be accounted for.

Enzyme migration along the DNA from a recognition site, before heteroduplex formation is initiated, offers an explanation for several features of eukaryote recombination. First, it could account for the occurrence of crossing over between the sites of alleles without either site apparently being included in hybrid DNA. Such crossing over is frequent in some genes (see Section 4). On the other hand, to account for the data on polarity in recombination, it seems necessary to postulate initiation of recombination at fixed sites that appear to be outside the genes.

Enzyme migration offers a solution to this paradox, by allowing initiation outside a gene to become manifest only within it. Moreover, there is evidence for extragenic initiation of intragenic crossing over. Lissouba *et al.* (1962) found that mutant 277 in spore-colour gene *46* of *Ascobolus immersus* showed crossing over in the interval between its site and that of mutant 63 to its left, provided that mutant 138, which mapped to the right of 277, was not also present. For crossing over to occur to the left of 277 it seemed to be necessary for 277 to be the right-hand member of those crossed. It is necessary for 63 and 138 to be right-hand members in order to show conversion, and for 277 to be a right-hand member in order to show postmeiotic segregation. The inference was that recombination has to be initiated at the right-hand end of this gene, not only to account for this polarity in conversion, as discussed in Section 12, but also to account for crossing over to the left of mutant 277 (see Whitehouse, 1967).

A second feature of eukaryote recombination that may be illuminated by the idea of enzyme migration before heteroduplex formation is the mutant specificity of crossing over between alleles. Such specificity was referred to in Section 4. As there is no hybrid DNA at either mutant site with such crossing over, it was not obvious how the mutant specificity came about. If a mutant had a molecular structure that interfered with heteroduplex formation, selection for recombinants might imply selection for meiotic cells in which enzyme migration moved heteroduplex initiation beyond the site of the mutant, as suggested above for mutant TM429 at *his-3* in *Neurospora*. Whether crossing over took place between the sites of the alleles would no doubt also depend on their distance apart (see Section 4). There is evidence from the work of Rizet and associates with spore-colour mutants of *Ascobolus immersus* that mutants which favour crossing over with an allele may be deletions (see Whitehouse, 1967). Such multisite mutants are just those that might be expected to interfere with heteroduplex formation.

Thirdly, enzyme migration before hybrid DNA is initiated allows recombination of distal origin to bring about conversion at a proximal site, and vice versa. This is illustrated by the dotted lines in Fig. 85(a) and by the proximal polarity (P1 in excess of P2) shown in the third row of Table 37, despite the evident distal origin of these events. These results may make a re-appraisal necessary of some data on polarity reversal.

Emerson and Yu-Sun (1967) discovered that two closely linked white-spored mutants, w-10 and w-78, of the Pasadena strain of *Ascobolus immersus* showed different frequencies and patterns of aberrant asci depending on to which of two wild strains, K and P, they were crossed. Lamb and Helmi (1978) investigated this situation further and identified control factors, K and P, responsible for the variation in behaviour. These factors were closely linked to the *w-10* locus, occasionally showing co-conversion with the spore-colour mutants. Lamb and Helmi detected a third control factor, 91, and isolated two more spore-colour mutants mapping at the *w-10* locus. The three control factors behaved as alleles. From their results, Lamb and Helmi (1978) concluded that the *K/P/91* site may be an initiation point for recombination. They suggested that if the initial step in recombination were the asymmetric invasion of one chromatid by a nucleotide chain from the other, the variations which they observed in conversion pattern of mutants at the *w-10* locus with different *K/P/91* genotypes might be an indirect consequence of variations in the invasion frequencies of each nucleotide chain of each chromatid. If this were true, it would be unnecessary to postulate that the control factors also affected mismatch repair. The *K/P/91* site in *Ascobolus* may therefore act in the same way as proposed for the *cog* site in *Neurospora*, an endonuclease recognizing different nucleotide sequences at the site with different efficiencies.

The M26 mutant in the *ade6* gene of *Schizosaccharomyces pombe*, investigated by Gutz (1971), was referred to in Sections 9 and 10. This mutant gave wild-type recombinants when crossed with distant alleles with a frequency approximately 13 times that shown by a closely linked allele, M375, when crossed with them. No other *ade6* mutant among the 394 available showed behaviour similar to M26. Tetrad analysis revealed that M26 showed conversion with a frequency of about 5% in crosses with wild type, conversion to wild type being about 12 times as frequent as to mutant. Three other *ade6* mutants tested gave aberrant ascus frequencies of 0.3–0.9%, with no significant excess of conversion in one direction over the other. Of 19 alleles mapped in relation to M26 on the basis of recombination frequencies, four were to its left and 15 to its right. The co-conversion shown by M26 with alleles and, through this, its effect on the conversion of the alleles, have already been mentioned (Sections 9 and 10). It was evident that the frequency of conversion of closely linked alleles was greatly increased in the presence of M26, and its predominant direction was determined by M26, being to mutant in *trans* and to wild type in *cis* crosses (Table 24). Gutz (1971) suggested that the DNA at the M26 site preferentially undergoes single-strand breakage, thereby initiating recombination at this site.

Goldman and Smallets (1979) investigated the effect of M26 when homozygous on conversion of the allele M210. They analysed 1005 unselected asci and found eight 3:1, five 1:3 and one 5:3 segregation at the site of M210. The frequency of aberrant segregation (1.4%) was double that observed for M210 when M26 was heterozygous, and the direction of conversion lacked the bias imparted to alleles by M26 as a result of co-conversion in the heterozygote (Table 24). It was evident that in the homozygote M26 acted independently as an initiation site in chromatids of each parentage to give an additive result, like *cis* and *trans* crosses superimposed. Goldman and Smallets (1979) concluded that the initiation of recombination was triggered by the M26 nucleotide sequence itself, and not by an interaction between this sequence and that of wild type.

The YS17 mutation in the *buff* spore-colour gene of *Sordaria brevicollis* investigated by MacDonald and Whitehouse (1979) was referred to in Section 14. It behaved like M26 in *Schizosaccharomyces pombe*. YS17 showed conversion with a frequency of about 10%, (Plate 5) whereas 28 other *buff* mutants gave frequencies of 0.5% or less; over 98% of the conversion was to wild type; alleles showed a much increased frequency of aberrant segregation in its presence; and the aberrant segregation involved the same chromatid at each site (see Section 14). In all these respects it resembled M26. YS17 mapped near the distal end of *buff*, 25 alleles mapping proximal to it and three distal to it, and, as expected from its high conversion frequency, YS17 had an over-riding effect on the polarity in the gene. All these observations are in agreement with the hypothesis that YS17 acts as an initiation site for recombination.

If M26 in *ade6* of *Schizosaccharomyces pombe* and YS17 in *buff* of *Sordaria brevicollis* are mutations that act as recombination initiators, as the evidence suggests, they must act as a recipient of a nucleotide chain from the other parent, not a donor to it. This is because they both show high frequency conversion to wild type rather than to mutant. The same conclusion was reached with the *cog* site in *Neurospora* from evidence of a different kind.

To sum up, the evidence about the initiation of recombination in eukaryotes points to a single-strand break in one chromatid, leading initially to a heteroduplex in one chromatid only, as predicted by the Meselson–Radding model. The molecule recognized initially seems to be become the recipient of a strand from the other parent, as expected if the nuclease nicks the homologous DNA molecule and not that which it recognized. Migration of the nuclease along the DNA from the recognition site, prior to the initiation of heteroduplex formation, offers an explanation of several features of the data.

16. CROSSOVER FREQUENCY AND POSITION IN RELATION TO ABERRANT SEGREGATION

The association between aberrant segregation and crossing over was established from crosses between mutants and wild type with flanking markers present (Section 5). The results of genetic analysis of asci with wild-type recombinants from crosses of alleles in the presence of flanking markers are available for four genes (Table 38). The data reveal that aberrant segregation of proximal and distal mutants may be associated with different frequencies of crossing over. This was first demonstrated by Fogel and Hurst (1967) with the *his1* locus in *Saccharomyces cerevisiae* and was found by Marcou (1969) with spore-colour gene *14* in *Podospora anserina*. It was also evident in data for random prototrophs obtained by various authors and shown graphically by Whitehouse and Hastings (1965). For example, in the *pan-2* gene of *Neurospora crassa* investigated by Case and Giles (1959), prototrophs obtained from crossing alleles towards the proximal end of the gene showed about 40% crossing over, while at the distal end the corresponding figure was about 75%. With random prototrophs, one cannot exclude those resulting from crossing over between the sites of the alleles, but the *pan-2* data show a gradual change in the crossover frequency from the proximal to the distal end of the gene, with no relation to the number of intervening mutant sites between the sites of those crossed: crossing over between the mutant sites would be expected to be more frequent with distant sites. It is likely that many of the prototrophs, both with *his1* and *pan-2*, result from hybrid DNA reaching one of the mutant sites but not the other. The inference would then be that hybrid DNA of proximal origin was associated with a different frequency of crossing over from hybrid DNA entering the same gene from the other end.

Table 38. Numbers of asci of various flanking marker genotypes obtained when there was aberrant segregation at a mutant site and normal 4:4 segregation (2:2 for yeast) at an allelic site. The non-crossover asci include those with an incidental crossover. In the data of Kitani and associates for *Sordaria fimicola*, incompletely analysed asci have been distributed among the possible genotypes in proportion to their frequency in the fully scored asci.

Species and gene	Reference	Type of aberrant segregation (C, conversion to wild-type; P, postmeiotic segregation)	Aberrant segregation at proximal site				Aberrant segregation at distal site			
			Non-crossover† (a)	Proximal or median crossover† (b)	Distal crossover† (c)	Percentage crossover (uncorrected in brackets)	Non-crossover† (a)	Proximal crossover† (c)	Median or distal crossover† (b)	Percentage crossover (uncorrected in brackets)
Saccharomyces cerevisiae, his1	Fogel and Hurst (1967)	C	474 + 42	270 − 10	103 − 32	(44) 39	49 + 3	7	77 − 3	(63) 61
	Savage and Hastings (1981)	C	530 + 17	285 − 5	115 − 12	(43) 41	66 + 3	10 − 1	87 − 2	(60) 58
Podospora anserina, spore colour gene 14	Marcou (1969)	C	40 + 3	24	3 − 3	(40) 36	78 + 5	7 − 5	12	(20) 14
Sordaria fimicola, grey	Kitani and Olive (1969)	P	251	137	31	40	146	24	81	42
	Kitani and Whitehouse (1974)	C	72	60	10	49	34	5	13	35
Sordaria brevicollis, buff	MacDonald and Whitehouse (1979)	C‡	83 + 3	14 − 2	8 − 1	(21) 18	105 + 6	20 − 1	37 − 5	(35) 31
	Sang and Whitehouse (1982)	P	44 + 3	7 − 2	10 − 1	(28) 23	120 + 8	9 − 7	21 − 1	(20) 15
	Theivendirarajah and Whitehouse (1982)	C	84 + 7	30 − 4	26 − 3	(40) 35	94 + 7	7 − 5	55 − 2	(40) 35

†The numbers after a plus or minus sign are the corrections made on account of incidental crossovers. These numbers are based on the number of non-crossover asci with an incidental crossover on (a) either side, (b) the proximal side and (c) the distal side of the gene, but with allowance for the contrary effect of incidental crossovers in conjunction with associated ones (see section 5). In the *Sordaria fimicola* data these two corrections cancel out.

‡These crosses involved mutant YS17, which was responsible for all the conversion observed.

A further complication is shown by the *buff* gene in *Sordaria brevicollis*, where Sang and Whitehouse (1982) found that postmeiotic segregation at a distal site was associated with less crossing over than conversion at this site (Table 38). This difference was less evident (and not significant with the existing data) at proximal sites. This discovery agrees with the observation by Bond (1973) that random wild-type recombinants from pairwise crosses of *buff* mutants are associated with about 28% crossing over if they occur singly and about 45% if they occur in pairs, and that the single wild types have more P2 flanking marker genotypes (implying an event at the distal site), relative to P1, than the paired wild types.

The frequency of crossing over associated with aberrant segregation shows great diversity, ranging from about 15% for random prototrophs from crosses of alleles mapping near the proximal end of *met-7* in *Neurospora* (Murray, 1969) to about 75% at the distal end of *pan-2* (see above, though crossing over between the sites of alleles may contribute to this figure). About 15% crossing over has also been found for conversion at a distal site in gene *14* in *Podospora* (Marcou, 1969: see Table 38) and for postmeiotic segregation under some circumstances at *buff* in *Sordaria brevicollis* (Tables 19 and 38). Most investigations with *Neurospora* have shown significantly less than 50% crossing over: see graphs in review by Whitehouse and Hastings (1965). Ascus analyses with both species of *Sordaria* have also given values below 50% (see Tables 18, 19 and 38). For *Saccharomyces cerevisiae* a figure of 50% is often quoted, but does not seem to be justified by the evidence. The data of Fogel and Hurst (1967) for *his1* indicate an overall frequency of 42% crossing over in association with aberrant segregation and the results obtained by Savage and Hastings (1981) gave a similar figure (Table 38). From genetic analysis of unselected asci from crosses of a *Sup6* mutant and wild type, DiCaprio and Hastings (1976) found that conversion at *Sup6* showed 37% associated crossing over, or 43% if co-conversions with *cdc14* were included. A random spore analysis by prototroph selection from numerous crosses of *Sup6* mutants gave an average of 39% flanking marker recombination. All these frequencies for yeast were based on large samples and differed significantly from 50%.

Asci showing aberrant segregation at one site and normal 4:4 segregation at an allelic site provide information about the position of crossovers associated with the aberrant segregation, as well as their frequency: see Table 38. The information is not as detailed as one would wish because conversion means the loss of the mutant as a genetic marker. It is for this reason that two of the three crossover positions are grouped together in the table. More precise localization of crossing over is possible, however, with postmeiotic segregation, provided the hybrid DNA is asymmetric. This is illustrated by the data of Kitani and associates for postmeiotic segregation at one site and normal 4:4 segregation at the other site in crosses between *grey* mutants of *Sordaria fimicola*. Some of the asci in this investigation were not fully scored owing to incomplete spore germination. These asci have been allocated to each alternative genotype in accordance with the numbers of known genotype. The data reveal that hybrid DNA is largely asymmetric when the aberrant segregation is at the distal site of the pair (Table 39(iii)),

Table 39. Data of Kitani and Olive (1969) and Kitani and Whitehouse (1974) from crosses between *grey* spore-colour mutants in *Sordaria fimicola* in the presence of flanking markers. The numbers are given of asci with postmeiotic segregation at one mutant site and normal 4:4 segregation at the other. The 3:5 segregations could be recognized (except where the ascus phenotype was the same as a non-recombinant ascus) because the spores of the mutants crossed differed in colour. Incompletely analysed asci have been distributed among the alternative genotypes open to them in proportion to the frequency of those genotypes in that cross.

Mutant site showing postmeiotic segregation	Mutants crossed	Class of postmeiotic segregation	Non-crossover			Crossover		
			Genotype	Number observed	Tetratype frequency	Proximal	Median	Distal
(i) Proximal	1—2—5—3 6 4 4b	Aberrant 4:4	—	4	0%	0 (prox.+med. brace)		0
	1—5 6	5:3	Tritype	33		12	1	8
			Tetratype	0		(b)†	(a)†	0
	2—5	3:5	Tritype	—		—	0	—
			Tetratype	0		(b)†	(a)†	1

(ii) Proximal	1—3
	   4
	   4b
		Aberrant 4:4	—	43		34 (prox.+med. brace)		3

	5—4 5—4b	5:3	Tritype	99	42%	58	32	13
			Tetratype	72		(b)†	(a)†	7
		3:5	Tritype	—	(60%)‡	—	7	—
			Tetratype	7		(b)†	(a)†	0
(iii) Distal	As (i) and (ii)	Aberrant 4:4	—	11		2	11 (med.+dist. brace)	
		5:3	Tritype	92	2%	22	2 (c)†	49 (b)†
			Tetratype	2		0	(c)†	(b)†
		3:5	Tritype	41	0%	0	5 (c)†	14 (b)†
			Tetratype	0		0	(c)†	(b)†

† Same genotype as corresponding tritype with (a) proximal, (b) median, (c) distal crossover.

‡ A cross of mutants 1 and 4 in *cis* configuration gave 2 tritype and 3 tetratype.

but predominantly symmetric when at the proximal site, provided that the distal mutant is towards the right-hand (distal) end of the gene (Table 39(ii)). With the distal mutant further to the left, the hybrid DNA at the proximal site is mostly asymmetric (Table 39(i)). The degree of symmetry is shown by the frequencies of aberrant 4:4 segregation and of tetratype non-crossover 5:3 and 3:5 segregation. The ambiguity in scoring crossover position if the hybrid DNA is symmetric is indicated in a footnote to the table.

The data on crossover position contradict the Meselson–Radding model, which predicts that the crossover will be at the far end of the hybrid DNA, that is, the opposite end from the initiation site (Fig. 82). It is likely that hybrid DNA will end between the mutant sites in a majority of the asci with aberrant segregation at one site and normal 4:4 segregation at the other. In other words, if the parental genotype was

$$\frac{M\ 1 + N}{m + 2\ n},$$

where 1 and 2 are the alleles and M/m and N/n are the flanking markers, the crossover ascus genotypes expected with asymmetric hybrid DNA generating a 5:3 or 3:5 segregation will be

$$(1)\ \begin{array}{ccc} M & 1 & +N \\ M & 1/+ & 2\ n \\ m & + & +N \\ m & + & 2\ n \end{array} \quad \text{and} \quad (2)\ \begin{array}{ccc} M & 1 & +N \\ M & 1 & 2\ n \\ m & 1/+ & +N \\ m & + & 2\ n \end{array},$$

or

$$(3)\ \begin{array}{ccc} M\ 1 & + & N \\ M\ 1 & +/2 & n \\ m\ + & + & N \\ m\ + & 2 & n \end{array} \quad \text{and} \quad (4)\ \begin{array}{ccc} M\ 1 & 2 & n \\ M\ 1 & 2 & n \\ m\ + & +/2 & N \\ m\ + & 2 & n \end{array},$$

depending whether entry of hybrid DNA is proximal ((1) and (2)) or distal ((3) and (4)). These correspond to the crossover position described as median in Table 39. Asci of these genotypes, however, were found to be rare when the hybrid DNA was asymmetric (Table 39(i) and (iii)). Instead, a majority of the asci had the crossover near the supposed initiation site, that is, on the proximal side of the gene when the aberrant segregation was at the proximal site, and on the distal side with distal aberrant segregation. In other words, the genotypes favoured were

$$\begin{array}{ccc} M & 1 & +N \\ M & + & 2\ n \\ m & 1/+ & +N \\ m & + & 2\ n \end{array} \quad \text{and} \quad \begin{array}{ccc} M\ 1 & + & N \\ M\ 1 & + & n \\ m\ + & +/2 & N \\ m\ + & 2 & n \end{array}$$

for 5:3 segregation at the proximal and distal sites, respectively. As can be seen from the table, these were about 20 times as frequent as the genotypes (1) and (3) above predicted by the Meselson–Radding model. Comparable results have been obtained for the *buff* gene in *Sordaria brevicollis*: no examples have been found of the genotypes (1)–(4) above that would be expected from asymmetric hybrid DNA reaching one mutant site and terminating in a crossover, in a sample of about 100 crossover asci (Sang and Whitehouse, 1982; Theivendirarajah and Whitehouse, 1982).

Further evidence for the rarity of median crossing over is provided by the rarity or absence of reciprocal recombination of alleles in many crosses of fungal mutants, particularly if their sites are near together. Mismatch correction in the appropriate direction in any of genotypes (1)–(4) above, expected with asymmetric hybrid DNA and a median crossover, will give rise to reciprocal recombination of alleles.

There is another unexpected crossover position, evident in both the postmeiotic segregation and the conversion data in Tables 38 and 39. A substantial fraction of the crossovers are separated from the mutant site showing aberrant segregation by the site showing normal 4:4 segregation. Thus, in *his1* in *Saccharomyces cerevisiae* about a quarter of the crossovers associated with conversion at a proximal site are situated distal to the other mutant site (Table 38) and in the *grey* locus in *Sordaria fimicola* 30% of the crossovers associated with a 5:3 segregation at a distal site occur on the proximal side of the gene (Table 39(iii)). These crossovers are not incidental to the aberrant segregation as they involve the same chromatids and, moreover, undetected incidental crossovers have been allowed for (see footnote to Table 38).

One explanation for these non-adjacent crossovers is that the hybrid DNA extended beyond the second mutant site but the mismatch was corrected back to give the normal 4:4 segregation that was observed. This explanation is favoured by Savage and Hastings (1981) with *his1* in yeast. Quantitative evidence in support of normal 4:4 segregation being restored at one of the sites in two-point crosses was obtained by Rossignol and Haedens (1978) with closely linked mutants at the *b2* spore-colour locus in *Ascobolus immersus*. Another explanation for the non-adjacent crossovers is that the aberrant segregation and the crossover are discrete, though associated, events. The normal 4:4 segregation at the second site would then be primary and not the result of restoration after hybrid DNA formation.

Information about the frequency of occurrence of hybrid DNA at the second site is possible with mutants that show postmeiotic segregation, if the hybrid DNA is confined to one chromatid. If the parental genotype was

$$\frac{M\ 1 + N}{m + 2\ n},$$

using the same notation as above, hybrid DNA in the *MN* chromatid extending to both sites, followed by mismatch correction to wild type at one of them, will give rise to either (1) $M + +/2\ N$ or (2) $M\ 1/+ + N$. If both mutants show conversion to wild type equally often when in hybrid DNA, genotypes (1) and (2) will be equally frequent. A similar conclusion applies to the genotypes with flanking marker crossing over, and to those with hybrid DNA in the *mn* chromatid, where the corresponding non-crossover genotypes are (3) $m + +/2 n$ and (4) $m 1/+ + n$. Hybrid DNA confined to one site, however, can give rise to genotypes (2) and (3) only.

Applying this analysis to data for the *buff* gene in *Sordaria brevicollis*, it appears that hybrid DNA extends to the second site in approximately 10% of the recombinant asci (Sang and Whitehouse, 1982; Theivendirarajah and Whitehouse, 1982). The frequency of non-adjacent crossovers, however, is approximately 50% (Table 38). A similar discrepancy was found by MacDonald and Whitehouse (1979) with the YS17 *buff* mutant. Thus, there seems to be a real possibility that two associated events may be responsible for some of the non-adjacent crossovers.

The idea of conversion and a separately initiated crossover close by was proposed by Fincham (1974) to account for his results with the *amination-1* (*am-1*) gene in *Neurospora crassa*. In pairwise crosses of *am-1* mutants in the presence of flanking markers, one of the non-crossover flanking marker genotypes (P1) was more frequent among random prototrophs than the other (P2) in almost every cross. This could be accounted for by initiation of recombination predominantly on the proximal side of the gene. The two recombinant flanking marker genotypes (R1 and R2), however, among the prototrophs were almost equal in frequency. The occurrence of P1 > P2 but R1 = R2 raised the question of how the R2 genotype ($M + + n$) arose. As pointed out in Section 5, the occurrence of R2 wild types is not predicted with the simplest assumptions about the recombination mechanism (see for example, Table 21). This genotype has usually been attributed to independent correction to wild type at the two allelic mutant sites, on the assumption that the hybrid DNA extended to both sites (Fig. 86(i)). If R1 and R2 are equally frequent, the presumption would be that the heteroduplex extended to both mutant sites in every meiotic cell that gave rise to prototrophs, the R1 genotypes ($m + + N$) arising in the same way as R2 but with donor and recipient molecules reversed. But if hybrid DNA regularly covers both sites the frequency of P1 prototrophs is expected to equal P2, as these would arise in a similar way. Thus, P1 ($M + + N$) would be generated by the events shown in Fig. 86(i) omitting the isomerization, and P2 ($m + + n$) similarly with donor and recipient interchanged.

In order to account for equality of R1 and R2 but inequality of P1 and P2, Fincham (1974) suggested that P1 and P2 arose, predominantly at least, from hybrid DNA extending to one allelic site only, but that crossovers were often initiated separately close by. If this initiation occurred on the opposite side of the gene from the event that gave rise to conversion, an R2 genotype would result (Fig. 86(ii)). If the crossover was on the same side of the gene as the conversion, an R1 genotype would be produced. Equality of R1 and R2 would then be attributed to an equal chance of the crossover event occurring to the left or to the right of the conversion event. This hypothesis will thus account for the other

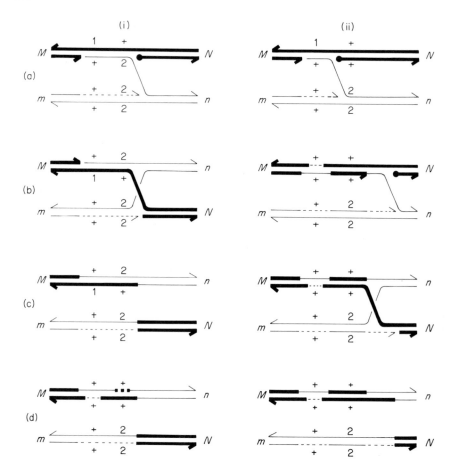

Fig. 86.   Hypotheses based on that of Meselson and Radding (1975) to explain the origin of prototrophs with the R2 flanking marker genotype ($M + + n$) in a cross between alleles 1 and 2 in the presence of flanking markers M/m and N/n. The parental genotype was

$$\frac{M\ 1 + N}{m + 2\ n}.$$

Line thicknesses indicate parentage.

(i) *Independent mismatch correction.* (a) A strand from the *mn* chromatid has entered the *MN* molecule to give a heteroduplex that extends across both mutant sites. Repair synthesis has occurred in *mn* (broken line) and exonucleolytic breakdown in *MN* (●). (b) The polymerase-nuclease has left its substrate and isomerization has taken place. (c) The crossed strands have been cut and nicks have been sealed. (d) Independent mismatch correction to wild type has taken place at the mutant sites. This correction might occur at any time after (a).

(ii) *Separately initiated crossover.* (a) A strand from the *mn* chromatid has given a heteroduplex in *MN* at the proximal site only. (b) The crossing strand has been cut, mismatch correction to wild type has taken place and nicks have been sealed. A second heteroduplex region has been formed on the distal side of the gene. (c) Isomerization has taken place. (d) The crossed strands have been cut and nicks have been sealed.

discrepancy between observation and the predictions of the Meselson–Radding model, namely, that many of the crossovers seem to occur at or near the initiation site of the event that gave rise to aberrant segregation.

The idea that crossovers may be separately initiated near the site of aberrant segregation is discussed further in Section 19.

17. RECOMBINATION ENZYMES

Extensive biochemical studies have been made by Herbert Stern and associates with *Lilium* pollen mother cells cultured *in vitro*. Hotta, Ito and Stern (1966) found that about 0.3% of the total DNA of the nucleus of *L. longiflorum* was synthesized at zygotene and pachytene. The pachytene synthesis, unlike that at zygotene, was found by Hotta and Stern (1971a) not to be inhibited by hydroxyurea. They inferred that it was of the nature of repair synthesis. This would be in keeping with the DNA synthesis postulated during the asymmetric phase of the Meselson–Radding model, and also required for mismatch repair.

The reannealing kinetics of the DNA synthesized at pachytene were investigated in *L. henryi* by Smyth and Stern (1973). The reannealing of denatured molecules was measured by chromatography on hydroxyapatite. They found that the pachytene-labelled DNA reannealed faster than the total DNA. From the rate, they concluded that the sequences traced must be repeated 2400–3000 times in the nucleus. Very highly repeated sequences, such as occur in centromeric heterochromatin in many organisms, were not labelled at pachytene, and surprisingly, unique sequences were not labelled appreciably. Smyth and Stern (1973) estimated that not more than 9% of the label was in unique sequences. They concluded that the pachytene repair was localized to specific regions with moderate repeats of nucleotide sequence. They suggested that the initiation sites for recombination outside the genes were in moderately repeated DNA, the hybrid DNA extending from these sites into the unique sequences of the genes. Nevertheless, it is surprising that so much of the pachytene repair synthesis should be in repeated sequences.

Howell and Stern (1971) investigated the enzyme activities peculiar to pachytene. The cultured pollen mother cells develop synchronously and provided excellent materal for this purpose. *Lilium* cultivars Cinnabar and Bright Star, hybrids of *L. speciosum* and *L. henryi*, were used. An endonuclease with a sharp pH optimum at 5.2 was partially purified from pachytene cells. It was found to cause single-strand breaks in duplex DNA, giving 3′-phosphoryl and 5′-hydroxyl end-groups at nicks. The endonuclease activity was zero at leptotene, rose during zygotene to reach a peak in early pachytene and then fell to zero again by the end of pachytene. Two DNA repair activities were also found to be present at meiotic prophase and to have maximum activity in late zygotene and early pachytene. These were a polynucleotide kinase that can phosphorylate 5′-hydroxyl end-groups produced by the endonuclease, and a polynucleotide ligase that can join the resulting 5′-phosphate to a 3′-hydroxyl end-group. A non-specific phosphatase, present throughout meiotic prophase, could dephosphorylate the 3′ ends formed by the endonuclease. The activity specifically at pachytene of the endonuclease and the repair enzymes is in precise agreement with current ideas about the recombination mechanism. Moreover, evidence was obtained by Hotta and Stern (1974) that nicking and repair of chromosomal DNA take place at pachytene. They analysed the sedimentation behaviour of DNA extracted from *Lilium* pollen mother cells at various stages of meiosis. Degradation of DNA during its preparation was minimized by extracting it from pollen mother cell protoplasts, the cell walls having been removed enzymically. Denatured DNA from pachytene cells showed sedimentation peaks at 62 s and 104 s in an alkaline glycerol gradient. The 62 s peak, unlike that at 104 s, was not found at other stages of meiosis. There was no corresponding peak in the sedimentation profile of duplex DNA. Evidently single-stranded breaks occur at pachytene and are soon repaired, as the 62 s peak in single-stranded DNA is transitory. The hybrid cultivar Black Beauty shows little or no chromosome pairing at meiosis and few or no chiasmata. It was found also to lack the 62 s peak in the sedimentation profile of the denatured DNA from pachytene cells. This observation gave strong support for the hypothesis that the nicking at pachytene is related to recombination. It is known that chromosome doubling in Black Beauty to give a tetraploid plant allows normal pairing and chiasma formation. Hotta and Stern (1974) pointed out that this might mean that DNA nicking at pachytene took place only in paired chromosomes, the defect in diploid Black Beauty being regulatory and not genetic.

The number of sites in the chromosomes where pachytene nicking takes place seems to be grossly in excess of the number required for recombination. Hotta and Stern (1974) pointed out that the 62 s and 104 s sedimentation peaks for the single-stranded DNA corresponded to molecular weights of 47×10^6 and 170×10^6 daltons, respectively, implying an average of 3.6 nicks in each strand of 170×10^6 daltons molecular weight. They assumed that half the early pachytene DNA was present as fragments of molecular weight 47×10^6 daltons. If the total DNA of the *Lilium* pollen mother cell has a molecular weight of 66×10^{12} daltons (see Howell and Stern, 1971), it follows that there should be about 700 000 sites of nicking in each pachytene nucleus. Yet *Lilium* has about 36 chiasmata per meiosis. This estimate of the number of sites of single-strand breaks in pachytene DNA was comparable to the estimated number of sites of pachytene DNA synthesis. Hotta and Stern (1971a) used bromodeoxyuridine as a density label for the pachytene synthesis and found that the position of these heavy molecules on a caesium chloride density gradient was unaffected by shearing the total DNA into fragments of about 500 000 daltons molecular weight. They inferred that the labelled regions were probably appreciably smaller than this. In other words, the DNA synthesized at pachytene was distributed in short stretches at many sites. Howell and Stern (1971) estimated that 0.1% of the total DNA underwent repair replication at pachytene, implying synthesis of DNA of molecular weight 66×10^9 daltons. Assuming that the individual DNA segments that undergo synthesis at pachytene have a molecular weight of not more than 100 000 daltons, it was evident that there were at least 660 000 regions of repair replication per nucleus.

Evidence that the sites of nicking of pachytene DNA are not randomly distributed was obtained by Hotta and Stern (1974) from an analysis of the weight distribution of s values in the 35–80 s region of the alkaline gradient. The scatter about the mean value was found to be 65% of a theoretical plot for randomly nicked molecules. Evidently the nicks are more regularly spaced than a random distribution. This conclusion, and the similarity in the number of sites for nicking and for synthesis, support the hypothesis that the nicking occurs in the regions of repeated sequence known to be involved in the synthesis. They favoured the idea that the stimulus for repair replication was activation of the endonuclease, this activation, as already suggested, being brought about by the chromosome pairing.

Hotta and Stern (1971b) isolated a protein from the nuclei of *Lilium* pollen mother cells that showed DNA binding activity. The nuclei were disrupted by sonication and freed from nucleohistones by fractionation. The protein was found to have a very high affinity for single-stranded DNA, and to have properties similar to the gene-*32* protein of phage T4, catalysing the renaturation of DNA. In its presence 60% of denatured *Lilium* DNA became duplex in 2 h at 25 °C. The protein was not detected in somatic tissues and was present at meiosis only at zygotene and pachytene, its activity reaching a peak in early pachytene. Hotta and Stern (1971b) suggested that the binding protein facilitated complementary pairing between DNA strands. This activity might be required at zygotene for matching regions specifically involved in the pairing of homologues, and at pachytene for matching regions involved in recombination. A DNA-binding protein with properties similar to that of *Lilium* was found by Hotta and Stern (1971c) in spermatocyte nuclei at meiotic prophase obtained from several mammals (*Rattus*, *Bos*, *Homo*). They suggested that the presence of such a protein was likely to prove a universal feature of recombination.

A detailed study of the biochemistry of pachytene in *Mus musculus* was made by Hotta, Chandley and Stern (1977a,b). A close similarity with *Lilium* was found, not only as regards the occurrence of a DNA-binding protein, but also in the formation and repair of nicks, the occurrence of repair synthesis, and the localization of this synthesis predominantly in regions of moderately repeated nucleotide sequence. Like *Lilium*, about half the DNA of the pachytene nucleus in the mouse gave a peak at 62 s in an alkaline gradient. This implied that where nicking occurred the spacing of the nicks was the same as in *Lilium*, that is, such as to give a single-strand molecular weight of about 48×10^6 daltons. On the other hand, the mouse nucleus has only one-twentieth of the DNA content of that of *Lilium*, so there were evidently correspondingly fewer nicks. Another difference was that the mouse pachytene synthesis occurred predominantly in sequences repeated about 400 times – a lower figure than was found with *Lilium*. These are minor differences, however, and the outstanding features of the results were the remarkable similarities. The authors concluded that the chemistry of events associated with recombination was virtually identical in *Lilium* and *Mus*, had evidently been conserved over immense periods of time and was probably universal in eukaryotes. The genetic data, discussed earlier, support this conclusion.

Jacobson *et al.* (1975) investigated the nicking of DNA in *Saccharomyces cerevisiae* at meiosis. Diploid cells heterozygous for pairs of alleles at both the *lys2* and *trp5* loci were used, so that recombination could be monitored. The DNA was obtained intact by lysis of spheroplasts derived from the yeast cells. The sedimentation pattern of the DNA was examined in neutral and alkaline sucrose gradients at various times after transfer of the cells to sporulation medium. This transfer inhibits the mitotic budding and induces the premeiotic DNA synthesis and meiosis. Single-strand breaks were found to arise in the DNA with the onset of meiosis. They occurred in both the template and newly synthesized (^{3}H-labelled) strands, following the premeiotic DNA synthesis. The authors estimated that two to six nicks arose per 10^9 daltons molecular weight of DNA. The DNA content of the nucleus at the time of recombination is equivalent to a molecular weight of about 36×10^9 daltons, so, unlike *Lilium* and *Mus*, the estimated number of nicks is not appreciably in excess of the requirements of recombination.

In an attempt to find recombination-deficient mutants in *Ustilago maydis*, Holliday and Halliwell (1968) looked for deoxyribonuclease deficiency. After mutagen treatment cells were plated on agar containing DNA. Subsequently the colonies were rescued by replica plating and the DNA in the medium precipitated with hydrochloric acid. Colonies with reduced extracellular nuclease activity were revealed by their lack of a halo of clear agar. A double mutant deficient in extracellular DNase activity and probably in another nuclease was obtained in this way. Badman (1972) used the same technique and obtained two mutants deficient in extracellular DNase activity. She then used these mutants in a further mutation experiment, scoring this time for intracellular DNase deficiency. This was made possible by rinsing the plates with toluene to make the cell membranes permeable and then incubating at 32 °C for 6 h before precipitating the DNA in the agar with acid. Five mutants deficient in intracellular DNase were isolated in this way. The mutants, which were recessive, mapped at two loci, *nuc-1* and *nuc-2*, causing extracellular and intracellular DNase deficiency, respectively. The effects of the *nuc-1 nuc-2* double mutant homozygote on both mitotic and meiotic recombination were investigated using strains heterozygous for the linked *ad-1* and *me-1* mutants to score crossing over, and heterozygous for two non-complementing mutations at the nitrate reductase (*nar-1*) locus to score conversion. Badman (1972) found that conversion was abolished, but crossing over still took place, though the distribution of mitotic crossovers was abnormal.

Holloman and Holliday (1973) purified 8600-fold an intracellular nuclease, DNase I, extracted from wild-type *Ustilago maydis*. They also detected, but did not purify, a second intracellular enzyme, DNase II. They found that a *nuc-1* mutant, lacking extracellular DNase activity, had normal levels of DNase I and II, while a *nuc-1 nuc-2* double mutant had low levels of both enzymes. The DNase I was highly active on denatured DNA. Studies by Holloman (1973) with such DNA labelled with ^{32}P at the 5' end or with ^{3}H at the 3' end indicated that the enzyme may degrade from both ends. The enzyme was found also to be capable of introducing single-strand breaks into duplex DNA. The explanation for the low levels of both intracellular nucleases in the double mutant was not clear, but the possibility that the conversion deficiency was caused by the lack of DNase I was tested by Ahmad, Holloman and Holliday (1975) using heteroduplex molecules of phage SPP1.

The first clear demonstration of mismatch correction in heteroduplex DNA was made by Spatz and Trautner (1970) with phage SPP1 of *Bacillus subtilis*. The strands of the DNA of this phage differ sufficiently in density to be separable in a caesium chloride gradient. Mutants were induced with hydroxylamine, which is known to cause $G \cdot C$ to $A \cdot T$ transitions. Reciprocal heteroduplex molecules were prepared by mixing the heavy strand of wild type with the light strand of the mutant, and vice versa, and incubating at 70 °C to allow the strands to anneal. One heteroduplex would have a $G \cdot T$ mismatch and the other $A \cdot C$. *B. subtilis* cells were then infected with the heteroduplex molecules. This transfection, like transformation, is possible at a certain stage of the cell cycle. A single burst analysis was then carried out, that is, the phage progeny released from individual cells were scored for the mutant character. The numbers of cells giving wild type only, a mixture of wild-type and mutant particles, or mutant only, were counted. With mutant no. 161, for example, the relative frequencies of these three classes of cells were 0.69, 0.23 and 0.08 when the wild-type strand was heavy, and 0.05, 0.15 and 0.80 when the wild-type strand was light. These results implied that mismatch correction had occurred in a majority of the cells, with the heavy strand favoured. With some mutants the opposite result was obtained, the light strand being favoured. By inducing mutations in the separated strands of the phage, Spatz and Trautner (1970) established the purine–pyrimidine orientation of mutants relative to the heavy and light strands. They found that mismatch correction favoured purine excision with some of the mutants and pyrimidine excision with others. They concluded that the excision was influenced by features of the DNA other than the mismatch.

Ahmad, Holloman and Holliday (1975) tested the effect of the *Ustilago maydis* DNas I on SPP1 DNA heteroduplex for mutant 161. After 15 min exposure to the enzyme, competent cells of *Bacillus subtilis*, that is, susceptible to transfection, were added and incubated for 30 min at 37 °C. A single-strand break leads to a failure of the SPP1 DNA to infect the host. Infection was measured by plating the cells and scoring the number of infective centres. There was some inactivation of homoduplex SPP1 DNA (either +/+ or 161/161) used as a control, but the heteroduplex molecules showed a three- to sevenfold greater inactivation. Heteroduplexes containing a deletion (4307) of about 3% of the SPP1 genome were inactivated by DNase I much more rapidly than the single base mismatch of mutant 161. It was evident that the *Ustilago* DNase I preferentially inactivates DNA containing mismatched bases. Ahmad *et al.* (1975) suggested that the enzyme binds to an end of the DNA molecule and then diffuses along it, introducing nicks where base pairing is disturbed. They attributed the inactivation of homoduplex molecules to imperfect base pairing. The discovery that mismatched bases provide a substrate for DNase I gave strong support to the hypothesis that conversion results from mismatch correction, since DNase I deficiency was known to be associated with conversion deficiency. Further evidence that mismatched bases provide a substrate for DNase I action was obtained by Pukkila (1978). She used linear molecules of *Escherichia coli* phage φ80 DNA into which had been inserted a host gene for a transfer RNA. Pairs of heteroduplexes were prepared from molecules differing by single base substitutions in the *E. coli* tRNA gene. It was found that the enzyme introduced a nick near certain mismatched bases but not others. Those cut were an $A \cdot A$ and $T \cdot T$ pair and a $G \cdot A$ and $C \cdot T$ pair, but not the third kind of transversion: $G \cdot G$ and $C \cdot C$. An $A \cdot C$ and $G \cdot T$ pair of mismatches was cut at one position in the molecule but not at another, implying that the neighbouring base sequence was also relevant. Pukkila also found that the enzyme cleaved circular duplex DNA molecules of phage PM-2 containing supercoils, producing relaxed circular and linear molecules.

Johnston and Nasmyth (1978) discovered that a temperature-sensitive *Saccharomyces cerevisiae* cell cycle mutant at the *cdc9* locus was deficient in DNA ligase. At the restrictive temperature (37 °C) the mutant was defective in DNA replication and in the repair of DNA damaged by ultraviolet irradiation. The DNA synthesized at that temperature was found to contain many single-strand breaks. Game, Johnston and von Borstel (1979) found that holding a diploid homozygous for the *cdc9* mutant at 37 °C and then plating cells at the permissive temperature (23 °C) gave rise to increased intra- and intergenic recombination. They proposed that the single-strand breaks may lead to recombination. Fabre and Roman (1979) used a diploid heterozygous in *trans* for the temperature-sensitive *cdc9* mutant and for an allelic mutant. They found that X-rays induced conversion to *cdc9*+ and that for low doses there was a linear increase in frequency of convertants with increasing dose, but the yield was 10 times greater if the cultures were kept at 22 °C for 6 h compared with cells incubated at 35 °C immediately after irradiation. They concluded that the *cdc9*-controlled ligase was involved in induced conversion as well as DNA replication.

Disomic strains, that is, haploid with one extra chromosome, have been exploited in *S. cerevisiae* to obtain recombination-deficient mutants. By using a strain heterozygous for two allelic auxotrophic mutants in the disomic chromosome, recombination-deficient progeny can be selected, following mutagen treatment, by their reduced frequency of prototrophs. Roth and Fogel (1971) used *leu2* alleles in a strain disomic for chromosome III and also heterozygous for mating-type and selected mutants deficient in meiotic recombination, and Rodarte-Ramón and Mortimer (1972) used *arg4* alleles in a strain disomic for chromosome VIII and selected mutants deficient in X-ray- and ultraviolet-induced mitotic recombination. Mutants deficient in conversion were obtained: see Rodarte-Ramón (1972) and Fogel and Roth (1974). A comparable technique was used by Parag and Parag (1975) working with *Aspergillus nidulans* carrying a duplicated segment, one in the normal position on chromosome I and one translocated to chromosome II. These segments were found to recombine with one another. The duplication included the *adE* gene, so, by using a strain heterozygous for two *adE* alleles, progeny deficient in adenine prototrophs could be selected. The use of these selective techniques in disomic or duplication strains that are otherwise haploid has enabled recessive mutants to be obtained that may be deficient in recombination enzymes.

The limited information available about enzymes involved in eukaryote recombination is in excellent agreement with current ideas about the mechanism, apart from the riddles posed by the large number of sites of nicking at pachytene in *Lilium* and *Mus* and the extent to which the associated repair synthesis is in non-unique nucleotide sequences.

18. CONTROL OF CROSSOVER FREQUENCY AND DISTRIBUTION

The occurrence in most organisms of interference between crossovers within chromosome arms was described in Section 1. Stadler (1959) investigated whether conversion events without crossing over showed interference with crossovers. He selected wild-type recombinants at the *cysteine* (*cys*) locus in the left arm of linkage group VI of *Neurospora crassa*, and scored them for flanking markers. He had two markers on one side of *cys*, and compared their recombination frequency when there was conversion at *cys* with and without associated crossing over. The results are given in Table 40. It was evident that non-crossover recombination events did not show interference with crossovers. Similar results were obtained by Fogel and Hurst (1967) with asci showing conversion at the *histidine 1* locus in *Saccharomyces cerevisiae* and by Carlson (1971) with recombinants at the *r* locus (*rudimentary* wings) in *Drosophila melanogaster* (Table 40). It is likely therefore that the lack of interference shown by non-crossover events is of general occurrence. Although the cause of interference has not been established, Stadler's discovery reveals an important difference between crossovers and non-crossovers which any hypothesis of the interference mechanism must take account of.

The absence of interference between crossovers and non-crossovers explains how intragenic recombination involving more than two chromatids can arise. This has often been recorded: see, for example, Table 12(g) and Section 11. Events apparently involving more than two chromatids at the same site have also been observed cytologically at meiosis in *Locusta migratoria* by Tease (1978) with 5-bromodeoxyuridine labelling followed by Hoechst fluorochrome treatment.

The synaptonemal complex or *synapton* which is formed at zygotene and persists until diplotene, provides a structural framework for the pairing of homologous chromosomes (review, Rasmussen and Holm, 1980). In male *Drosophila melanogaster* and in the *c(3)G* mutant the synapton is lacking and there is no crossing over. In female *Bombyx mori* (Silkworm), on the other hand, the synapton is present but crossing over is lacking. It appears that synapton formation is necessary for meiotic crossing over but that, as might be expected, there are other requirements too. On the other hand, by transferring yeast cells from meiotic to vegetative culture medium selective for recombinants. Olson and Zimmermann (1978a) found that meiotic recombination could occur before the synapton had formed. They concluded that recombination in *Saccharomyces*

Table 40. Data on interference between recombination events of different kinds. On the *Drosophila* linkage map, B, f, g, r, and tc stand for Bar eyes, forked bristles, garnet eye-colour, rudimentary wings and tiny chaetae, respectively.

Organism and reference	Wild-type recombinants at C			Recombination frequencies in the marked intervals				
	Event at C	Number†	BD interval	A Outer left-hand marker	B Inner left-hand marker	C Gene in which recombinants selected	D Inner right-hand marker	E Outer right-hand marker
Neurospora crassa (Stadler, 1959)				—	lys-5	cys	ylo	ad-1
					5.5	18.3	11.1	1
	—	282	Recombinant					3.9 ⎱ $x^2 = 16.8$
	—	361	Parental					13.3 ⎰ $n = 1$, $P \ll 0.01$
Saccharomyces cerevisiae (Fogel and Hurst, 1967)				—	thr 3	his 1	arg 6	trp 2
				34	2.4	9.9	19.3	
	Conversion	452	Recombinant				12.1 ⎱ $x^2 = 23.2$	
	Conversion	528	Parental				19.3 ⎰ $n = 1$, $P \ll 0.01$	
	Reciprocal recombination	97	Recombinant				10.8	
Drosophila melanogaster (Carlson, 1971)				g	tc	r	f	B
				7.2	2.9	2.2	0.3	9
	—	1765	Recombinant	0.06 ⎱ $x^2 = 108.3$			0 ⎱ $x^2 = 3.3$	
	—	1189	Parental	6.2 ⎰ $n = 1$, $P \ll 0.01$			0.25 ⎰ $n = 1$, $P = 0.1$–0.05	

†The numbers in Fogel and Hurst's data refer to asci with wild-type spores.

cerevisiae does not require the synapton. They also inferred that it was not involved in X-ray-induced mitotic recombination, since no synapton was found (Olson and Zimmermann, 1978*b*).

Carpenter (1975) discovered in *Drosophila melanogaster* structures associated with the synapton to which she gave the name *recombination nodule*. These were more or less spherical, about 100 nm in diameter, densely staining, and located adjacent to the central element of the synapton, as seen in electron micrographs of oocytes at pachytene (Fig. 87). She found that the number of nodules per nucleus and their locations along the bivalents corresponded with crossover numbers and distribution. The number and position of the nodules were established from serial sections of entire meiotic nuclei. Confirmation of a relationship between nodules and crossing over has been obtained with several organisms (for references see Carpenter (1979*b*) and Rasmussen and Holm (1980)).

Carpenter (1979*a*) described a second type of recombination nodule in *Drosophila*. These were ellipsoidal rather than spherical and measured about 100 nm x 35 nm, though the length was variable. They may be precursors of the spherical nodules as they arise earlier in pachytene than the spherical ones and disappear earlier, in mid-pachytene. The maximum number of ellipsoidal nodules per cell was found to exceed that of spherical nodules, so if these represent a developmental sequence, not all ellipsoidal nodules can become spherical. Carpenter (1979*a*) recorded the positions of nodules along the chromosome arms and found that spherical nodules showed interference with one another, as expected if they correspond to crossovers. On the other hand, the ellipsoidal nodules appeared to be randomly distributed and did not show interference with one another, though there were indications of interference between ellipsoidal and spherical. The latter observation conflicts with the hypothesis that ellipsoidal nodules correspond to non-crossover recombination events, as these show no

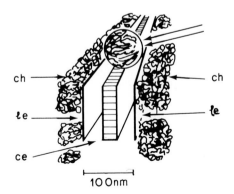

Fig. 87.   Diagram to show the typical position of a recombination nodule (double arrow) in relation to the synapton: ce, central element; ch, chromatin; le, lateral element. (Reproduced by permission of Dr A. T. C. Carpenter from A. T. C. Carpenter, *Proc. natn. Acad. Sci. U.S.A.* **72**, 3187 (1975).)

interference with crossovers (Table 38). Carpenter (1979*a*) discussed two hypotheses relating the nodules to recombination:

(1) *Ellipsoidal nodules are precursors of spherical nodules.* The sites of recombination would then be chosen before the outcome as a crossover or non-crossover was determined, some nodules remaining ellipsoidal and others, spaced out to accommodate interference, becoming spherical.

(2) *Ellipsoidal nodules are not precursors of spherical nodules.* The crossover or non-crossover outcome at a site would then be determined at or before the time of formation of the nodules, the two kinds of nodules corresponding at all times to the two kinds of event.

If, as seems likely, the nodules comprise the enzymes and DNA-binding protein required for recombination, these requirements are expected to be the same whether the outcome is a crossover or a non-crossover. Moreover, according to the Meselson–Radding hypothesis, the outcome will be determined relatively late, after the polymerase-nuclease has left its substrate (see Section 14). Thus, hypothesis (1) above, with some of the ellipsoidal nodules developing into spherical ones, seems to fit current ideas better than hypothesis (2), with its decision between crossover and non-crossover at an early stage.

In agreement with the conclusion that the nodules are the sites of recombination, and that this involves DNA synthesis, Carpenter (1981) found that both ellipsoidal and spherical nodules of *D. melanogaster* showed significant labelling compared with elsewhere in the nucleus, following incubation with [^{3}H]thymidine and preparation of electron microscope autoradiographs of serial sections.

Carpenter (1979*b*) investigated recombination nodules in *Drosophila* females homozygous for recessive mutants giving reduced frequencies of crossing over. The mutants were at three loci, *mei-9*, *mei-41* and *mei-218*, and were known to give crossover frequencies of about 10%, 50% and 10% respectively, of normal. Moreover, *mei-41* and *mei-218* mutants give relatively fewer distal crossovers than normal. This distribution in fact corresponds more closely to the physical length of the euchromatic regions of the chromosomes, distal sites being relatively favoured in wild-type *Drosophila*. With *mei-9* mutants the crossover distribution is like that in wild type. Two categories of *Drosophila* recombination mutants, of which these are representative, had been recognized by Sandler *et al.* (1968): those that affect crossover distribution by altering preconditions for exchange, and those that affect the probability of exchange without affecting the preconditions and so do not influence crossover distribution. From investigation of a number of recombination-defective mutants isolated from the wild, Sandler *et al.* favoured such a two-step process with sites for exchange being established first, followed by the exchange process itself.

Carpenter (1979*b*) examined one *mei-9* mutant and two each at the *mei-41* and *mei-218* loci. She found that both kinds of nodules were of normal frequency and distribution in the *mei-9* homozygote, but greatly reduced in numbers and of abnormal morphology in the *mei-41* and *mei-218* homozygotes. In these two mutants the number and distribution of spherical nodules matched those of crossovers. From the *mei-9* results it was evident that the defect causing the crossover deficiency was not in the formation of nodules. It was also apparent that spherical nodules did not form after crossing over had taken place, since the number of nodules was normal, despite the low crossover frequency. The implication was that at least some of the steps in recombination were mediated by the nodules. Carpenter argued that if ellipsoidal nodules were precursors of spherical, this change of morphology must occur when events were switched to the crossover pathway, but before the crossover actually took place. The defect caused by the *mei-9* mutant would then be manifest after spherical nodule formation but before crossing over occurred. The *mei-9* locus is also active in somatic cells, playing a part in DNA repair, for Boyd, Golino and Setlow (1976) found that a *mei-9* mutant was deficient in the excision of ultraviolet-induced pyrimidine dimers. The inference from the *mei-9* results that the nodules are directly involved in recombination was supported by the presence of abnormal nodules in the other mutants. Carpenter favoured the hypothesis that all aspects of recombination were mediated by the nodules.

In discussion of interference and recombination nodules the current terminology is misleading, with references to conversion and to non-reciprocal exchange as alternatives to crossing over, whereas they occur in association with it as well as without it. I have suggested (Whitehouse, 1973*b*) the term *crossunder* for 'non-crossover recombination event'.

As mentioned above, the *mei-41* and *mei-218* mutants in *Drosophila* alter the crossover distribution from a favouring of distal sites in wild type to a more or less random distribution in the mutants. If it is true that the sites of recombination are chosen before their outcome (crossover or crossunder) is determined, the relative frequencies of crossovers and crossunders at particular loci will differ in a *mei-41* or *mei-218* background from those found in wild type. It might be possible to test this at the *r* locus (*rudimentary* wings) or at the *rosy* (*ry*) eye-colour locus.

Mutants at the *mei-2* locus in *Podospora anserina* were found by Simonet (1973) to increase the frequency of crossing over in proximal regions of the chromosomes and to decrease it in distal regions. The effect thus resembled that of the *mei-41* and *mei-218* mutants in *Drosophila*. It would be interesting to compare the flanking marker behaviour of wild-type recombinants from crosses of alleles at various loci in *Podospora* in a wild-type and in a *mei-2* background. Simonet found a fourfold increase in recombination near the proximally located locus *14* in the *mei-2* background compared with wild type. Moreover, conversion to wild type at the site of mutant 230 at locus *14* was associated with more crossing over: in wild type, 17 crossunders, one crossover (6%); in *mei-2*, 13 crossunders, five crossovers (28%). This difference, though not significant, is in the direction expected if part of the control of crossover frequency and distribution is made by a control of how many recombination events become crossovers and how many remain as non-crossovers.

The distribution of crossovers in the X-chromosome of female *Drosophila melanogaster* at meiosis was found by Sandler (1954, 1957) to be profoundly altered by certain structural changes within it. The normal crossover frequency gave a unimodal distribution, 90% of the cells having either one or two crossovers, these alternatives being about equally frequent. If an inversion involving the greater part of the chromosome transferred part of the proximal heterochromatin to a distal position, then the cross-over distribution in the homozygote for the inversion was bimodal, cells with no crossovers being as frequent as those with two and together accounting for 90%. Another well documented phenomenon in *Drosophila* is the effect of heterozygosity for structural changes, particularly inversions. These often increase the frequency of crossing over in the other chromosomes (review: Lucchesi and Suzuki, 1968). It would be interesting to investigate the effects of heterozygosity for structural changes, and of transfer of heterochromatin to a distal site, on flanking marker behaviour of convertants, for example, at the *r* or *ry* loci. Interaction between different chromosomes in the determination of crossover frequency is also evident when they are structurally normal. Mather (1936) reported a negative correlation between chiasma frequencies in different bivalents in a majority of 27 insects and flowering plants examined, implying competition between bivalents during chiasma formation.

Studies on chiasma frequency and distribution in *Secale cereale* (Rye) and other plants have revealed an elaborate control system. Rees and Naylor (1960) found that even within a rye anther there was a gradient of increasing chiasma frequency from tip to base. One of the F_2 progeny of a hybrid between two species of *Secale* was found by G. H. Jones and Rees (1964) and Jones (1967) to be a recessive mutant which had a random distribution of chiasmata, both as regards number and position, instead of the normal situation in *Secale* in which there is little variation in chiasma frequency per bivalent and most of the chiasmata are localized distally. From a wider study of rye genotypes, Jones (1974a) found a high degree of correlation in the amounts of chiasma variation between and within pollen mother cells. The underlying common basis in this variation was identified as the error component of variance. Jones inferred that the rye genotypes differed primarily in their degree of control over chiasma formation. Increasing variation in chiasma frequency between bivalents was correlated with increasing proportions of non-distal chiasmata. Jones suggested that the mutant with randomly distributed chiasmata was defective in the establishment of exchange positions rather than in exchange itself. He concluded that recombination in *Secale* is probably a two-step process comparable to that proposed by Sandler *et al.* (1968) for *Drosophila*, exchange position being established before exchange itself. The occurrence in *Secale* of genotypes ranging in a continuous series from distal localization of chiasmata to random distribution indicated a complex genetic control of the distribution of sites of crossing over, with the control evidently capable of fine adjustment.

It has often been claimed that the spacing out of chiasmata implies that they form sequentially. Jones (1974b) argued that there was no evidence for this. In his rye mutant with randomly distributed chiasmata, both localization of chiasmata and interference between chiasmata have broken down, indicating that a single control governs both phenomena. Sequential formation, however, implies separate controls, since interference cannot operate until the first chiasma has formed. The correlated response in diverse rye genotypes of the chiasma frequency variation between bivalents and of the proportion of non-distal chiasmata showed that the relationship between localization and interference was of general occurrence in *Secale*. The occurrence of a similar association of reduced interference with reduced localization in meiotic mutants of *Drosophila* and *Podospora* points to the widespread occurrence of a single genetic control of the positions of all the chiasmata in a bivalent.

Supernumerary chromosomes, that is, non-essential chromosomes additional to the normal set, can contribute to the fine control of chiasma frequency and distribution, R. N. Jones and Rees (1967) found that in rye, with increasing numbers of supernumerary chromosomes, there was increasing variation in chiasma frequencies between pollen mother cells within plants, and also increasing variation between bivalents within pollen mother cells. Similar results have been obtained with other species of flowering plants, mean chiasma frequency some-times being affected (review: Jones, 1975).

The most suitable organism for investigating to what extent control of crossover frequency and distribution operates at the level of total recombination events, and to what extent at the level of crossovers versus crossunders, seems to be *Podospora anserina*, where spore-colour loci and mutants modifying crossover distribution are available (Marcou *et al.*, 1979). From the similarities referred to above in the control of variation in crossover frequency and distribution in *Drosophila*, *Podospora* and *Secale*, it is likely that conclusions derived from *Podospora* would apply widely in eukaryotes.

The idea of G. H. Jones (1974*a*,*b*) of a single control of the positions of all the chiasmata in a bivalent, in any organism showing localization of chiasmata or interference, is not necessarily in conflict with the occurrence of independent control of recombination at specific sites found by Catcheside (1977) and associates in *Neurospora crassa* (see Sections 12 and 15). This is because the two types of control act at such different levels. Distal localization of chiasmata might still allow considerable flexibility as to which initiation site for recombination was involved in individual cells. Furthermore, as already discussed, localization of chiasmata and interference may be mediated at the level of crossover versus crossunder rather than at the level of initiation of recombination. Control of recombination in highly specific localized regions, analogous to that demonstrated in *Neurospora*, has been found in the basidiomycete *Schizophyllum commune* by Simchen and Stamberg (1969). In both species there was natural variation in the controls, which were detected by comparing strains obtained from different wild sources. Control at the level of the chromosome was discovered by Parker (1975) in the flowering plant *Hypochaeris radicata*. From a French population he isolated a recessive mutant in which chiasma formation failed in chromosome no. IV, while remaining at normal levels in the other three bivalents. The failure was not in chromosome pairing, which was normal.

To account for the low frequency of mitotic recombination compared with meiotic, Holliday (1968) suggested that one or more of the essential enzymes are repressed in mitotic cells. Agents which increase crossing over might then induce these enzymes by inactivating the repressor. Another possibility considered by

Holliday was that agents such as ultraviolet light that promote mitotic recombination (see Sections 2 and 4) cause damage to the DNA that can be repaired by a process involving recombination. This would be comparable to the recombination repair pathway known in *Escherichia coli* (see Chapter VI, Section 9). Subsequently, he obtained evidence (Holliday, 1971) that damage to DNA of a mitotic cell in *Ustilago maydis* derepresses one or more enzymes required for such a repair process involving recombination. He found that exposure of diploid cells to gamma rays induced recombination with high frequency. Recombination was measured by using a diploid heterozygous for two non-complementing mutants at the *nar-1* locus. This is the structural gene for nitrate reductase, an enzyme that is induced when wild-type cells are transferred to a culture medium containing nitrate as the sole source of nitrogen. In the *nar-1* double heterozygote the enzyme can be synthesized only if a wild-type recombinant has arisen. The product of the enzyme action, nitrite, accumulated because the cells were deficient in the next enzyme in the pathway. Thus, assay for nitrite provided a sensitive test of recombination frequency. The level of recombination with increasing dose of gamma rays was found to correspond to the number of cells able to multiply, rather than to the total number of cells. Holliday concluded that ability to multiply depended on an inducible DNA repair mechanism involving recombination.

Similar conclusions were reached by Fabre and Roman (1977) from experiments with *Saccharomyces cerevisiae*. They irradiated a haploid strain with ultraviolet light or X-rays and then paired it with a diploid strain heterozygous in *trans* for a pair of alleles at the *ade6* locus. The haploid carried both *ade6* mutants. They found that recombination in the diploid to give adenine prototrophs was stimulated by the irradiation of the haploid. They favoured the idea that recombination was repressed in somatic cells and that radiation released some factor necessary for recombination. Exposure of the ultraviolet-irradiated haploid to light, which would no doubt reduce the number of pyrimidine dimers through photoreactivation, was found to reduce the number of recombinants. Fabre and Roman (1977) concluded that damage to the DNA was responsible for the induction of recombination as part of a repair process.

Evidence for recombination repair is also provided by the occurrence of radiation-sensitive mutants defective in mitotic allelic recombination. These were reported in *Ustilago* by Holliday (1965), in *Aspergillus nidulans* by Shanfield and Käfer (1969) and Jansen (1970a,b), and in *Saccharomyces cerevisiae* by Hunnable and Cox (1971). One of the the the *Ustilago* mutants, *rec-1*, was investigated in detail by Holliday *et al.* (1976). It showed spontaneous mitotic recombination with relatively high frequency, but a reduced level of ultraviolet-induced recombination. They suggested that the *rec-1* gene had a regulatory function controlling the repression of recombination during the mitotic cell cycle. If the mutant had an altered repressor that bound only weakly to its operator, the spontaneous mitotic recombination would be accounted for. They suggested that the sensitivity to radiation might result from a failure to induce a recombination repair mechanism. Mutations of comparable phenotype have been described in *Aspergillus* by Parag and Parag (1975) and in *Saccharomyces* by Boram and Roman (1976). It thus seems likely that a repair process requiring recombination may be of general occurrence in fungi, and that damage to DNA of a mitotic cell derepresses recombination.

Repression of mitotic recombination seems to be associated with repression of sister-chromatid exchange. Such exchanges were first detected by Taylor, Woods and Hughes (1957) in their investigation of chromosome replication in *Vicia faba*. As described in Section 1, they used autoradiography, following ^{3}H-labelling of newly synthesized strands of DNA. They found that sister chromatids sometimes underwent exchange, revealed as a label switch in the autoradiograph. There has been much debate as to how far sister-chromatid exchanges are a consequence of the presence of the label needed to detect them (review: Wolff, 1977). It seems clear that many, perhaps a majority, do result from the labelling. This applies also to the harlequin staining described in Section 1, using a fluorescent dye and Giemsa stain after two rounds of replication in the presence of 5-bromo-deoxyuridine. Thus, Schvartzman *et al.* (1979) concluded that most of the sister-chromatid exchanges observed in this way in *Allium cepa* (Onion) root tips were induced by the bromodeoxyuridine. In marked contrast to this low spontaneous exchange rate, Chaganti, Schonberg and German (1974) discovered that in the rare human disease called Bloom's syndrome, which is caused by a single recessive gene, *bl*, sister-chromatid exchanges were frequent. Using harlequin staining, their frequency in lymphocytes in culture was found to be about 13 times the normal rate (Plate 6). This discovery is of particular interest because lymphocytes from *bl* homozygotes also show a high frequency of quadriradial chromosome configurations at mitosis (German, 1964). As pointed out in Section 2, these imply mitotic crossing over. It seems as if the recessive gene responsible for Bloom's syndrome leads to derepression of recombination both between homologous chromosomes and between sister chromatids. Tice, Windler and Rary (1978) studied a mixed culture of normal human fibroblast cells and those from a Bloom's patient. They found that the *bl bl* cells produced an agent capable of increasing the frequency of sister-chromatid exchanges in normal cells.

Rommelaere, Susskind and Errera (1973) found that ultraviolet irradiation increased the frequency of both mitotic crossing over and sister-chromatid exchange in cultured cells of *Cricetulus griseus* (Chinese Hamster). The mitotic crossing over was observed by inducing fusion between cells labelled with [^{3}H]thymidine and others labelled with [^{14}C]thymidine, and finding chromosomes carrying both labels. Schvartzman *et al.* (1979) favoured the Meselson–Radding hypothesis as the mechanism of sister-chromatid exchange, since they found exchanges in *Allium* root tips when only one of the four polynucleotide chains contained bromodeoxyuridine. They inferred from this that a break in only one strand was needed to initiate sister-chromatid exchange. (See also Section 7.)

On present evidence it is reasonable to suppose a common mechanism for recombination between sister chromatids and between homologous chromosomes, both at mitosis and meiosis. The differences seem to be in regulation, sister-chromatid exchange being repressed at all times and homologous exchange in mitotic cells. In fungi, homologous exchange, at least, is derepressed when the DNA is damaged, as is sister-chromatid exchange in mammalian cells.

19. CONCLUSIONS

A hypothesis of the mechanism of eukaryote recombination with many attractive features was put forward by Sobell (1972–75). It has not been discussed here because it predicts that recombination events with parental flanking markers and those with recombinant flanking markers will be equally frequent (Sobell, 1974). This prediction conflicts with much of the experimental data.

Stahl (1979) stated that recombination data for *Saccharomyces cerevisiae* put special demands on the Meselson–Radding model. There are two peculiarities of the yeast data. First, there is the equal frequency of conversion to wild type and to mutant, even for deletion mutants, with postmeiotic segregation usually rare and apparently not associated with a particular class of mutant (Section 11). These results would be expected on the Meselson–Radding hypothesis, however, if the yeast nuclease involved in mismatch repair, after recognizing the mismatch, cut one of the two strands without regard to the molecular nature of the mismatch. The conversion behaviour of deletions seems to call for the same explanation as for point mutants, because the two kinds of mutants showed co-conversion (see Table 29). The discovery by McKnight, Cardillo and Sherman (1981) of a 5000 base-pair deletion showing conversion only to wild type, while a closely linked insertion of similar length shows conversion equally often to wild type and to mutant (see Section 11), raises some interesting questions, more particularly since the two mutants showed frequent co-conversion.

The second peculiarity of yeast recombination is the rarity of aberrant 4:4 asci. But this can be accounted for, as Meselson and Radding themselves pointed out (Section 14), if the asymmetric phase of their model was unusually persistent through the polymerase-nuclease having a high affinity for its substrate. Nevertheless, Stahl (1979) preferred an explanation of the yeast data in which conversion was attributed to localized replication of the DNA occurring independently of the presence of the mutants. This hypothesis, however, cannot explain marker effects such as those that give rise to map expansion (Section 10) or those discussed in Section 11.

Esposito (1978) claimed that mitotic recombination in yeast occurs at the two-chromatid stage before DNA synthesis has occurred. He investigated spontaneous mitotic recombination in diploid *Saccharomyces cerevisiae* heterozygous in *trans* for two mutants at the *trp5* locus by plating on medium lacking tryptophan to select for recombinants. The diploid was also heterozygous for a mutant at the *ade5* locus, which is some distance distal to *trp5* in the same chromosome arm, and homozygous for a mutant at the *ade2* locus, which is in another chromosome. The presence of this mutant, as already pointed out (Section 4), leads to the accumulation of a red pigment, unless there is also homozygosity for the *ade5* mutant, which blocks synthesis of the red pigment. A majority of the tryptophan prototrophs formed red colonies, but a few were white or had red and white sectors. The genotypes of 31 of the latter were analysed. To obtain sectored tryptophan-independent colonies there has to be an event at *trp5* that generates two *trp5*$^+$ chromatids, one of which is associated with crossing over (leading to homozygosity for *ade5* and a white sector). An asymmetric distribution of hybrid DNA, such as is believed to occur in yeast, will not give rise to two wild-type chromatids. To explain his results, Esposito invoked crossing over at the two-chromatid stage, with independent mismatch correction to wild type in the associated hybrid DNA at each mutant site in *trp5*, and with the strand exchange resolved, not by cleavage, but by replication during the subsequent round of DNA synthesis. It seems equally likely, however, that the sectored prototrophs resulted from normal recombination at the four-chromatid stage, selection for sectoring implying selection for rare double events at *trp5*, one associated with crossing over and the other not, and with at least three chromatids involved altogether (cf. Table 12(g) and Sections 11 and 18).

Eukaryote recombination has proved to be a difficult phenomenon to explain because of its internal contradictions: recombination of alleles may be reciprocal or non-reciprocal; conversion may be associated with parental or with non-parental flanking marker genotypes; hybrid DNA may be distributed symmetrically or asymmetrically; mismatches may be repaired or remain uncorrected to give postmeiotic segregation. One of the merits of the Meselson–Radding model is its ability to explain the occurrence of all these alternatives in one and the same gene. As pointed out in Sections 15 and 16, however, several modifications to the model seem to be required.

First, it seems necessary to invert the initial steps, such that the site recognized for initiation becomes the recipient of a strand from the homologous molecule, not a donor to it. This requirement would be met if the endonuclease recognized a site on one DNA duplex and cut one strand in the homologous region of a second DNA duplex, with which it was closely paired.

Secondly, some of the data require transfer of the initiating enzyme along the DNA, before formation of a heteroduplex. This transfer might occur by nick migration, or by enzyme movement without nicking. The latter alternative is preferable, as it would allow the nicking to occur in the homologue, as suggested above.

Thirdly, it seems to be necessary to control the isomerization process to accommodate the elaborate control that is normally found of crossover frequency and distribution. No less than five lines of evidence point to a two-step process, as though initially isomerization was not allowed and subsequently, at a controlled number and distribution of the sites, the enzyme aggregate was reactivated to give a second recombination event in proximity to the first and this time associated with isomerization. The evidence for a two-step process was discussed in Sections 16 and 18. It is derived from investigations of (1) the recombination behaviour of meiotic mutants in *Drosophila melanogaster*, (2) the properties of recombination nodules in *Drosophila*, (3) the chiasma frequency and distribution in various *Secale* genotypes, (4) the flanking marker genotypes in *am-1* allelic recombinants in *Neurospora crassa*, and (5) the crossover positions in asci showing aberrant segregation at a mutant site and normal segregation at an allelic site. The diversity of these approaches and the similarity of their conclusions give support for the idea – the *crossunder hypothesis*, as it might be called – that randomly distributed non-crossover events (crossunders) may provide the 'nodes' – the preconditions for crossing over – postulated originally by Bridges (1915).

VIII.   General conclusions

The striking feature of the work on recombination surveyed in Chapters II–VII is the way in which ideas about the mechanism have converged. There is now general agreement that the process occurs by breakage and joining, though break and copy seems also to be a contributory factor in at least one instance. Breakage and joining was first demonstrated in phage λ (Chapter V, Section 2) and soon afterwards in T4 (Chapter IV, Section 7), in *Escherichia coli* following conjugation (Chapter VI, Section 3), in eukaryotes (Chapter VII, Section 1) and elsewhere. Break and copy was favoured at one time for phage f1 (Chapter III, Section 6(c)) but subsequently this idea was abandoned in favour of break and rejoin (Chapter III, Section 6(e)). Break and copy, however, seems necessary to account for the data on recombination by the Red system in the middle region of the genome of phage λ (Chapter V, Section 11).

The idea that homologous pairing may precede strand breakage has recently gained ground following the startling discoveries of the behaviour of the *recA* protein *in vitro* described in Chapter VI, Section 11. Pairing before breakage had been suggested by Moore (1972), who had investigated recombination in the fructose transport (*ftr*) gene in the basidiomycete *Coprinus cinereus* (formerly *C. lagopus*). Pairing of unbroken molecules was favoured by Champoux (1977). He mixed complementary single-stranded circles of the DNA of simian virus 40 in the presence of a topo-isomerase – a DNA untwisting enzyme isolated from the liver of *Rattus norvegicus* – and found that the circles completely renatured to form a covalently closed duplex ring. The evidence was the sedimentation rate in alkali, the buoyant density in caesium chloride gradients containing ethidium bromide, and the resistance to a nuclease specific for single-stranded DNA. Champoux concluded that the enzyme produced transient single-strand breaks sufficient to allow the strands to rewind completely. He suggested that the enzyme might promote the initiation of recombination by allowing reciprocal heteroduplex formation before strand breakage. The initial event would be the unpairing of the strands at homologous regions in two DNA molecules. Evidence for the occurrence *in vitro* of such interlocked duplexes with reciprocal heteroduplex regions was obtained by Potter and Dressler (1978). They mixed closed circular duplex DNA molecules of a plasmid with extracts from *E. coli* of *recA* + genotype, and observed under the electron microscope figure of eight molecules with an extended region of pairing at the node. These interwrapped double circles were resistant to denaturation by alkali, implying continuity of all four strands.

Proteins that bind to single-stranded DNA evidently have an important function in recombination through their ability to promote complementary pairing and hence heteroduplex formation. The gene-*32* protein of phage T4 was the first such protein to be described (Chapter IV, Section 9) and subsequently a similar protein was found in eukaryotic cells at the zygotene and pachytene stages of meiosis (Chapter VII, Section 17). In *E. coli* both the *ssb* protein and the *recA* protein bind to simplex DNA and play a part – the latter an essential part – in recombination (Chapter VI, Section 11).

There is evidence from the behaviour of gene-*32* mutants that in phage T4 the gene-*32* protein and enzymes involved in recombination form a multienzyme complex (Chapter IV, Section 9). Sang and Whitehouse (1979) suggested that the occurrence of an enzyme complex might explain the relationship they had found at the *buff* locus in *Sordaria brevicollis* between different aspects of the recombination process, namely, mismatch correction and flanking marker crossing over (Chapter VII, Section 16). The existence of recombination nodules in eukaryotes (Chapter VII, Section 18) also points to the occurrence of an enzyme aggregate.

Specific recognition sites for the initiation of recombination by an endonuclease have been proposed for eukaryotes (Chapter VII, Sections 12 and 15) and for the RecBC pathway in *E. coli* (Chapter VI, Section 8). The occurrence of enzyme migration along the DNA from these sites has been postulated to accommodate some of the data, and cutting in *trans* has been proposed for the initiation of eukaryote recombination (Chapter VII, Section 15). Such nicking of a closely associated homologous DNA molecule was first suggested for DNA repair in *E. coli* (Chapter VI, Section 11).

At one time the nicking of both recombining molecules at corresponding positions more or less simultaneously, as in Holliday's model, was the favoured hypothesis for the initiation of recombination. Latterly, however, an asymmetric start, as in the Meselson–Radding model, has been given strong support both in *E. coli* (Chapter VI, Section 11) and in eukaryotes (Chapter VII, Section 14). The chief exception to this as a general hypothesis is the integration of the λ genome into that of *E. coli* promoted by the λ Int recombination system (Chapter V, Section 9). The Int protein is a topo-isomerase and two molecules of it are believed to act simultaneously to cut and exchange strands of the same polarity in each of the two recombining molecules as the initial step of integration.

The idea that a single-strand break in a DNA molecule, or a free end to a strand, can promote recombination is well established. Michalke (1967) found that recombination frequency increased sharply towards the ends of the genome of phage T1 of *E. coli*. Genome ends have also been found to promote recombination in T4 (Chapter IV, Section 6), but the circularly permuted genome ends in T4 makes this more difficult to demonstrate. Single-strand breaks are well documented as a source of recombination in φX174 (Chapter III, Section 5(e)) and are believed to trigger repair by recombination following ultraviolet-induced damage to DNA in *E. coli* (Chapter VI, Section 9). Another possibility, however, is that the single-stranded region in the damaged molecule provides a binding site for *recA* protein, leading to homologous pairing and strand transfer to the damaged molecule (see Cunningham *et al.*, 1980). Ligase deficiency has been observed to promote recombination in T4 (Chapter IV, Section 11) and in yeast (Chapter VII, Section 17), again suggesting that single-strand breaks are recombinogenic. In keeping with these findings, Hotchkiss (1971) believed that in general 'the two parents enter unequally into the individual confrontation'. He favoured the idea that 'one parent strand essentially "attacks" the other parent'. A similar notion underlies the single-strand aggression model for the initiation of recombination in ϕX174 (Chapter III, Section 5(h)) and the asymmetric heteroduplex model for phage f1 (Chapter III, Section 6(e)), as well as the Meselson–Radding model.

Exonucleolytic action is evidently involved in many recombination systems, for example, the exonuclease controlled by genes *46* and *47* in T4 (Chapter IV, Section 8), the λ exonuclease in the Red pathway (Chapter V, Section 7) and exonucleases V, VIII and I in the *E. coli* RecBC, RecE and RecF pathways, respectively (Chapter VI, Sections 5–7). The function of the DNA degradation brought about by these enzymes is not clear, though one possibility is the strand assimilation shown by the λ exonuclease (Chapter V, Section 10). Concerted exonuclease and polymerase action on recipient and donor duplexes, respectively, is a central feature of the Meselson–Radding model.

The importance of supercoiling in promoting recombination was discovered with investigations *in vitro* of the *E. coli* RecA and λ Int systems (Chapter VI, Section 11, and Chapter V, Section 9). Supercoiling need not be restricted to the genomes of phages with small circular molecules, provided that the ends of a DNA segment are prevented from revolving, so supertwisting may have wide importance in recombination. Moreover, superhelicity need not be a legacy of the preceding DNA replication, since it is now known that DNA gyrase can impart supercoiling to a DNA molecule (Chapter V, Section 8). Perhaps one of the functions of the synapton is to allow supercoiling to occur or be maintained in segments of the DNA of eukaryotes. Recombination can evidently occur, however, in the absence of the synapton (see Chapter VII, Section 18).

Heteroduplex formation was originally proposed as one of several possible explanations for partial heterozygotes in phage T2 (Chapter IV, Section 3). Hybrid DNA is now believed to be of general occurrence as a recombination intermediate. Evidence for it has been obtained in transformation (Chapter II, Section 2) and in eukaryotic recombination (Chapter VII, Section 7), as well as in *E. coli* and its phages. The discovery of branch migration (Chapter IV, Section 8; Chapter III, Section 5(f)) explained the occurrence of long heteroduplex segments, such as have been observed in phage f1 (Chapter III, Section 6(d)) and in both Int- and Red-promoted recombination between genomes of phage λ (Chapter V, Sections 9 and 12, respectively). Cunningham *et al.* (1980) reported the formation *in vitro* of very long heteroduplex regions by *recA* protein of *E. coli*. This observation was based on electron microscopy of joint molecules derived from single-stranded circular DNA and linear homologous duplex DNA.

The occurrence of mismatch repair was established through transfection with artificially prepared heteroduplexes of the DNA of phage SPP1 of *Bacillus subtilis* (see Chapter VII, Section 17). Evidence pointing to such repair had already been obtained from study of integration efficiency in the pneumococcus (Chapter II, Section 3(a)) and from transfection with artificially prepared λ heteroduplexes (Chapter V, Section 13). The idea had previously been advanced that conversion in eukaryotes resulted from mismatch correction, and this has now gained much support (Chapter VII, Sections 7, 11 and 17).

There are some interesting variations in the details of mismatch repair in different organisms. In heteroduplexes with wild type of mutations in phage T4 involving deletion or addition of a few nucleotides, the looped-out strand seems to be favoured for excision (Chapter IV, Section 10), while in *Ascobolus* and *Sordaria* the converse applies (Chapter VII, Section 11). In yeast the longer and the shorter strands seem to be exised equally often. With substitution mutants, particular base mismatches are recognized and excised when the *hex* system is functioning in the pneumococcus (Chapter II, Section 3(b)). A similar situation, though with an influence of neighbouring base sequence, seems to apply with *Ustilago maydis* DNase I action (Chapter VII, Section 17). With phage SPP1 of *Bacillus subtilis* neighbouring base sequence seems largely to determine which strand is excised (Chapter VII, Section 17), and it is possible that *Ascobolus* and *Sordaria* are similar, since substitution mutants show much variation in the relative frequencies of different kinds of aberrant asci and do not fall into discrete classes in this respect (Chapter VII, Section 11).

There is evidence from several sources that excision in mismatch repair is extensive. In phage λ the excision appears to be, at least predominantly, in the 5' to 3' direction (Chapter V, Section 13), whereas in pneumococcal transformation and in eukaryote recombination excision seems to occur in both directions from the site of mismatch (Chapter II, Section 3(a), and Chapter VII, Section 10, respectively).

Mismatch repair in recombination does not seem to have any biological significance. The suggestion was made by Ahmad, Holloman and Holliday (1975) that its importance was in the repair of replication errors. Two investigations give support to this idea.

The *hex* system in the pneumococcus not only repairs mismatches in the heteroduplex produced during transformation, but also reduces the frequency of spontaneous mutations (Chapter II, Section 3(b)). Such mutations are likely to arise through errors in DNA replication. The implication of this finding, made by Tiraby and Sicard (1973*b*) and Tiraby and Fox (1973), was that the *hex* system led to fewer replication errors, presumably by excision of mismatched nucleotides. Ahmad *et al.* (1975) pointed out that random repair would frequently erase the wild type in favour of the mutant. They suggested that the discontinuous nature of DNA synthesis would provide strand ends, and so allow non-random repair, with excision initiated at an end. With reference to transformation, Guild and Shoemaker (1976) proposed that the *hex* system, after recognizing a mismatch, searched for nicks or gaps and excised that strand, thus providing an explanation for the preferential excision of donor markers. Claverys, Roger and Sicard (1980) pointed out that such an activity is just what is required if the enzyme corrects replication errors by excising the newly synthesized strand at sites of mismatch.

Investigations with *E. coli* also give support for the idea that mismatch repair may be primarily a means to reduce replication errors. The DNA adenine methylase (*dam*) gene of *E. coli* controls the methylation of adenine residues within the sequence 5' G-A-T-C 3' and the methylation occurs after replication. Thus, old and new strands could be distinguished by the presence or absence of methylation. Moreover, *dam*⁻ mutants show an increased frequency of spontaneous mutation. Radman *et al.* (1980) were thus led to test the idea that in mismatch repair there is preferential excision of the non-methylated strand in a methylated/non-methylated heteroduplex. Phage λ carrying different genetic markers was grown in wild-type and *dam*⁻ hosts, giving methylated (me⁺) and non-methylated (me⁻) DNA, respectively. Heteroduplexes were prepared by extracting the DNA, separating the *l* and *r* strands by the methods described earlier (Chapter V, Section 13) and annealing at 65 °C. Transfection of wild-type *E. coli* was carried out using λ heteroduplexes with one methylated and one unmethylated strand and heterozygous in *trans* for a mutation in the *cI* gene and an amber mutation (no. 3) in gene *P*. These are well separated markers. Over 600 plaques were analysed from each of the following heteroduplexes:

$$(a) \left\{ \frac{l \text{ me}^+ + P}{r \text{ me}^- \ cI +} \right\} \text{ and } (b) \left\{ \frac{l \text{ me}^- + P}{r \text{ me}^+ \ cI +} \right\}.$$

These differ only in the strand that is methylated. Heteroduplex (a) gave over 93% of plaques of genotype + *P*, and heteroduplex (b) gave over 94% of genotype *cI* +. When contaminations of the *r*-strand by the *l*-strand and vice versa were taken into account, Radman *et al.* (1980) concluded that the bias in the recovery of parental markers was at least 100-fold in favour of the methylated genotype. This result was not caused by loss of the non-methylated strand because non-methylated and methylated λ DNA were found to be equally active after transfection. When the experiment was repeated using host strains deficient in mismatch repair, the methylation-induced bias in marker recovery was abolished. It was evident that the methylation-induced bias was provoked by mismatch repair. The methylation, and the mismatch repair that is sensitive to it, appear to be part of a system for detecting and correcting errors in replication.

Mutants of *E. coli* deficient in mismatch repair were first reported by Nevers and Spatz (1975). They analysed single bursts following transfection with heteroduplex λ DNA heterozygous for a temperature-sensitive *cI* mutant and wild type. They found that mixed bursts were rare with wild-type host cells, indicating frequent mismatch repair, but with host cells carrying a *mutU* (formerly called *uvrE*) mutant, there were numerous mixed bursts, implying deficiency in mismatch repair. Interestingly, the *mutU* mutant is associated with a raised level of spontaneous mutation. Further mutants deficient in mismatch repair, revealed in a similar way by transfection with heteroduplex λ DNA, were isolated by Rydberg (1978) and Glickman and Radman (1980) and found also to show high spontaneous mutation rates. These mutants mapped at three other previously known mutator loci, *mutH*, *mutL* and *mutS*. The *mutL* and *mutS* mutants were found by Glickman and Radman (1980) to reduce recombination between amber mutants 3 and 80 at the P locus in λ without affecting intergenic recombination. This is the result expected if the host mutants are deficient in mismatch repair and if such repair is the primary source of allelic recombination.

Pukkila et al. (1982) obtained support for the idea that in E. coli post-replicative mismatch repair is directed by adenine methylation so as to lower the mutation rate. Heteroduplexes of phage λ DNA were prepared using as the source of methylated strands DNA which had been methylated in vitro with purified adenine methylase, such that the adenine residues in all the 5' G-A-T-C 3' sequences were methylated. This methylation to saturation was demonstrated by treatment with restriction enzyme Mbo I (from Moraxella bovis), which will cleave at this site only if the adenines of both strands are unmethylated. Pukkila et al. (1982) found in that in transfection experiments using heteroduplexes of a cI mutant and wild type and with one strand methylated and the other not, excision of the unmethylated strand (the right direction) rather than the methylated one (the wrong direction) at the site of mismatch was favoured by methylation to saturation. Thus, in a heteroduplex consisting of the cI mutant on an unmethylated l (or light) strand and wild type on a methylated r (or heavy) strand, unmixed cI bursts, implying correction in the wrong direction, amounted to only 4% of the total bursts with full methylation of the r strand, but to 26% when there was partial methylation. Pukkila et al. (1982) also found, however, that the two chemically distinct mismatches at the mutant site (cI in l, + in r, and vice versa) were repaired at widely different rates. Evidently the repair frequency of a mismatch is not determined solely by its distance from the methylation sites.

Unexpectedly, Pukkila et al. (1982) also discovered that, when both strands of the heteroduplex were fully methylated, mismatch repair at the site of the cI mutant did not take place. This finding may have implications for recombination of closely linked mutants: as pointed out above, there is evidence that mismatch repair is a primary source of allelic recombinants.

A constant feature of recombination data has been the occurrence of both crossover and non-crossover flanking marker genotypes when closely linked mutants show recombination, as reported, for example, in E. coli by Glickman and Radman (1980), in phage T4 (Chapter IV, Section 6), and in eukaryotes (Chapter VII, Section 5). In T4 the two kinds of event have been distinguished as a crossover heteroduplex and an insertion heteroduplex, and in eukaryotes as a crossover and a crossunder, respectively (Chapter VII, Section 18). According to the Meselson–Radding model these alternative outcomes for an event depend on isomerization – the rotation of arms on one side of the point of strand exchange relative to the other. It has been evident ever since Holliday (1964b) proposed a strand exchange as an intermediate in recombination that the way in which the strands were cut at the exchange point would determine the outcome (see Chapter VII, Section 7).

Recently, an enzyme coded by phage T4 has been characterized that appears to have the ability to resolve such a strand exchange. Gene 49 of T4 is involved in the packaging of DNA in the head of the virus. Infection with mutants in this gene results in accumulation of partially filled heads and a DNA precursor pool that sediments more rapidly than normal. Kemper and Garabett (1981) purified an endonuclease controlled by gene *49* and found that it degraded *in vitro* the rapidly sedimenting DNA isolated from cells infected with *49*-defective phage. The enzyme produced molecules about one-third the size of the T4 genome. Kemper, Garabett and Courage (1981) investigated the substrate specificity of the enzyme *in vitro*. They found that when helix-destabilizing proteins from *E. coli* or T4 (the gene-*32* product) were added to the reaction, the enzyme showed a marked preference for the rapidly sedimenting DNA. Duplex T4 DNA was a poor substrate, so the cleavage sites in the rapidly sedimenting DNA were evidently not specific nucleotide sequences. Cleavage of simplex molecules or gaps in duplex molecules occurred with the gene-*49* enzyme, but was inhibited by the addition of helix-destabilizing protein. Cleavage of the rapidly sedimenting DNA was not inhibited in this way. Kemper *et al.* (1981) concluded that single-stranded regions were not part of the cleavage sites. In agreement with this, the rapidly sedimenting DNA was found to be resistant to single-strand-specific endonuclease S1. Kemper *et al.* believe that the rapidly sedimenting DNA is generated by branched recombinant structures which, in the absence of gene-*49* enzyme, prevent the filling of the phage head. They suggested that the substrate for the endonuclease may be sites of strand exchange which, as Sigal and Alberts (1972) showed, need not contain any single-stranded DNA (see Chapter VII, Section 14).

The final step in recombination is believed to be the sealing of nicks by DNA ligase. An exception to this is Int-promoted recombination in phage λ where a coordinated nicking–swivelling–ligation event takes place and the *E. coli* DNA ligase is not required (Chapter V, Section 9). The evidence that ligase is involved in recombination includes the behaviour of T4 mutants defective in gene *30*, which codes for T4-induced polynucleotide ligase (Chapter IV, Sections 7 and 8); the finding that *E. coli* DNA ligase can seal the joint after strand assimilation associated with the action of λ exonuclease (Chapter V, Section 10); and the discovery that the *cdc9*-controlled ligase in yeast plays a part in X-ray-induced conversion (Chapter VII, Section 17).

The occurrence of multiple recombination events in proximity, leading to negative interference, has been widely reported. One of the earliest examples of this phenomenon was described by Bresch (1955) with phage T1 of *E. coli*. The most frequent cause of negative interference seems to be mismatch correction in hybrid DNA. Support for this explanation of negative interference is provided by the work of Glickman and Radman (1980) described above. They found that mutations at the *mutL* and *mutS* loci in *E. coli*, which cause a deficiency in mismatch repair, also reduced negative interference: as already mentioned, the frequency of allelic recombination was decreased without any significant effect on intergenic recombination. Another source of negative interference is the occurrence of insertion heteroduplexes, implying two breaks in one DNA strand in proximity (Chapter IV, Section 6).

A further source of negative interference is the initiation of a second recombination event, similar in kind to the first one and in proximity to it. Such clustering of discrete events, as distinct from mismatch repair secondary to hybrid DNA formation, has been postulated to occur in transformation (Chapter II, Section 4), in recombination in phage T4 (Chapter IV, Sections 6 and 11) and in eukaryotes (Chapter VII, Sections 16 and 19).

The similarities between the recombination mechanisms in diverse organisms are impressive. Although the process is not fully understood in any organism, sufficient is known to indicate that all may result from variations of the same kinds of event. *In vitro* techniques offer the promise of rapid advance in understanding some of these component steps in recombination. The *in vitro* methods are exemplified by the work on *recA* protein (Chapter VI, Section 11), on Int-promoted recombination in phage λ (Chapter V, Section 9) and on plasmid recombination described above (Potter and Dressler, 1978). These authors have devised a biochemical assay to detect recombination intermediates (Potter and Dressler, 1979). One set of input plasmid DNA molecules was labelled radioactively. A second set of homologous molecules was not radioactive but contained multiple copies of the *E. coli* lactose (*lac*) operator, introduced by genetic manipulation using a restriction eyzyme. Recombination intermediates were detected as radioactive molecules that bound to a nitrocellulose filter containing *lac* repressor, since this binds specifically to the *lac* operator. It was found, as expected, that the intermediates recovered had a figure of eight structure, as seen under the electron microscope. The assay is expected to be useful in purifying recombination enzymes.

IX. Non-homologous recombination by transposable elements

The control of high frequency mutation in *Zea mays* (Maize) had been investigated by McClintock (reviews 1965, 1968). She discovered that a genetic element which she called *Dissociation* (*Ds*) was responsible for the appearance of a recessive allele in a closely linked gene affecting a character of the endosperm and, furthermore, mutation back to the dominant allele occurred with high frequency during the development of the maize grain, leading to a variegated phenotype. The high frequency mutation took place only in the presence of another dominant genetic element, *Activator* (*Ac*), which was unlinked to *Ds*. McClintock found that *Ds* was frequently associated with chromosome breakage or other structural changes at its locus. Another peculiarity was that both *Ac* and *Ds* could be transferred to new positions in the linkage group, or even to a different chromosome altogether. With transfer of *Ds* to a new position, the original dominant phenotype was restored for the character determined by the gene at the old site, but a recessive mutant was generated for the gene at the new site and, again, this mutant showed high frequency somatic mutation in the presence of *Ac*. McClintock called *Ac* and *Ds* controlling elements, since they appeared to act at a certain stage of development and only within the nucleus.

Two additional systems of controlling elements are known in maize: *Dotted* (*Dt*) and *Suppressor-mutator* (*Spm*) – see review by Fincham and Sastry (1974). The three systems act independently.

Controlling elements with properties similar to those of maize are known in several flowering plants causing, for example, coloured spots on the corolla of *Antirrhinum majus* as a result of high frequency mutation (referred to in Chapter VII, Section 2) at the *pallida* (*pal*) locus (Plate 7): see Harrison and Fincham (1964). The high frequency mutation associated with controlling elements may occur in more than one direction, giving rise, for example, to pale spots as well as dark ones. Fincham and Harrison (1967) recorded multiple alleles arising in this way at the *pal* locus in *Antirrhinum*. Dooner and Nelson (1979) showed that a *Ds*-associated mutant of a structural gene (*bronze*, *bz*) for one of the enzymes catalysing anthocyanin biosynthesis in maize showed mutation, in the presence of *Ac*, to several different alleles of *bz*, which differed in the properties of the enzyme.

Controlling elements are also known in *Drosophila melanogaster*, causing high frequency mutation, for example, at the *white* (*w*) eye-colour locus: see review by

Green (1980). The *Drosophila* results show a close parallel to those obtained in maize.

It is evident that the occurrence of controlling elements is widespread in eukaryotes. This phenomenon has been greatly illuminated in an unexpected way, through investigations of mutants in *Escherichia coli* with unusual properties.

Saedler and Starlinger (1967) described mutants of the *galactose (gal)* operon of *E. coli* that from their reversion behaviour were evidently not deletions or base substitutions, and yet all three enzymes of galactose fermentation, coded by the operon, were lacking. A polar effect on genes distal to a mutated gene with respect to the promoter of an operon is well known and is associated with mutants causing polypeptide termination. But the *gal* mutants were not suppressed by amber or ochre suppressors, and the polar effect was stronger than that associated with amber or ochre mutants. Jordan, Saedler and Starlinger (1968) demonstrated that the mutants resulted from the insertion of a segment of DNA into the operon. They showed that phage λ DNA carrying such strongly polar *gal* mutations had a higher buoyant density in a caesium chloride gradient than otherwise identical phage carrying a wild-type galactose operon, and similar results were obtained by Shapiro (1969). Furthermore, the insertions were not random segments of the *E. coli* genome but specific nucleotide sequences. This was demonstrated by Hirsch, Starlinger and Brachet (1972), who tested the homology of insertions by annealing separated strands of phage λ carrying them, and examining the molecules under the electron microscope. Non-homology was revealed as single-stranded loops. They found that the mutations comprised the insertion, in either orientation, of one or other of two specific nucleotide sequences, which they called insertion sequences *1* and *2* (IS*1* and IS*2*). These consisted of about 800 and 1400 nucleotide pairs, respectively, with no detectable homology between them. The insertions occurred at many different positions in the operon, possibly at random sites. Similar results were obtained by Fiandt, Szybalski and Malamy (1972).

Subsequently, more complex transposable elements were recognized and given the name *transposon* (review: Cohen, 1976). Transposons (Tn), unlike insertion sequencies, contain additional genes unrelated to insertion and consequently are larger, usually of more than 2000 nucleotide pairs. Plasmids, unlike transposable elements, are self-replicating and so can multiply independently of the host genome. Plasmids, however, may contain insertion elements. The movement of transposons carrying genes for resistance of antibiotics offered an explanation of the well known transfer of such resistance from one plasmid to another and thence from one bacterial species to another. The movement of transposable elements in the *E. coli* genome or its plasmids was found to be independent of the *recA* gene product. Transposons within plasmids were found often to end with inverted repeats, that is, the duplex DNA at the two ends had identical nucleotide sequences but one had been inverted relative to the other. It was suggested that the reverse repeat might be recognized by the enzyme – a transposase – postulated to bring about transposition. IS*2* has been found to carry a promoter at

Plate 7.   Inflorescence of *Antirrhinum majus* homozygous for *pallida-recurrens*, a mutation at the *pallida* locus. The high frequency mutation to *pallida*⁺ associated with *pallida-recurrens* gives rise to coloured spots and streaks on the corolla. (Courtesy of B. J. Harrison.)

which transcription can be initiated, with the consequence that an adjacent gene would be turned on or off according to the orientation of the insertion sequence.

The parallels between the behaviour of transposable elements in *E. coli* and controlling elements in maize and other eukaryotes are striking and have been discussed by Fincham and Sastry (1974) and Nevers and Saedler (1977). The initial mutation would be caused by the insertion of a transposable element into a gene, either the coding part or the neighbouring regulatory sites. The high frequency mutation would be the consequence of the excision of the transposable element again, and the various mutant end-products would depend on whether the excision was precise, restoring the wild type, or whether too few or too many nucleotides were excised. Chromosome breakage would be the consequence of a failure to rejoin after excision. The need for *Ac* in the *Ac–Ds* system would be accounted for if it supplied the excision function for *Ds*.

An important discovery relevant to the mechanism of transposition in *E. coli* was made by Calos, Johnsrud and Miller (1978) and Grindley (1978). They found that a nine base-pair segment of host DNA was duplicated when IS*1* was inserted, with the result that the insertion element was bounded by this small direct repeat. The nucleotide sequence of the repeat was found to be different at each insertion site, in keeping with the idea that insertion is at random positions, or nearly so. Several other insertion elements were found also to be spanned by a nine base-pair repeat of host DNA and others by a five base-pair repeat (for references see Grindley and Sherratt, 1979). Hypotheses for transposition that incorporated replication of a short segment of host DNA at the insertion site were proposed by Grindley and Sherratt (1979), Shapiro (1979) and Arthur and Sherratt (1979): see Fig. 88.

Research in this field was the subject of the Cold Spring Harbor Symposium in 1980 (*Cold Spring Harbor Symposia on Quantitative Biology*, Vol. 45, published 1981). A few of the outstanding discoveries are summarized below.

IS*1* has a 35 base-pair nucleotide sequence at one end which is also present, but inverted, at the other end. When a plasmid (A) carrying IS*1* combines with a different plasmid (B) not carrying it – a process called *co-integration* – it is found that IS*1* as well as the nine base-pair segment of B at the insertion site are duplicated. The resulting molecule has the structure:

$$\cdots 9\text{bp} - \text{IS}1 - \text{plasmid A} - \text{IS}1 - 9\text{bp} - \text{plasmid B} \cdots$$

where the dotted lines indicate joining to form a circle. Sequence analysis revealed that IS*1* was directly repeated, that is, it was not inverted. Insertion could occur at many different sites in plasmid B, so the nine base-pair repeat varied in sequence from one co-integrate to another.

Transposon *3* (Tn*3*) consists of nearly 5000 nucleotide pairs and contains three genes,

$$- tnpA - tnpR - bla -$$

which code for a transposase, a repressor molecule, and β-lactamase, respectively.

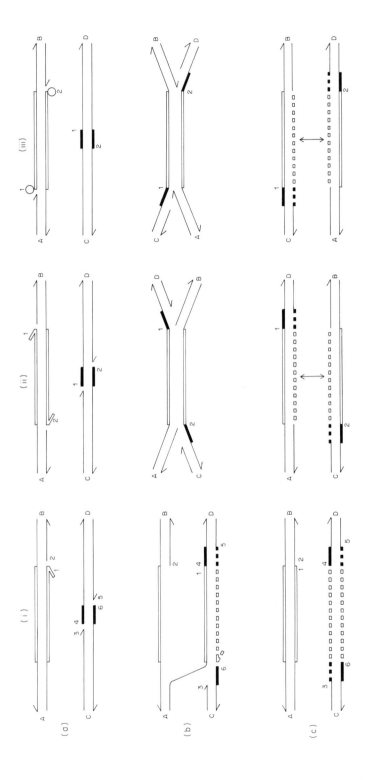

The last-mentioned enzyme confers resistance to ampicillin. There is a 38 base-pair sequence at one end of Tn3 which is inverted at the other end, and it has been shown that the ends are required for transposition. The product of *tnpR* regulates the frequency of transposition and is necessary for the formation of normal transposition products. Co-integrate molecules formed by Tn3 have properties similar to those described above for IS1, except that the host DNA sequence that is replicated at the integration site is of five base pairs instead of nine. In the models proposed by Shapiro (1979) and Arthur and Sherratt (1979), co-integrates are an intermediate in transposition (Fig. 88(ii) and (iii)). Resolution of co-integrates requires the product of gene *tnpR* and it also requires the presence of an internal region of the transposon between *A* and *R* genes. It is believed that the *tnpR* product mediates site-specific recombination in this region. The steps would be as follows. Plasmid A carrying Tn3 would combine with plasmid B, as indicated for IS1, to give a co-integrate of structure

(b) · · · 5bp — Tn3 — plasmid A — Tn3 — 5bp — plasmid B · · · (b)

where (b) and (b) are joined. Reciprocal recombination would then occur between

---

Fig. 88.   The diagram illustrates models for the insertion of transposable DNA into a recipient genome. AB is the donor molecule carrying an insertion element, indicated by white rectangles. CD is the recipient molecule with the target site (five or nine base pairs) indicated by black rectangles. Half-arrowheads indicate 3' ends. Broken lines imply newly synthesized strands.

(i) *The model of Grindley and Sherratt (1979).* (a) One strand of the donor molecule is nicked at the 3' end of the insertion element, and staggered cuts are made at the 5' ends of the target site in the recipient molecule. The numbers identify strand ends. (b) The 3' end of the insertion element is ligated to one of the 5' ends of the target site. DNA replication then proceeds from end no. 5, copying target sequence and insertion element. Subsequently, ligation of the newly synthesized strand to end no. 6 takes place, and further replication occurs in the opposite direction from end no. 3, using the newly synthesized strand as template. (c) Finally, further nicking and ligation restores the donor molecule to its original condition, the replication of the insertion element being conservative.

(ii) *The model of Shapiro (1979).* (a) Staggered cuts are made at the 3' ends of the insertion element and the 5' ends of the target site. (Shapiro pointed out that there was no compelling reason for choosing these ends rather than the other ones (see (iii) below) but the two pairs of cleavages must have opposite polarities.) (b) Ligation occurs of the two ends labelled 1 and of the two ends labelled 2. (c) DNA replication, initiated in the recipient molecule (CD), fills the gaps, giving a co-integrate containing two copies of insertion element and target sequence. Unlike the model of Grindley and Sherratt (1979), the insertion element is thus replicated semi-conservatively. The co-integrate is resolved by site-specific reciprocal recombination within the insertion element (double-headed arrow).

(iii) *The model of Arthur and Sherratt (1979).* (a) A nuclease, shown as a circle, nicks the 5' ends of the transposable element and becomes covalently bound to these ends, thereby conserving the energy of the broken phosphodiester bond. (Arthur and Sherratt considered the alternative possibility of nicking the 3' ends also to be feasible: see (ii)(a) above.) (b) Each 5' end, with nuclease attached, attacks the recipient genome, making staggered nicks and ligating to the 3' ends of the target site (positions 1 and 2). (c) As (ii)(c) above, except that with the initial nicking reversed the replication is initiated in the donor molecule (AB).

the two copies of Tn*3* at the position marked X below:

$$\text{Tn}3$$

(b) $\cdots$ 5bp — *tnpA* — *tnpR* — *bla* — plasmid A — $\cdots$ (a)

X

(a) $\cdots$ *tnpA* — *tnpR* — *bla* — 5bp — plasmid B $\cdots$ (b)

where (a) and (a) are joined, and (b) and (b) are joined. The reciprocal exchange would restore the original plasmid A carrying Tn*3*,

(a) $\cdots$ Tn*3* — plasmid A $\cdots$ (a),

and also give rise to plasmid B carrying it,

(b) $\cdots$ 5bp — Tn*3* — 5bp — plasmid B $\cdots$ (b).

Site-specific recombination within Tn*3* was first proposed by Gill *et al.* (1978) on the basis of the phenotype of deletion mutants of Tn*3*.

These results agree well with the models proposed by Shapiro (1979) and Arthur and Sherratt (1979) (see Fig. 88(ii) and (iii)). They suggested that there is (1) single-strand cleavage in opposite strands at each end of the transposable element, and at each end of the five or nine base-pair host sequence; (2) cross-ligation between host sequence and transposable element at the points of cleavage, generating replication forks at each end of the transposable element; and (3) semi-conservative replication of each strand of the transposable element and the five or nine base-pair host sequence to which it is now joined. Resolution of the co-integrate would be by site-specific reciprocal recombination, as already discussed.

Transposition of an insertion element to a new site within the same DNA molecule will generate either a deletion or an inversion of a segment of the plasmid DNA, if the process follows the steps proposed by Shapiro (1979) or Arthur and Sherratt (1979). Thus a deletion (segment BC) would arise if B were joined to C in Fig. 88, and an inversion (segment BD) if B were joined to D. The orientation of the inversion will reverse if recombination occurs between the two copies of the insertion element that span it.

The movements of transposable elements in *E. coli* described above involve the insertion of the element at a new site without its loss from the original site. This implies replication of the element, as in the models (Fig. 88). In transposition of controlling elements in maize, the insertion element appears to be removed from its previous site. Such excision accounts for high frequency mutation.

A bacteriophage of *E. coli* called Mu has been the subject of much research as, although it can integrate in the host genome as prophage, it differs greatly from λ in its method of integration. It has now become apparent that Mu is in effect a giant transposon, though unlike other transposable elements, Mu is self-replicating. It can integrate at many different places, apparently at random, in the *E. coli* genome. In transposition, the ends of Mu DNA are used in a similar way to those of other transposable elements and five base pairs of host DNA at the integration site are replicated in the course of transposition, as in Tn*3*. Co-integrates can be formed by Mu, but there is no evidence for their resolution in the way that has been established for Tn*3*. There may be a separate pathway for insertion of Mu, distinct from the co-integration pathway.

The recombination system involved in the transposition of insertion elements differs fundamentally from normal recombination in that it does not depend on homology of nucleotide sequence between the two recombining molecules. This is evident from the random – or near random – insertion sites; the ability of insertion elements to be integrated in the host DNA in either orientation; and the duplication of a diversity of host five or nine base-pair sequences when insertion occurs. The presumption, therefore, is that the transposase recognizes the inverted repeat at the ends of the transposable element, but not the nucleotide sequence of the target site – the host sequence that will become duplicated.

Transposable elements seem likely to be of general occurrence in living organisms. They evidently have implications for many different phenomena. The high frequency mutation which it appears they cause has already been discussed. Controlled inversion of a DNA segment is responsible for phase variation of flagellar antigens in *Salmonella*, and for variation in the host range of phage Mu, where a segment called G of 3000 nucleotide pairs is flanked by short inverted duplications. Inversion of the G segment is mediated by a Mu gene called *Gin*. Controlled transpositions are involved in mating-type switching in yeasts, in antigenic variation in trypanosomes, and in rearrangements in mammalian antibody genes during lymphocyte development, thereby accounting for the great diversity of antibody molecules that an individual can produce. As already indicated, the non-homologous recombination brought about by transposable elements differs basically in mechanism from homologous recombination, and has evidently been favoured by natural selection for quite different reasons.

Glossary

Aberrant 4 :4
Segregation of a mutation and wild type at meiosis such that two of the four products show postmeiotic segregation (see Chapter VII, Section 4).

Aberrant segregation
Segregation of a mutation and wild type at meiosis in a ratio other than a simple 2:2 (see Chapter VII, Section 4).

Amber mutant
A mutation that has generated the triplet UAG in the messenger RNA leading to premature polypeptide termination.

Amber suppressor
A mutation that suppresses an amber mutant, probably by a mutation in a gene for a transfer RNA such that an amino acid is inserted instead of polypeptide termination.

Annealing
The association of complementary single strands of DNA to form the duplex molecule.

Auxotroph
An individual with a growth requirement. (Cf. *Prototroph*.)

Branch migration
A change of pairing partner in duplex DNA as a result of the release of hydrogen bonding between base pairs and the formation of new bonds (see Chapter III, Section 5(f), and Chapter IV, Section 8).

Chiasma
The node where paired chromosomes remain in contact in late prophase and metaphase of the first division of meiosis (see Chapter VII, Section 1).

Chromatid
A daughter chromosome still held to its sister at the centromere.

Cis
The same chromosome or genome, not the homologous one of opposite parentage. The term is used particularly for a double heterozygote of genotype

$$\frac{a\,b}{+\,+},$$

where *a* and *b* are closely linked mutations. (Cf. *Trans*.)

Co-conversion
Conversion of two or more closely linked mutations to the same parental genotype (see Chapter VII, Section 9).

Coincidence
A measure of the likelihood of a second crossover in proximity to a first one (see Chapter V, Section 4, and Chapter VII, Section 1).

Co-integration
The joining of two plasmids into a single molecule through the action of an insertion element.

Competence
The state of a bacterial cell when it will take up extracellular DNA in transformation or transfection.

Conversion
Non-reciprocal recombination generating 3 : 1 ratios for a character difference in a tetrad (see Chapter VII, Section 4).

Cross-link
A covalent bond joining the two strands of duplex DNA.

Crossover
Progeny with a non-parental genotype; and the process that gives rise to such recombinants of linked mutants. In prokaryotes the term is commonly used for any process that generates recombinants, whereas in eukaryotes it is usually restricted to those arising by reciprocal exchange (crossing over) in contrast to non-reciprocal events (conversion and postmeiotic segregation).

Crossover heteroduplex
The joining of duplex segments of DNA of differing parentage by means of a heteroduplex region. Flanking markers would thus have a crossover genotype (see Chapter IV, Section 6). (Cf. *Insertion heteroduplex*.)

Crossunder
The term used here for a 'non-crossover recombination event', that is, recombination associated with a parental genotype for flanking markers (see Chapter VII, Section 18).

Deletion mutant
A mutation resulting from the loss of more than one nucleotide, and hence unable to recombine with two or more mutations that can recombine with one another.

D-loop
A triple-stranded region in DNA containing two paired strands and an unpaired strand, the latter covalently joined at both ends to the rest of the duplex DNA (see Chapter VI, Section 11).

Duplex
The double-stranded structure of DNA.

Endonuclease
An enzyme that cuts one or both strands of an intact molecule of nucleic acid.

Excision
The process of cutting out a segment of one strand of a duplex DNA molecule, for example, if it contains a mismatched base or a pyrimidine dimer.

Exonuclease
An enzyme that degrades nucleic acid from a nick or gap, or an end.

Five-prime (5′) end
The end of a polynucleotide with a 5′-phosphate (or hydroxyl) radical. (Cf. *Three-prime (3′) end*.)

Frameshift mutant
A mutation resulting from the addition or deletion of one or two (but not three) nucleotides, with the result that the genetic code is misread for the remainder of the gene.

Gap
A cut in one strand of duplex DNA from which nucleotides have been removed. (Cf. *Nick*.)

Genome
Used here for the 'chromosome' of bacteria and viruses.

Heteroduplex
Duplex DNA where the two strands are of different parentage.

Heterokaryon
A binucleate or multinucleate cell containing nuclei of more than one genotype.

Homoduplex
Duplex DNA in which the two strands are precisely complementary, without any mismatched bases.

Hybrid DNA
DNA in which the two strands of the duplex are derived from different parents, that is, heteroduplex DNA.

Insertion element
A general term for insertion sequences and transposons.

Insertion heteroduplex
The insertion of a segment of one strand of the DNA of one parent into a duplex DNA molecule of the other parent in place of its corresponding segment. Flanking markers would thus have a non-crossover genotype (see Chapter IV, Section 6). (Cf. *Crossover heteroduplex*.)

Insertion sequence
Transposable DNA containing no known genes unrelated to transposition. (Cf. *Transposon*.)

Integration
The incorporation of a phage genome into that of its host, or of transforming DNA into that of the recipient cell.

Interference
A non-random distribution of recombination events. Events more widely spaced than random show positive interference, less widely negative.

Ligase
An enzyme that seals a phosphodiester link at a nick in duplex nucleic acid to produce an intact molecule.

Locus
The place on the genome or chromosome occupied by a gene. (Cf. *Site.*)

Lysis
Breakdown of a bacterial cell as a result of phage attack.

Lysogen
A bacterial cell carrying a prophage and hence liable to undergo lysis if the phage is released.

Map expansion
A frequency of wild-type recombinants of a pair of mutants exceeding the sum of similar frequencies for intervening pairs (see Chapter VII, Section 10).

Marker
A genetic difference between two strains which has arisen as a result of mutation.

Minus (−)
Placed after the name of a gene, this indicates that a mutation has occurred rendering it non-functional (or partially so). (Cf. *Plus.*)

Mismatch
Non-complementary bases in duplex DNA as a result of heteroduplex formation from molecules differing by a mutation. Mismatched bases may also arise by mutation or by replication errors.

Mis-sense mutant
A substitution mutant that leads to the insertion of a different amino acid in the polypeptide.

Modification
An alteration to DNA, usually by methylation of specific bases, to prevent attack by a restriction enzyme.

Nick
A cut in one strand of duplex DNA without removal of any nucleotides. (Cf. *Gap.*)

Ochre mutant
A mutation that has generated the triplet UAA in the messenger RNA leading to premature polypeptide termination.

Ochre suppressor
A mutation that suppresses an ochre mutant, probably by a mutation in a gene for a transfer RNA such that an amino acid is inserted instead of polypeptide termination.

Octad
The group of eight spores derived from a single fungal ascus.

Opal mutant
A mutation that has generated the triplet UGA in the messenger RNA leading to premature polypeptide termination.

Opal suppressor
A mutation that suppresses an opal mutant, probably by a mutation in a gene for a transfer RNA such that an amino acid is inserted instead of polypeptide termination.

P1, P2
The flanking marker genotypes of wild-type recombinants from a cross between closely linked mutants, P1 being the parental genotype of the proximal mutant and P2 the distal (see Chapter VII, Section 3). (Cf. *R1, R2.*)

Permissive
Applied to the situation when a conditional mutant (for example, an amber mutant, or a temperature-sensitive mutant) shows the non-mutant phenotype (for example, in an amber-suppressor strain, or at low temperature). (Cf. *Restrictive.*)

Plaque
The clear area in a bacterial colony resulting from cell breakdown through phage attack.

Plasmid
A self-replicating DNA molecule that exists in the cytoplasm of a cell and, potentially at least, may be integrated into the host genome.

Plus (+)
Placed opposite or after the name of a gene or a mutant, this indicates the normal (that is, wild-type or non-mutant) condition. (Cf. *Minus.*)

Point mutant
A mutation arising from the substitution, addition or deletion of one nucleotide.

Polar mutation
A mutation in one gene of an operon that affects other genes situated distally with respect to the promoter.

Polymerase
An enzyme that synthesizes nucleic acid, using the complementary strand as template.

Postmeiotic segregation
Segregation of a character difference at the first mitosis after meiosis, revealed as sectored colonies in yeast with four spores per ascus, or as 5:3, 3:5 or aberrant 4:4 segregation in eight-spored asci (see Chapter VII, Section 4).

Promoter
The site of initiation of transcription of a gene or an operon.

Prophage
The existence of bacteriophage in a cryptic form with the phage genome integrated in that of its host.

Prototroph
An individual lacking a growth requirement. (Cf. *Auxotroph.*)

R1, R2
The flanking marker genotypes of wild-type recombinants from a cross between closely linked mutants, R1 corresponding to an exchange between the mutant sites and R2 to exchanges in all three marked intervals (see Chapter VII, Section 3). (Cf. *P1, P2.*)

Restriction
The enzyme system of bacteria that restricts phage attack by cutting both strands of the DNA of the virus.

Restrictive
Applied to the situation when a conditional mutant (for example, an amber mutant or a temperature-sensitive mutant) shows the mutant phenotype. (Cf. *Permissive.*)

Simplex
Single-stranded, in contrast to duplex.

Single burst
Lysis of a bacterial cell in isolation from other cells.

Sister-chromatid exchange
An exchange, by breakage and rejoining, between sister chromatids. Each chromatid is believed to have one duplex DNA molecule as its axis (see Chapter VII, Section 18).

Site
The place on the genome or chromosome occupied by a mutation. (Cf. *Locus.*)

Spheroplast
A cell devoid of its cell wall and enclosed only by the inner membrane.

Stop mutant
A mutation (also called a nonsense mutation) that has generated one of the three triplets (UAG, amber; UAA, ochre; UGA, opal) in the messenger RNA leading to premature polypeptide termination.

Stop-mutant suppressor
A mutation that suppresses a stop mutant, probably by a mutation in a gene for a transfer RNA such that an amino acid is inserted instead of polypeptide termination.

Strand assimilation
The progressive incorporation of a strand from one DNA duplex into another (see Chapter V, Section 10).

Substitution mutant
A mutation resulting from the substitution of one base for another.

Supercoiled (superhelical, supertwisted)
Duplex DNA with a different number of coils from that expected from the double helix. Naturally occurring supercoiled DNA is underwound, with fewer coils than would be shown by an unrestrained duplex such as a nicked circular molecule (see Chapter V, Section 9, and Chapter VI, Section 11).

Synapton
The term used here for the synaptonemal complex, the organelle present at pachytene of meiosis and associated with homologous pairing of chromosomes (see Chapter VII, Section 18).

Three-prime (3') end
The end of a polynucleotide with a 3'-hydroxyl (or phosphoryl) radical. (Cf. *Five-prime (5') end.*)

Trans
The homologous chromosome or genome, usually of opposite parentage. The term is used particularly for a double heterozygote of genotype

$$\frac{a\,+}{+\,b},$$

where *a* and *b* are closely linked mutations. (Cf. *Cis*.)

Transduction
The transfer of genetic information from one bacterial cell to another by means of a phage. In generalized transduction random segments of the bacterial genome are transferred. In specialized transduction segments close to the site of integration of the prophage are transferred.

Transfection
Infection of a cell with isolated viral nucleic acid leading to the production of viral particles.

Transformation
The transfer of genetic information by means of extracellular DNA.

Transition
A mutation in which one purine is substituted for the other, and similarly with the pyrimidines, that is, $A \cdot T$ is replaced by $G \cdot C$, or vice versa. (Cf. *Transversion*.)

Transposon
Transposable DNA containing genes unrelated to transposition. (Cf. *Insertion sequence*.)

Transversion
A mutation in which purine and pyrimidine are transposed, that is, $A \cdot T$ is replaced by $T \cdot A$, or $G \cdot C$ by $C \cdot G$, or $A \cdot T$ by $C \cdot G$, or vice versa. (Cf. *Transition*.)

References

References with three or more authors are placed in chronological order.

Abdel-Monem, M., and Hoffman-Berling, H. (1976). Enzymic unwinding of DNA. 1. Purification and characterization of a DNA-dependent ATPase from *Escherichia coli. Eur. J. Biochem.* **65**, 431–440.

Abel, W. O. (1967). Analyse der Interferenz unter verschiedenen Temperaturbedingungen bei *Sphaerocarpos*. I. Chromatidinterferenz. *Molec. gen. Genet.* **99**, 49–61.

Ahmad, A., Holloman, W. K., and Holliday, R. (1975). Nuclease that preferentially inactivates DNA containing mismatched bases. *Nature, Lond.* **258**, 54–56.

Alberts, B. M. (1970). Function of gene 32-protein, a new protein essential for the genetic recombination and replication of T4 bacteriophage DNA. *Fedn Proc. Fedn Am. Socs exp. Biol.* **29**, 1154–1163.

Alberts, B. M., and Frey, L. (1970). T4 bacteriophage gene 32: a structural protein in the replication and recombination of DNA. *Nature, Lond.* **227**, 1313–1318.

Alberts, B. M., Amodio, F. J., Jenkins, M., Gutmann, E. D., and Ferris, F. L. (1969). Studies with DNA–cellulose chromatography. I. DNA-binding proteins from *Escherichia coli. Cold Spring Harb. Symp. quant. Biol.* **33**, 289–305.

Amati, P., and Meselson, M. (1965). Localized negative interference in bacteriophage λ. *Genetics* **51**, 369–379.

Angel, T., Austin, B., and Catcheside, D. G. (1970). Regulation of recombination at the *his-3* locus in *Neurospora crassa. Aust. J. biol. Sci.* **23**, 1229–1240.

Anraku, N., and Lehman, I. R. (1969). Enzymic joining of polynucleotides. VII. Role of the T4-induced ligase in the formation of recombinant molecules. *J. molec. Biol.* **46**, 467–479.

Anraku, N., and Tomizawa, J. (1965*a*). Molecular mechanisms of genetic recombination in bacteriophage. III. Joining of parental polynucleotides of phage T4 in the presence of 5-fluorodeoxyuridine. *J. molec. Biol.* **11**, 501–508.

Anraku, N., and Tomizawa, J. (1965*b*). Molecular mechanisms of genetic recombination of bacteriophage. V. Two kinds of joining of parental DNA molecules. *J. molec. Biol.* **12**, 805–815.

Anraku, N., Anraku, Y., and Lehman, I. R. (1969). Enzymic joining of polynucleotides. VIII. Structure of hybrids of parental T4 DNA molecules. *J. molec. Biol.* **46**, 481–492.

Arthur, A., and Sherratt, D. (1979). Dissection of the transposition process: a transposon-encoded site-specific recombination system. *Molec. gen. Genet.* **175**, 267–274.

Avery, O. T., MacLeod, C. M., and McCarty, M. (1944). Studies on the chemical nature of the substance inducing transformation of pneumococcal types. Induction of transformation by a desoxyribonucleic acid fraction isolated from pneumococcus Type III. *J. exp. Med.* **79**, 137–158.

Baas, P. D., and Jansz, H. S. (1972*a*). Asymmetric information transfer during φX174 DNA replication. *J. molec. Biol.* **63**, 557–568.

Baas, P. D., and Jansz, H. S. (1972*b*). φX174 replicative from DNA replication, origin and direction. *J. molec. Biol.* **63**, 569–576.

368

Bachmann, B. J., and Low, K. B. (1980). Linkage map of *Escherichia coli* K-12, edition 6. *Microbiol. Rev.* **44**, 1–56.

Badman, R. (1972). Deoxyribonuclease-deficient mutants of *Ustilago maydis* with altered recombination frequencies. *Genet. Res.* **20**, 213–229.

Baker, R., and Tessman, I. (1967). The circular genetic map of phage Sl3. *Proc. natn. Acad. Sci. U.S.A.* **58**, 1438–1445.

Baker, R., Doniger, J., and Tessman, I. (1971). Roles of parental and progeny DNA in two mechanisms of phage Sl3 recombination. *Nature new Biol.* **230**, 23–25.

Barbour, S. D., Nagaishi, H., Templin, A., and Clark, A. J. (1970). Biochemical and genetic studies of recombination proficiency in *Escherichia coli*, II. *Rec*$^+$ revertants caused by indirect suppression of *rec*- mutations. *Proc. natn. Acad. Sci. U.S.A.* **67**, 128–135.

Barrell, B. G., Air, G. M., and Hutchison, C. A. (1976). Overlapping genes in bacteriophage φX174. *Nature, Lond.* **264**, 34–41.

Bateson, W. (1909). *Mendel's Principles of Heredity*, Cambridge University Press, Cambridge.

Bateson, W., Saunders, E. R., and Punnett, R. C. (1905). Experimental studies in the physiology of heredity. *Rep. Evol. Comm. R. Soc.* **2**, 1–55, 80–99.

Bell, L., and Byers, B. (1979). Occurrence of crossed strand-exchange forms in yeast DNA during meiosis. *Proc. natn. Acad. Sci. U.S.A.* **76**, 3445–3449.

Benbow, R. M., Hutchison, C. A., Fabricant, J. D., and Sinsheimer, R. L. (1971). Genetic map of bacteriophage φX174. *J. Virol.* **7**, 549–558.

Benbow, R. M., Eisenberg, M., and Sinsheimer, R. L. (1972). Multiple length DNA molecules of bacteriophage φX174. *Nature new Biol.* **237**, 141–144.

Benbow, R. M., Zuccarelli, A. J., Davis, G. C., and Sinsheimer, R. L. (1974*a*). Genetic recombination in bacteriophage φX174. *J. Virol.* **13**, 898–907.

Benbow, R. M., Zuccarelli, A. J., and Sinsheimer, R. L. (1974*b*). A role for single-strand breaks in bacteriophage φX174 genetic recombination. *J. molec. Biol.* **88**, 629–651.

Benbow, R. M., Zuccarelli, A. J., and Sinsheimer, R. L. (1975). Recombinant DNA molecules of bacteriophage φX174. *Proc. natn. Acad. Sci. U.S.A.* **72**, 235–239.

Benz, W. C., and Berger, H. (1973). Selective allele loss in mixed infections with T4 bacteriophage. *Genetics* **73**, 1–11.

Berger, H., and Benz, W. C. (1975). Repair of heteroduplex DNA in bacteriophage T4. In *Molecular Mechanisms for the Repair of DNA* (P. C. Hanawalt and R. B. Setlow, eds), pp. 149–154, Plenum Press, New York.

Berger, H., and Warren, A. J. (1969). Effects of deletion mutants on high negative interference in T4D bacteriophage. *Genetics* **63**, 1–5.

Berger, H., Warren, A. J., and Fry, K. E. (1969). Variations in genetic recombination due to amber mutations in T4D bacteriophage. *J. Virol.* **3**, 171–175.

Bernstein, H. (1969). Repair and recombination in phage T4. I. Genes affecting recombination. *Cold Spring Harb. Symp. quant. Biol.* **33**, 325–331.

Birky, C. W. (1978). Transmission genetics of mitochondria and chloroplasts. *A. Rev. Genet.* **12**, 471–512.

Blanco, M., Levine, A., and Devoret, R. (1975). *lexB*: a new gene governing radiation sensitivity and lysogenic induction in *Escherichia coli* K-12. In *Molecular Mechanisms for the Repair of DNA* (P. C. Hanawalt and R. B. Setlow, eds), pp. 379–382, Plenum Press, New York.

Bodmer, W. (1966). Integration of deoxyribonuclease-treated DNA in *Bacillus subtilis* transformation. *J. gen. Physiol.* **49**, no. 6, Suppl., 233–258.

Bodmer, W. F., and Ganesan, A. T. (1964). Biochemical and genetic studies of integration and recombination in *Bacillus subtilis* transformation. *Genetics* **50**, 717–738.

Bole-Gowda, B. N., Perkins, D. D., and Strickland, W. N. (1962). Crossing-over and interference in the centromere region of linkage group I of *Neurospora*. *Genetics* **47**, 1243–1252.

Bond, D. J. (1973). Polarity of gene conversion and post-meiotic segregation at the *buff* locus in *Sordaria brevicollis. Genet. Res.* **22**, 279–289.

Boon, T., and Zinder, N. D. (1970). Genetic recombination in bacteriophage f$_1$: transfer of parental DNA to the recombinant. *Virology* **41**, 444–452.

Boon, T., and Zinder, N. D. (1971). Genotypes produced by individual recombination events involving bacteriophage f$_1$. *J. molec. Biol.* **58**, 133–151.

Boram, W. R., and Roman, H. (1976). Recombination in *Saccharomyces cerevisiae*: a DNA repair mutation associated with elevated mitotic gene conversion. *Proc. natn. Acad. Sci. U.S.A.* **73**, 2828–2832.

Boyd, J. B., Golino, M. D., and Setlow, R. B. (1976). The *mei-9^a* mutant of *Drosophila melanogaster* increases mutagen sensitivity and decreases excision repair. *Genetics* **84**, 527–544.

Boyle, J. M., and Symonds, N. (1969). Radiation-sensitive mutants of T4D. I. T4*y*: a new radiation-sensitive mutant; effect of the mutation on radiation survival, growth and recombination. *Mutation Res.* **8**, 431–439.

Brenner, S., Streisinger, G., Horne, R. W., Champe, S. P., Barnett, L., Benzer, S., and Rees, M. W. (1959). Structural components of bacteriophage. *J. molec. Biol.* **1**, 281–292.

Bresch, C. (1955). Zum Paarungsmechanismus von Bakteriophagen. *Z. Naturf.* **10b**, 545–561.

Breschkin, A. M., and Mosig, G. (1977*a*). Multiple interactions of a DNA-binding protein *in vivo*. I. Gene 32 mutations of phage T4 inactivate different steps in DNA replication and recombination. *J. molec. Biol.* **112**, 279–294.

Breschkin, A. M., and Mosig, G. (1977*b*). Multiple interactions of a DNA-binding protein *in vivo*. II. Effects of host mutations on DNA replication of phage T4 gene 32 mutants. *J. molec. Biol.* **112**, 295–308.

Bresler, S. E., and Lanzov, V. A. (1967). Mechanism of genetic recombination during bacterial conjugation of *Escherichia coli* K-12. II. Incorporation of the donor DNA fragment into the recombinant chromosome. *Genetics* **56**, 117–124.

Bridges, C. B. (1915). A linkage variation in *Drosophila. J. exp. Zool.* **19**, 1–12.

Broker, T. R. (1973). An electron microscopic analysis of pathways for bacteriophage T4 DNA recombination. *J. molec. Biol.* **81**, 1–16.

Broker, T. R., and Lehman, I. R. (1971). Branched DNA molecules: intermediates in T4 recombination. *J. molec. Biol.* **60**, 131–149.

Brooks, K., and Clark, A. J. (1967). Behaviour of λ bacteriophage in a recombination deficient strain of *Escherichia coli. J. Virol.* **1**, 283–293.

Brown, N. L., and Smith, M. (1977). The sequence of a region of bacteriophage φX174 DNA coding for parts of genes *A* and *B. J. molec. Biol.* **116**, 1–30.

Buchanan, R. E., Holt, J. G., and Lessel, E. F. (eds) (1966). *Index bergeyana*. E. & S. Livingstone, Edinburgh and London.

Burnet, F. M. (1927). The relationship between heat-stable agglutinogens and sensitivity to bacteriophage in the *Salmonella* group. *Br. J. exp. Path.* **8**, 121–129.

Burnet, F. M., and McKie, M. (1930*a*). Balanced salt action as manifested in bacteriophage phenomena. *Aust. J. exp. Biol. med. Sci.* **7**, 183–198.

Burnet, F. M., and McKie, M. (1930*b*). The electrical behaviour of bacteriophages. *Aust. J. exp. Biol. med. Sci.* **7**, 199–209.

Cairns, J. (1961). An estimate of the length of the DNA molecule of T2 bacteriophage by autoradiography. *J. molec. Biol.* **3**, 756–761.

Calef, E. (1957). Effect on linkage maps of selection of crossovers between closely linked markers. *Heredity* **11**, 265–279.

Calef, E., and Licciardello, G. (1960). Recombination experiments on prophage host relationships. *Virology* **12**, 81–103.

Calos, M. P., Johnsrud, L., and Miller, J. H. (1978). DNA sequence at the integration sites of the insertion element IS*1. Cell* **13**, 411–418.

Campbell, A. M. (1962). Episomes. *Adv. Genet.* **11**, 101–145.

Campbell, A. M. (1965). The steric effect in lysogenization by bacteriophage lambda. II. Chromosomal attachment of the b_2 mutant. *Virology* **27**, 340–345.

Carlson, P. S. (1971). A genetic analysis of the *rudimentary* locus of *Drosophila melanogaster*. *Genet. Res.* **17**, 53–81.

Carlson, P. S. (1974). Mitotic crossing-over in a higher plant. *Genet. Res.* **24**, 109–112.

Carpenter, A. T. C. (1975). Electron microscopy of meiosis in *Drosophila melanogaster* females. II. The recombination nodule – a recombination-associated structure at pachytene? *Proc. natn. Acad. Sci. U.S.A.* **72**, 3186–3189.

Carpenter, A. T. C. (1979a). Synaptonemal complex and recombination nodules in wild-type *Drosophila melanogaster* females. *Genetics* **92**, 511–541.

Carpenter, A. T. C. (1979b). Synaptonemal complex and recombination nodules in recombination-deficient mutants of *Drosophila melanogaster*. *Chromosoma* **75**, 259–292.

Carpenter, A. T. C. (1981). EM autoradiographic evidence that DNA synthesis occurs at recombination nodules during meiosis in *Drosophila melanogaster* females. *Chromosoma* **83**, 59–80.

Carroll, R. B., Neet, K. E., and Goldthwait, D. A. (1972). Self-association of gene-32 protein of bacteriophage T4. *Proc. natn. Acad. Sci. U.S.A.* **69**, 2741–2744.

Case, M. E., and Giles, N. H. (1959). Recombination mechanisms at the *pan-2* locus in *Neurospora crassa*. *Cold Spring Harb. Symp. quant. Biol.* **23**, 119–135.

Case, M. E., and Giles, N. H. (1964). Allelic recombination in *Neurospora*: tetrad analysis of a three-point cross within the *pan-2* locus. *Genetics* **49**, 529–540.

Cassuto, E., and Radding, C. M. (1971). Mechanism for the action of λ exonuclease in genetic recombination. *Nature new Biol.* **229**, 13–16; **230**, 128.

Cassuto, E., Lash, T., Sriprakash, K. S., and Radding, C. M. (1971). Role of exonuclease and β protein of phage λ in genetic recombination. V. Recombination of λ DNA *in vitro*. *Proc. natn. Acad. Sci. U.S.A.* **68**, 1639–1643.

Cassuto, E., Mursalim, J., and Howard-Flanders, P. (1978). Homology-dependent cutting in *trans* of DNA in extracts of *Escherichia coli*: an approach to the enzymology of genetic recombination. *Proc. natn. Acad. Sci. U.S.A.* **75**, 620–624.

Cassuto, E., West, S. C., Mursalim, J., Conlon, S., and Howard-Flanders, P. (1980). Initiation of genetic recombination: homologous pairing between duplex DNA molecules promoted by *recA* protein. *Proc. natn. Acad. Sci. U.S.A.* **77**, 3962–3966.

Castellazzi, M., George, J., and Buttin, G. (1972). Prophage induction and cell division in *E. coli*. I. Further characterization of the thermosensitive mutation *tif-1* whose expression mimics the effect of UV irradiation. *Molec. gen. Genet.* **119**, 139–152.

Castellazzi, M., Morand, P., George, J., and Buttin, G. (1977). Prophage induction and cell devision in *E. coli*. V. Dominance and complementation analysis in partial diploids with pleiotropic mutations (*tif*, *recA*, *zab* and *lexB*) at the *recA* locus. *Molec. gen. Genet.* **153**, 297–310.

Catcheside, D. G. (1966). A second gene controlling allelic recombination in *Neurospora crassa*. *Aust. J. biol. Sci.* **19**, 1039–1046.

Catcheside, D. G. (1975). Occurrence in wild strains of *Neurospora crassa* of genes controlling genetic recombination. *Aust. J. biol. Sci.* **28**, 213–225.

Catcheside, D. G. (1977). *The Genetics of Recombination*, Edward Arnold, London.

Catcheside, D. G., and Angel, T. (1974). A *histidine-3* mutant, in *Neurospora crassa*, due to an interchange. *Aust. J. biol. Sci.* **27**, 219–229.

Cavalli-Sforza, L. L., and Jinks, J. L. (1956). Studies on the genetic system of *Escherichia coli* K-12. *J. Genet.* **54**, 87–112.

Chaganti, R. S. K., Schonberg, S., and German, J. (1974). A manyfold increase in sister chromatid exchanges on Bloom's syndrome lymphocytes. *Proc. natn. Acad. Sci. U.S.A.* **71**, 4508–4512.

Champoux, J. J. (1977). Renaturation of complementary single-stranded DNA circles:

complete rewinding facilitated by the DNA untwisting enzyme. *Proc. natn. Acad. Sci. U.S.A.* **74**, 5328–5332.

Champoux, J. J., and Dulbecco, R. (1972). An activity from mammalian cells that untwists superhelical DNA – a possible swivel for DNA replication. *Proc. natn. Acad. Sci. U.S.A.* **69**, 143–146.

Chase, M., and Doermann, A. H. (1958). High negative interference over short segments of the genetic structure of bacteriophage T4. *Genetics* **43**, 332–353.

Chilton, M.-D. (1967). Transforming activity in both complementary strands of *Bacillus subtilis* DNA. *Science, N.Y.* **157**, 817–819.

Chilton, M.-D., and Hall, B. D. (1968). Transforming activity in single-stranded DNA from *Bacillus subtilis*. *J. molec. Biol.* **34**, 439–451.

Chovnick. A. (1958). Aberrant segregation and pseudoallelism at the garnet locus in *Drosophila melanogaster*. *Proc. natn. Acad. Sci. U.S.A.* **44**, 333–337.

Chovnick, A. (1961). The garnet locus in *Drosophila melanogaster*. I. Pseudoallelism. *Genetics* **46**, 493–507.

Chovnick, A., Ballantyne, G. H., Baillie, D. L., and Holm, D. G. (1970). Gene conversion in higher organisms: half-tetrad analysis of recombination within the rosy cistron of *Drosophila melanogaster*. *Genetics* **66**, 315–329.

Chovnick, A., Ballantyne, G. H., and Holm, D. G. (1971). Studies on gene conversion and its relationship to linked exchange in *Drosophila melanogaster*. *Genetics* **69**, 179–209.

Clark, A. J. (1967). The beginning of a genetic analysis of recombination proficiency. *J. cell. comp. Physiol.* **70**, Suppl. 1, 165–180.

Clark, A. J. (1973). Recombination deficient mutants of *E. coli* and other bacteria. *A. Rev. Genet.* **7**, 67–86.

Clark, A. J., and Margulies, A. D. (1965). Isolation and characterization of recombination-deficient mutants of *Escherichia coli* K-12. *Proc. natn. Acad. Sci. U.S.A.* **53**, 451–459.

Claverys, J. P., Roger, M., and Sicard, A. M. (1980). Excision and repair of mismatched base pairs in transformation of *Streptococcus pneumoniae*. *Molec. gen. Genet.* **178**, 191–201.

Clowes, R. C., and Moody, E. E. M. (1966). Chromosomal transfer from 'recombination-deficient' strains of *Escherichia coli* K-12. *Genetics* **53**, 717–726.

Cohen, S. N. (1976). Transposable genetic elements and plasmid evolution. *Nature, Lond.* **263**, 731–738.

Cole, R. S. (1973). Repair of DNA containing interstrand crosslinks in *Escherichia coli*: sequential excision and recombination. *Proc. natn. Acad. Sci. U.S.A.* **70**, 1064–1068.

Craig, N. L., and Roberts, J. W. (1980). *E. coli* recA protein-directed cleavage of phage λ repressor requires polynucleotide. *Nature, Lond.* **283**, 26–30.

Cunningham, R. P., Shibata, T., DasGupta, C., and Radding, C. M. (1979). Single strands induce recA protein to unwind duplex DNA for homologous pairing. *Nature, Lond.* **281**, 191–195.

Cunningham, R. P., DasGupta, C., Shibata, T., and Radding, C. M. (1980). Homologous pairing in genetic recombination: *recA* protein makes joint molecules of gapped circular DNA and closed circular DNA. *Cell* **20**, 223–235.

Curtis, M. J., and Alberts, B. (1976). Studies on the structure of intracellular bacteriophage T4 DNA. *J. molec. Biol.* **102**, 793–816.

Darlington, A. J., and Bodmer, W. F. (1968). Events occurring at the site of integration of a DNA molecule in *Bacillus subtilis* transformation. *Genetics* **60**, 681–684.

Das, A., Court, D., and Adhya, S. (1976). Isolation and characterization of conditional lethal mutants of *Escherichia coli* defective in transcription termination factor rho. *Proc. natn. Acad. Sci. U.S.A.* **73**, 1959–1963.

DasGupta, C., Shibata, T., Cunningham, R. P., and Radding, C. M. (1980). The topology of homologous pairing promoted by *recA* protein. *Cell* **22**, 427–446.

Davies, R. W., Schreier, P. H., and Büchel, D. E. (1977). Nucleotide sequence of the attachment site of coliphage lambda. *Nature, Lond.* **270**, 757–760.

Davis, K. J., and Symonds, N. (1974). The pathway of recombination in phage T4. A genetic study. *Molec. gen. Genet.* **132**, 173–180.

Davis, R. W., and Parkinson, J. S. (1971). Deletion mutants of bacteriophage lambda. III. Physical structure of att^ϕ. *J. molec. Biol.* **56**, 403–423.

Delbrück, M., and Bailey, W. T. (1947). Induced mutations in bacterial viruses. *Cold Spring Harb. Symp. quant. Biol.* **11**, 33–37.

Delius, H., Mantell, N. J., and Alberts, B. (1972). Characterization by electron microscopy of the complex formed between T4 bacteriophage gene 32-protein and DNA. *J. molec. Biol.* **67**, 341–350.

Demerec, M. (1926). Reddish – a frequently 'mutating' character in *Drosophila virilis*. *Proc. natn. Acad. Sci. U.S.A.* **12**, 11–16.

Demerec, M. (1928). Mutable characters of *Drosophila virilis*. I. Reddish-alpha body character. *Genetics* **13**, 359–388.

Demerec, M., and Fano, U. (1945). Bacteriophage-resistant mutants in *Escherichia coli*. *Genetics* **30**, 119–136.

Deonier, R. C., and Davidson, N. (1977). The sequence organisation of the integrated F plasmid in two Hfr strains of *Escherichia coli*. *J. molec. Biol.* **107**, 207–222.

De Vries, H. (1903). *Befruchtung und Bastardierung*, Leipzig. (English translation by C. S. Gager entitled 'Fertilization and hybridization' as pp. 217–263 of *Intracellular Pangenesis Including a Paper on Fertilization and Hybridization*, Chicago, 1910).

DiCaprio, L., and Hastings, P. J. (1976). Gene conversion and intragenic recombination at the *SUP6* locus and the surrounding region in *Saccharomyces cerevisiae*. *Genetics* **84**, 697–721.

Doerfler, W., and Hogness, D. S. (1968). Gene orientation in bacteriophage lambda as determined from the genetic activities of heteroduplex DNA formed *in vitro*. *J. molec. Biol.* **33**, 661–678.

Doermann, A. H., and Boehner, L. (1963). An experimental analysis of bacteriophage T4 heterozygotes. I. Mottled plaques from crosses involving six *r*II loci. *Virology* **21**, 551–567.

Doermann, A. H., and Hill, M. B. (1953). Genetic structure of bacteriophage T4 as described by recombination studies of factors influencing plaque morphology. *Genetics* **38**, 79–90.

Doermann, A. H., and Parma, D. H. (1967). Recombination in bacteriophage T4. *J. cell. comp. Physiol.* **70**, Suppl. 1, 147–164.

Doniger, J., Warner, R. C., and Tessman, I. (1973). Role of circular dimer DNA in the primary recombination mechanism of bacteriophage S13. *Nature new Biol.* **242**, 9–12.

Dooner, H. K., and Nelson, O. E. (1979). Heterogeneous flavonoid glucosyltransferases in purple derivatives from a controlling element-suppressed *bronze* mutant in maize. *Proc. natn. Acad. Sci. U.S.A.* **76**, 2369–2371.

Drake, J. W. (1966). Heteroduplex heterozygotes in bacteriophage T4 involving mutations of various dimensions. *Proc. natn. Acad. Sci. U.S.A.* **55**, 506–512.

Dubnau, D., and Cirigliano, C. (1972). Fate of transforming DNA following uptake by competent *Bacillus subtilis*. IV. The endwise attachment and uptake of transforming DNA. *J. molec. Biol.* **64**, 31–46.

Dubnau, D., and Davidoff-Abelson, R. (1971). Fate of transforming DNA following uptake by competent *Bacillus subtilis*. I. Formation and properties of the donor-recipient complex. *J. molec. Biol.* **56**, 209–221.

Ebersold, W. T., and Levine, R. P. (1959). A genetic analysis of linkage group I of *Chlamydomonas reinhardi*. *Z. VererbLehre* **90**, 74–82.

Echols, H., and Gingery, R. (1968). Mutants of bacteriophage λ defective in vegetative genetic recombination. *J. molec. Biol.* **34**, 239–249.

Echols, H., and Green, L. (1979). Some properties of site-specific and general recombination inferred from Int-initiated exchanges by bacteriophage lambda. *Genetics* **93**, 297–307.

Echols, H., Gingery, R., and Moore, L. (1968). Integrative recombination function of bacteriophage λ: evidence for a site-specific recombination enzyme. *J. molec. Biol.* **34**, 251–260.

Edgar, R. S., and Lielausis, I. (1964). Temperature-sensitive mutants of bacteriophage T4D: their isolation and genetic characterization. *Genetics* **49**, 649–662.

Edgar, R. S., and Steinberg, C. M. (1958). On the origin of high negative interference over short segments of the genetic structure of bacteriophage T4. *Virology* **6**, 115–128.

Edgar, R. S. Denhardt, G. H., and Epstein, R. H. (1964). A comparative genetic study of conditional lethal mutations of bacteriophage T4D. *Genetics* **49**, 635–648.

Emerson, S. (1969). Linkage and recombination at the chromosome level. In *Genetic Organization* (E. W. Caspari and A. W. Ravin, eds), Vol. 1, pp. 267–360, Academic Press, New York and London.

Emerson, S., and Yu-Sun, C. C. C. (1967). Gene conversion in the Pasadena strain of *Ascobolus immersus*. *Genetics* **55**, 39–47.

Emmerson, P. T. (1968). Recombination deficient mutants of *Escherichia coli* K12 that map between *thyA* and *argA*. *Genetics* **60**, 19–30.

Emmerson, P. T. and West. S. C. (1977). Identification of protein X of *Escherichia coli* as the *recA⁺/tif⁺* gene product. *Molec. gen. Genet.* **155**, 77–85.

Enea, V., and Zinder, N. D. (1976). Heteroduplex DNA: a recombinational intermediate in bacteriophage f1. *J. molec. Biol.* **101**, 25–38.

Enea, V., Vovis, G. F., and Zinder, N. D. (1975). Genetic studies with heteroduplex DNA of bacteriophage f1. Asymmetric segregation, base correction and implications for the mechanism of genetic recombination. *J. molec. Biol.* **96**, 495–509.

Enquist, L. W., and Skalka, A. (1973). Replication of bacteriophage λ DNA dependent on the function of host and viral genes. I. Interaction of *red*, *gam* and *rec*. *J. molec. Biol.* **75**, 185–212.

Enquist, L. W., and Weisberg, R. A. (1976). The red plaque test: a rapid method for identification of excision defective variants of bacteriophage lambda. *Virology* **72**, 147–153.

Enquist, L. W., and Weisberg, R. A. (1977). A genetic analysis of the *att–int–xis* region of coliphage lambda. *J. molec. Biol.* **111**, 97–120.

Enquist, L. W., Nash, H., and Weisberg, R. A. (1979). Strand exchange in site-specific recombination. *Proc. natn. Acad. Sci. U.S.A.* **76**, 1363–1367.

Ephrussi-Taylor, H. (1960). On the biological functions of deoxyribonucleic acid. *Symp. Soc. gen. Microbiol.* **10**, 132–154.

Ephrussi-Taylor, H. (1961). Etude de la recombinaison à l'échelle moléculaire dans la transformation bactérienne. *J. Chim. phys.* **58**, 1090–1099.

Ephrussi-Taylor, H. (1966). Genetic recombination in DNA-induced transformation of pneumococcus. IV. The pattern of transmission and phenotypic expression of high and low-efficiency donor sites in the *amiA* locus. *Genetics* **54**, 211–222.

Ephrussi-Taylor, H., and Gray, T. C. (1966). Genetic studies of recombining DNA in pneumococcal transformation. *J. gen. Physiol.* **49**, no. 6, Suppl., 211–231.

Ephrussi-Taylor, H., Sicard, A. M., and Kamen, R. (1965). Genetic recombination in DNA-induced transformation of pneumococcus. I. The problem of relative efficiency of transforming factors. *Genetics* **51**, 455–475.

Esposito, M. S. (1971). Postmeiotic segregation in *Saccharomyces*. *Molec. gen. Genet.* **111**, 297–299.

Esposito, M. S. (1978). Evidence that spontaneous mitotic recombination occurs at the two-strand stage. *Proc. natn. Acad. Sci. U.S.A.* **75**, 4436–4440.

Fabre, F., and Roman, H. (1977). Genetic evidence for inducibility of recombination competence in yeast. *Proc. natn. Acad. Sci. U.S.A.* **74**, 1667–1671.

Fabre, F., and Roman, H. (1979). Evidence that a single DNA ligase is involved in replication and recombination in yeast. *Proc. natn. Acad. Sci. U.S.A.* **76**, 4586–4588.

Fareed, G. C., and Richardson, C. C. (1967). Enzymatic breakage and joining of deoxyribonucleic acid. II. The structural gene for polynucleotide ligase in bacteriophage T4. *Proc. natn. Acad. Sci. U.S.A.* **58**, 665–672.

Faulds, D., Dower, N., Stahl, M. M., and Stahl, F. W. (1979). Orientation-dependent recombination hotspot activity in bacteriophage λ. *J. molec. Biol.* **131**, 681–695.

Fiandt, M., Syzbalski, W., and Malamy, M. H. (1972). Polar mutations in *lac, gal* and phage λ consist of a few DNA sequences inserted with either orientation. *Molec. gen. Genet.* **119**, 223–231.

Fidianián, H. M., and Ray, D. S. (1972). Replication of bacteriophage M13. VII. Requirement of the gene 2 protein for the accumulation of a specific RFII species. *J. molec. Biol.* **72**, 51–63.

Fincham, J. R. S. (1974). Negative interference and the use of flanking markers in fine-structure mapping in fungi. *Heredity* **33**, 116–121.

Fincham, J. R. S. (1976). Recombination in the *am* gene of *Neurospora crassa* – a new model for conversion polarity and an explanation for a marker effect. *Heredity* **36**, 81–89.

Fincham, J. R. S., and Baron, A. J. (1977). The molecular basis of an osmotically reparable mutant of *Neurospora crassa* producing unstable glutamate dehydrogenase. *J. molec. Biol.* **110**, 627–642.

Fincham, J. R. S., and Harrison, B. J. (1967). Instability at the *pal* locus in *Antirrhinum majus*. II. Multiple alleles produced by mutation of one original unstable allele. *Heredity* **22**, 211–224.

Fincham, J. R. S., and Holliday, R. (1970). An explanation of fine structure map expansion in terms of excision repair. *Molec. gen. Genet.* **109**, 309–322.

Fincham, J. R. S., and Sastry, G. R. K. (1974). Controlling elements in maize. *A. Rev. Genet.* **8**, 15–50.

Fink, G. R. (1974). Properties of gene conversion of deletions in *Saccharomyces cerevisiae*. In *Mechanisms in Recombination* (R. F. Grell, ed.) pp. 287–293. Plenum Press, New York and London.

Fink, G. R., and Styles, C. A. (1974). Gene conversion of deletions in the *his4* region of yeast. *Genetics* **77**, 231–244.

Finnerty, V. G., Duck, P., and Chovnick, A. (1970). Studies on genetic organization in higher organisms. II. Complementation and fine structure of the *maroon-like* locus of *Drosophila melanogaster*. *Proc. natn. Acad. Sci. U.S.A.* **65**, 939–946.

Fischer-Fantuzzi, L., and Calef, E. (1964). A type of λ prophage unable to confer immunity. *Virology* **23**, 209–216.

Fogel, S., and Hurst, D. D. (1967). Meiotic gene conversion in yeast tetrads and the theory of recombination. *Genetics* **57**, 455–481.

Fogel, S., and Mortimer, R. K. (1969). Informational transfer in meiotic gene conversion. *Proc. natn. Acad. Sci. U.S.A.* **62**, 96–103.

Fogel, S., and Mortimer, R. K. (1970). Fidelity of meiotic gene conversion in yeast. *Molec. gen. Genet.* **109**, 177–185.

Fogel, S., and Roth, R. (1974). Mutations affecting meiotic gene conversion in yeast. *Molec. gen. Genet.* **130**, 189–201.

Fogel, S., Hurst, D. D., and Mortimer, R. K. (1971). Gene conversion in unselected tetrads from multipoint crosses. *Stadler Symp.* **1/2**, 89–110.

Fogel, S., Mortimer, R., Lusnak, K., and Tavares, F. (1979). Meiotic gene conversion: a signal of the basic recombination event in yeast. *Cold Spring Harb. Symp. quant. Biol.* **43**, 1325–1341.

Fornili, S. L., and Fox, M. S. (1977). Electron microscope visualization of the products of *Bacillus subtilis* transformation. *J. molec. Biol.* **113**, 181–191.

Foss, H. M., and Stahl, F. W. (1963). Circularity of the genetic map of bacteriophage T4. *Genetics* **48**, 1659–1672.

Fox, M. S. (1960). Fate of transforming deoxyribonucleate following fixation by transformable bactera. II. *Nature, Lond.* **187**, 1004–1006.

Fox. M. S. (1966). On the mechanism of integration of transforming deoxyribonucleate. *J. gen. Physiol.* **49**, no. 6, Suppl., 183–196.

Fox, M. S., and Allen, M. K. (1964). On the mechanism of deoxyribonucleate integration in pneumococcal transformation. *Proc. natn. Acad. Sci. U.S.A.* **52**, 412–419.

Fox, M. S., and Hotchkiss, R. D. (1957). Initiation of bacterial transformation. *Nature, Lond.* **179**, 1322–1325.

Fox, M. S., and Hotchkiss, R. D. (1960). Fate of transforming deoxyribonucleate following fixation by transformable bacteria. I. *Nature. Lond.* **187**, 1002–1003.

Francke, B., and Ray, D. S. (1971). Formation of the parental replicative form DNA of bacteriophage φX174 and initial events in its replication. *J. molec. Biol.* **61**, 565–586.

Franklin, N. C., Dove, W. F., and Yanofsky, C. (1965). The linear insertion of a prophage into the chromosome of *E. coli* shown by deletion mapping. *Biochem. biophys. Res. Commun.* **18**, 910–923.

Friedberg, E. C., and King, J. J. (1971). Dark repair of ultraviolet-irradiated deoxyribonucleic acid by bacteriophage T4: purification and characterization of a dimer-specific phage-induced endonuclease. *J. Bact.* **106**, 500–507.

Game, J. C., Johnston, L. H., and von Borstel, R. C. (1979). Enhanced mitotic recombination in a ligase-defective mutant of the yeast *Saccharomyces cerevisiae*. *Proc. natn. Acad. Sci. U.S.A.* **76**, 4589–4592.

Gelbart, W. M., McCarron, M., Pandey, J., and Chovnick, A. (1974). Genetic limits of the xanthine dehydrogenase structural element within the *rosy* locus in *Drosophila melanogaster*. *Genetics* **78**, 869–886.

Gellert, M., Mizuuchi, K., O'Dea, M. H., and Nash, H. A. (1976). DNA gyrase: an enzyme that introduces superhelical turns into DNA. *Proc. natn. Acad. Sci. U.S.A.* **73**, 3872–3876.

German, J. (1964). Cytological evidence for crossing-over *in vitro* in human lymphoid cells. *Science, N.Y.* **144**, 298–301.

Ghikas, A., and Lamb, B. C. (1977). The detection, in unordered octads, of 6+ :2m and 2+ :6m ratios with postmeiotic segregation, and of aberrant 4:4s, and their use in corresponding-site interference studies. *Genet. Res.* **29**, 267–278.

Gilbert, W., and Dressler, D. (1969). DNA replication: the rolling circle model *Cold Spring Harb. Symp. quant. Biol.* **33**, 473–484.

Giles, N. H. (1952). Studies on the mechanism of reversion in biochemical mutants of *Neurospora crassa*. *Cold Spring Harb. Symp. quant. Biol.* **16**, 283–313.

Giles, N. H. (1956). Forward and back mutation at specific loci in *Neurospora*. *Brookhaven Symp. Biol.* **8**, 103–125.

Gill, R., Heffron, F., Dougan, G., and Falkow, S. (1978). Analysis of sequences transposed by complementation of two classes of transposition-deficient mutants of Tn*3*. *J. Bact.* **136**, 742–756.

Gillen, J. R., Karu, A. E., Nagaishi, H., and Clark, A. J. (1977). Characterization of the deoxyribonuclease determined by lambda reverse as exonuclease VIII of *Escherichia coli*. *J. molec. Biol.* **113**, 27–41.

Gingery, R., and Echols, H. (1967). Mutants of bacteriophage λ unable to integrate into the host chromosome. *Proc. natn. Acad. Sci. U.S.A.* **58**, 1507–1514.

Gingold, E. B., and Ashworth, J. M. (1974). Evidence for mitotic crossing-over during the parasexual cycle of the cellular slime mould *Dictyostelium discoideum*. *J. gen. Microbiol.* **84**, 70–78.

Glassberg, J., Meyer, R. R., and Kornberg, A. (1979). Mutant single-strand binding protein of *Escherichia coli*: genetic and physiological characterization. *J. Bact.* **140**, 14–19.

Glickman, B. W., and Radman, M. (1980). *Escherichia coli* mutator mutants deficient in methylation-instructed mismatch correction. *Proc. natn. Acad. Sci. U.S.A.* **77**, 1063–1067.

Godson, G. N. (1974). Origin and direction of φX174 double- and single-stranded DNA synthesis. *J. molec. Biol.* **90**, 127–141.

Goldman, S. L., and Smallets, S. (1979). Site specific induction of gene conversion: the effects of homozygosity of the *ade6* mutant M26 of *Schizosaccharomyces pombe* on meiotic gene conversion. *Molec. gene. Genet.* **173**, 221–225.

Goldmark, P. J., and Linn, S. (1970). An endonuclease activity from *Escherichia coli* absent from certain *rec⁻* strains. *Proc. natn. Acad. Sci. U.S.A.* **67**, 434–441.

Goldmark, P. J., and Linn, S. (1972). Purification and properties of the *recBC* DNase of *Escherichia coli* K-12. *J. biol. Chem.* **247**, 1849–1860.

Goldthwait, D., and Jacob, F. (1964). Sur le mécanisme de l'induction du développement du prophage chez les bactéries lysogènes. *C. r. hebd. Séanc. Acad. Sci., Paris* **259**, 661–664.

Goodgal, S. H., and Postel, E. H. (1967). On the mechanism of integration following transformation with single-stranded DNA of *Hemophilus influenzae. J. molec. Biol.* **28**, 261–273.

Gordon, C. N., Rush, M. G., and Warner, R. C. (1970). Complex replicative form molecules of bacteriophages φX174 and S13 *su*105. *J. molec. Biol.* **47**, 495–503.

Gottesman, M. E., and Yarmolinsky, M. B. (1968). Integration-negative mutants of bacteriophage lambda. *J. molec. Biol.* **31**, 487–505.

Gottesman, M. M., Gottesman, M. E., Gottesman, S., and Gellert, M. (1974). Chracterization of bacteriophage λ reverse as an *Escherichia coli* phage carrying a unique set of host-derived recombination functions. *J. molec. Biol.* **88**, 471–487.

Gottesman, S., and Gottesman, M. E. (1975*a*). Elements involved in site-specific recombination in bacteriophage lambda. *J. molec. Biol.* **91**, 489–499.

Gottesman, S., and Gottesman, M. (1975*b*). Excision of prophage λ in a cell-free system. *Proc. natn. Acad. Sci. U.S.A.* **72**, 2188–2192.

Gray, T. C., and Ephrussi-Taylor, H. (1967). Genetic recombination in DNA-induced transformation of pneumococcus. V. The absence of interference, and evidence for the selective elimination of certain donor sites from the final recombinants. *Genetics* **57**, 125–153.

Green, M. M. (1960). Double crossing-over or gene conversion at the white loci in *Drosophila melanogaster? Genetics* **45**, 15–18.

Green, M. M. (1980). Transposable elements in *Drosophila* and other Diptera. *A. Rev. Genet.* **14**, 109–120.

Greenstein, M., and Skalka, A. (1975). Replication of bacteriophage lambda DNA: *in vivo* studies of the interaction between the viral *gamma* protein and the host *recBC* DNase. *J. molec. Biol.* **97**, 543–559.

Griffith, F. (1928). Significance of pneumococcal types. *J. Hyg., Camb.* **27**, 113–159.

Grindley, N. D. F. (1978). IS*1* insertion generates duplication of a nine base pair sequence at its target site. *Cell* **13**, 419–426.

Grindley, N. D. F., and Sherratt, D. J. (1979). Sequence analysis at IS*1* insertion sites: models for transposition. *Cold Spring Harb. Symp. quant. Biol.* **43**, 1257–1261.

Gross, J., and Englesberg, E. (1959). Determination of the order of mutational sites governing L-arabinose utilization in *Escherichia coli* B/r by transduction with phage Plbt. *Virology* **9**, 314–331.

Gross, J., Grunstein, J., and Witkin, E. M. (1971). Inviability of *recA⁻* derivatives of the DNA polymerase mutant of DeLucia and Cairns. *J. molec. Biol.* **58**, 631–634.

Guarneros, G., and Echols, H. (1970). New mutants of bacteriophage λ with a specific defect in excision from the host chromosome. *J. molec. Biol.* **47**, 565–574.

Gudas, L. J., and Mount, D. W. (1977). Identification of the *recA* (*tif*) gene product of *Escherichia coli. Proc. natn. Acad. Sci. U.S.A.* **74**, 5280–5284.

Gudas, L. J., and Pardee, A. B. (1976). DNA synthesis inhibition and the induction of protein X in *Escherichia coli*. *J. molec. Biol.* **101**, 459–477.

Guerrini, F. (1969). On the asymmetry of λ integration sites. *J. molec. Biol.* **46**, 523–542.

Guerrini, F., and Fox, M. S. (1968*a*). Genetic heterozygosity in pneumococcal transformation. *Proc. natn. Acad. Sci. U.S.A.* **59**, 429–436.

Guerrini, F., and Fox, M. S. (1968*b*). Effects of DNA repair in transformation-heterozygotes of Pneumococcus. *Proc. natn. Acad. Sci. U.S.A.* **59**, 1116–1123.

Guild, W. R., and Shoemaker, N. B. (1974). Intracellular competition for a mismatch recognition system and marker-specific rescue of transforming DNA from inactivation by ultraviolet irradiation. *Molec. gen. Genet.* **128**, 291–300.

Guild, W. R., and Shoemaker, N. B. (1976). Mismatch correction in pneumococcal transformation: donor length and *hex*-dependent marker efficiency. *J. Bact.* **125**, 125–135.

Gurney, T., and Fox, M. S. (1968). Physical and genetic hybrids formed in bacterial transformation. *J. molec. Biol.* **32**, 83–100.

Gutz, H. (1971). Site specific induction of gene conversion in *Schizosaccharomyces pombe*. *Genetics* **69**, 317–337.

Hall, J. D., and Howard-Flanders, P. (1975). Temperature-sensitive *recA* mutant of *Escherichia coli* K-12: deoxyribonucleic acid metabolism after ultraviolet irradiation. *J. Bact.* **121**, 892–900.

Hamlett, N., and Berger, H. (1975). Mutations altering genetic recombination and repair of DNA in bacteriophage T4. *Virology* **63**, 539–567.

Harm, W. (1963). Mutants of phage T4 with increased sensitivity to ultraviolet. *Virology* **19**, 66–71.

Harris, W. J., and Barr, G. C. (1969). Some properties of DNA in competent *Bacillus subtilis*. *J. molec Biol.* **39**, 245–255.

Harris, W. J., and Barr, G. C. (1971). Mechanism of transformation in *B. subtilis*. *Molec. gen. Genet.* **113**, 331–344.

Harrison, B. J., and Carpenter, R. (1977). Somatic crossing-over in *Antirrhinum majus*. *Heredity* **38**, 169–189.

Harrison B. J., and Fincham, J. R. S. (1964). Instability at the *pal* locus in *Antirrhinum majus*. I. Effects of environment on frequencies of somatic and germinal mutation. *Heredity* **19**, 237–258.

Hartman, N., and Zinder, N. D. (1974*a*). The effect of B specific restriction and modification of DNA on linkage relationships in f1 bacteriophage. I. Studies on the mechanism of B restriction *in vivo*. *J. molec. Biol.* **85**, 345–356.

Hartman, N., and Zinder, N. D. (1974*b*). The effect of B specific restriction and modification of DNA on linkage relationships in f1 bacteriophage. II. Evidence for a heteroduplex intermediate in f1 recombination. *J. molec. Biol.* **85**, 357–369.

Hastings, P. J., and Whitehouse, H. L. K. (1964). A polaron model of genetic recombination by the formation of hybrid deoxyribonucleic acid. *Nature, Lond.* **201**, 1052–1054.

Hayes, W. (1966*a*). Sex factors and viruses. *Proc. R. Soc. B* **164**, 230–245.

Hayes, W. (1966*b*). Some controversial aspects of bacterial sexuality. *Proc. R. Soc. B* **165**, 1–19.

Hayes, W. (1968). *The Genetics of Bacteria and Their Viruses*, 2nd edn, Blackwell Scientific Publications, Oxford.

Herman, R. K. (1965). Reciprocal recombination of chromosome and F-merogenote in *Escherichia coli*. *J. Bact.* **90**, 1664–1668.

Hershey, A. D. (1947). Spontaneous mutations in bacterial viruses. *Cold Spring Harb. Symp. quant. Biol.* **11**, 67–77.

Hershey, A. D., and Chase, M. (1952). Genetic recombination and heterozygotes in bacteriophage. *Cold Spring Harb. Symp. quant. Biol.* **16**, 471–479.

Hershey, A. D., and Rotman, R. (1948). Linkage among genes controlling inhibition of lysis in a bacterial virus. *Proc. natn. Acad. Sci. U.S.A.* **34**, 89–96.

Hershey, A. D., and Rotman, R. (1949). Genetic recombination between host-range and plaque-type mutants of bacteriophage in single bacterial cells. *Genetics* **34**, 44–71.

Hershey, A. D., Burgi, E., and Streisinger, G. (1958). Genetic recombination between phages in the presence of chloramphenicol. *Virology* **6**, 287–288.

Hexter, W. M. (1958). On the nature of the garnet locus in *Drosophila melanogaster. Proc. natn. Acad. Sci. U.S.A.* **44**, 768–771.

Hirsch, H.-J., Starlinger, P., and Brachet, P. (1972). Two kinds of insertions in bacterial genes. *Molec. gen. Genet.* **119**, 191–206.

Hofer, F., Hollenstein, H., Janner, F., Minet, M., Thuriaux, P., and Leupold, U. (1979). The genetic fine structure of nonsense suppressors in *Schizosaccharomyces pombe. Curr. Genet.* **1**, 45–61.

Hofschneider, P. H. (1963). Untersuchungen über 'kleine' *E. coli* K12 Bakteriophagen. *Z. Naturf.* **18b**, 203–210.

Hogness, D. S., Doerfler, W., Egan, J. B., and Black, L. W. (1967). The position and orientation of genes in λ and λdg DNA. *Cold Spring Harb. Symp. quant. Biol.* **31**, 129–138.

Holliday, R. (1961). Induced mitotic crossing-over in *Ustilago maydis. Genet. Res.* **2**, 231–248.

Holliday, R. (1962). Mutation and replication in *Ustilago maydis. Genet. Res.* **3**, 472–486.

Holliday, R. (1964a). The induction of mitotic recombination by mitomycin C in *Ustilago* and *Saccharomyces. Genetics* **50**, 323–335.

Holliday, R. (1964b). A mechanism for gene conversion in fungi. *Genet. Res.* **5**, 282–304.

Holliday, R. (1965). Radiation sensitive mutants of *Ustilago maydis. Mutation Res.* **2**, 557–559.

Holliday, R. (1966). Studies on mitotic gene conversion in *Ustilago. Genet. Res.* **8**, 323–337.

Holliday, R. (1968). Genetic recombination in fungi. In *Replication and Recombination of Genetic Material* (W. J. Peacock and R. D. Brock, eds), pp. 157–174, Australian Academy of Science, Canberra.

Holliday, R. (1971). Biochemical measure of the time and frequency of radiation-induced allelic recombination in *Ustilago. Nature new Biol.* **232**, 233–236.

Holliday, R., and Dickson, J. M. (1977). The detection of post-meiotic segregation without tetrad analysis in *Ustilago maydis. Molec. gen. Genet.* **153**, 331–335.

Holliday, R., and Halliwell, R. E. (1968). An endonuclease-deficient strain of *Ustilago maydis. Genet. Res.* **12**, 95–98.

Holliday, R., Halliwell, R. E., Evans, M. W., and Rowell, V. (1976). Genetic characterization of *rec-1*, a mutant of *Ustilago maydis* defective in repair and recombination. *Genet. Res.* **27**, 413–453.

Holloman, W. K. (1973). Studies on a nuclease from *Ustilago maydis*. II. Substrate specificity and mode of action of the enzyme. *J. biol. Chem.* **248**, 8114–8119.

Holloman, W. K., and Holliday, R. (1973). Studies on a nuclease from *Ustilago maydis*. I. Purification, properties and implication in recombination of the enzyme. *J. biol. Chem.* **248**, 8107–8113.

Holloman, W. K., and Radding, C. M. (1976). Recombination promoted by superhelical DNA and the *recA* gene of *Escherichia coli. Proc. natn. Acad. Sci. U.S.A.* **73**, 3910–3914.

Holloman, W. K., Wiegand, R., Hoessli, C., and Radding, C. M. (1975). Uptake of homologous single-stranded fragments by superhelical DNA: a possible mechanism for initiation of genetic recombination. *Proc. natn. Acad. Sci. U.S.A.* **72**, 2394–2398.

Horii, T., Ogawa, T., and Ogawa, H. (1980). Organization of the *recA* gene of *Escherichia coli. Proc. natn. Acad. Sci. U.S.A.* **77**, 313–317.

Horii, Z.-I., and Clark, A. J. (1973). Genetic analysis of the RecF pathway to genetic

379

recombination in *Escherichia coli* K12: isolation and characterization of mutants. *J. molec. Biol.* **80**, 327–344.

Horiuchi, K., and Zinder, N. D. (1972). Cleavage of bacteriophage f1 DNA by the restriction enzyme of *Escherichia coli* B. *Proc. natn. Acad. Sci. U.S.A.* **69**, 3220–3224.

Horiuchi, K., Vovis, G. F., Enea, V., and Zinder, N. D. (1975). Cleavage map of bacteriophage f1: location of the *Escherichia coli* B-specific modification sites. *J. molec. Biol.* **95**, 147–165.

Hosoda, J. (1976). Role of genes *46* and *47* in bacteriophage T4 reproduction. III. Formation of joint molecules in biparental recombination. *J. molec. Biol.* **106**, 277–284.

Hotchkiss, R. D. (1971). Toward a general theory of genetic recombination in DNA. *Adv. Genet.* **16**, 325–348.

Hotta, Y., and Stern, H. (1971*a*). Analysis of DNA synthesis during meiotic prophase in *Lilium*. *J. molec. Biol.* **55**, 337–355.

Hotta, Y., and Stern, H. (1971*b*). A DNA-binding protein in meiotic cells of *Lilium*. *Devl. Biol.* **26**, 87–99.

Hotta, Y., and Stern, H. (1971*c*). Meiotic protein in spermatocytes of mammals. *Nature new Biol.* **234**, 83–86.

Hotta, Y., and Stern, H. (1974). DNA scission and repair during pachytene in *Lilium*. *Chromosoma* **46**, 279–296.

Hotta, Y., Ito, M., and Stern, H. (1966). Synthesis of DNA during meiosis. *Proc. natn. Acad. Sci. U.S.A.* **56**, 1184–1191.

Hotta, Y., Chandley, A. C., and Stern, H. (1977*a*). Biochemical analysis of meiosis in the male mouse. II. DNA metabolism at pachytene. *Chromosoma* **62**, 255–268.

Hotta, Y., Chandley, A. C., and Stern, H. (1977*b*). Meiotic crossing-over in lily and mouse. *Nature, Lond.* **269**, 240–242.

Howard-Flanders, P., and Boyce, R. P. (1966). DNA repair and genetic recombination: studies of mutants of *Escherichia coli* defective in these processes. *Radiation Res.*, Suppl. 6, 156–184.

Howard-Flanders, P., and Theriot, L. (1966). Mutants of *Escherichia coli* K-12 defective in DNA repair and in genetic recombination. *Genetics* **53**, 1137–1150.

Howell, S. H., and Stern, H. (1971). The appearance of DNA breakage and repair activities in the synchronous meiotic cycle of *Lilium*. *J. molec. Biol.* **55**, 357–378.

Hradecna, Z., and Szybalski, W. (1969). Electron micrographic maps of deletions and substitutions in the genomes of transducing coliphages λdg and λbio. *Virology* **38**, 473–477.

Huberman, J. A., Kornberg, A., and Alberts, B. M. (1971). Stimulation of T4 bacteriophage DNA polymerase by the protein product of T4 gene 32. *J. molec. Biol.* **62**, 39–52.

Hunnable, E. G., and Cox, B. S. (1971). The genetic control of dark recombination in yeast. *Mutation Res.* **13**, 297–309.

Hurst, D. D., and Fogel, S. (1964). Mitotic recombination and heteroallelic repair in *Saccharomyces cerevisiae*. *Genetics* **50**, 435–458.

Hurst, D. D., Fogel, S., and Mortimer, R. K. (1972). Conversion-associated recombination in yeast. *Proc. natn. Acad. Sci. U.S.A.* **69**, 101–105.

Ihler, G., and Rupp, W. D. (1969). Strand-specific transfer of donor DNA during conjugation in *E. coli*. *Proc. natn. Acad. Sci. U.S.A.* **63**, 138–143.

Inouye, M., and Pardee, A. B. (1970). Changes of membrane proteins and their relation to deoxyribonucleic acid synthesis and cell devision of *Escherichia coli*. *J. biol. Chem.* **245**, 5813–5819.

Iyer, V. N., and Ravin, A. W. (1962). Integration and expression of different lengths of DNA during the transformation of pneumococcus to erythromycin resistance. *Genetics* **47**, 1355–1368.

Jacob, F., and Wollman, E. L. (1954). Etude génétique d'un bactériophage tempéré d'*Escherichia coli*. I. Le système génétique du bactériophage λ. *Annls Inst. Pasteur, Paris* **87**, 653–673.

Jacob, F., and Wollman, E. L. (1958). Genetic and physical determinations of chromosomal segments in *Escherichia coli*. *Symp. Soc. exp. Biol.* **12**, 75–92.

Jacob, F., and Wollman, E. L. (1961). *Sexuality and the Genetics of Bacteria*, Academic Press, New York and London.

Jacobson, G. K., Pinõn, R., Esposito, R. E., and Esposito, M. S. (1975). Single-strand scissions of chromosomal DNA during commitment to recombination at meiosis. *Proc. natn. Acad. Sci. U.S.A.* **72**, 1887–1891.

James, A. P. (1955). A genetic analysis of sectoring in ultraviolet-induced variant colonies of yeast. *Genetics* **40**, 204–213.

James, A. P., and Lee-Whiting, B. (1955). Radiation-induced genetic segregations in vegetative cells of diploid yeast. *Genetics* **40**, 826–831.

Jansen, G. J. O. (1970*a*). Survival of *uvsB* and *uvsC* mutants of *Aspergillus nidulans* after UV-irradiation. *Mutation Res.* **10**, 21–32.

Jansen, G. J. O. (1970*b*). Abnormal frequencies of spontaneous mitotic recombination in *uvsB* and *uvsC* mutants of *Aspergillus nidulans*. *Mutation Res.* **10**, 33–41.

Janssens, F. A. (1909). Spermatogénèse dans les Batraciens. V. La théorie de la chiasmatypie. Nouvelles interprétation des cinèses de maturation. *Cellule* **25**, 387–411.

Jessop, A. P., and Catcheside D. G. (1965). Interallelic recombination at the *his*-1 locus in *Neurospora crassa* and its genetic control. *Heredity* **20**, 237–256.

Johnson, P. H., and Sinsheimer, R. L. (1974). Structure of an intermediate in the replication of bacteriophage φX174 deoxyribonucleic acid: the initiation site for DNA replication. *J. molec. Biol.* **83**, 47–61.

Johnston, L. H., and Nasmyth, K. A. (1978). *Saccharomyces cerevisiae* cell cycle mutant *cdc*9 is defective in DNA ligase. *Nature, Lond.* **274**, 891–893.

Jones, G. H. (1967). The control of chiasma distribution in rye. *Chromosoma* **22**, 69–90.

Jones, G. H. (1971). The analysis of exchanges in tritium-labelled meiotic chromosomes. II. *Stethophyma grossum*. *Chromosoma* **34**, 367–382.

Jones, G. H. (1974*a*). Correlated components of chiasma variation and the control of chiasma distribution in rye. *Heredity* **32**, 375–387.

Jones, G. H. (1974*b*). Is chiasma determination sequential? *Nature, Lond.* **250**, 147–148.

Jones, G. H., and Rees, H. (1964). Genotypic control of chromosome behaviour in rye. VIII. The distribution of chiasmata within pollen mother cells. *Heredity* **19**, 719–730.

Jones, R. N. (1975). B-chromosome systems in flowering plants and animal species. *Int. Rev. Cytol.* **40**, 1–100.

Jones, R. N., and Rees, H. (1967). Genotypic control of chromosome behaviour in rye. XI. The influence of *B* chromosomes on meiosis. *Heredity* **22**, 333–347.

Jordan, E., Saedler, H., and Starlinger, P. (1968). *O*° and strong polar mutations in the *gal* operon are insertions. *Molec. gen. Genet.* **102**, 353–363.

Kaiser, A. D. (1955). A genetic study of the temperate coliphage λ. *Virology* **1**, 424–443.

Kaiser, A. D., and Masuda, T. (1970). Evidence for a prophage excision gene in λ. *J. molec. Biol.* **47**, 557–564.

Kaiser, K., and Murray, N. E. (1979). Physical characterisation of the 'Rac prophage' in *E. coli*. K12. *Molec. gen. Genet.* **175**, 159–174.

Kakar, S. N. (1963). Allelic recombination and its relation to recombination of outside markers. *Genetics* **48**, 957–966.

Katz, E. R., and Kao, V. (1974). Evidence for mitotic recombination in the cellular slime mold *Dictyostelium discoideum*. *Proc. natn. Acad. Sci. U.S.A.* **71**, 4025–4026.

Kellenberger, G., Zichichi, M. L., and Weigle, J. (1960). Mutations affecting the density of bacteriophage λ. *Nature, Lond.* **187**, 161–162.

Kellenberger, G., Zichichi, M. L., and Weigle, J. (1961). Exchange of DNA in the recombination of bacteriophage λ. *Proc. natn. Acad. Sci. U.S.A.* **47**, 869–878.

Kellenberger, G., Zichichi, M. L., and Epstein, H. T. (1962). Heterozygosis and recombination of bacteriophage λ. *Virology* **17**, 44–55.

Kemper, B., and Garabett, M. (1981). Studies on T4-head maturation. 1. Purification and characterization of gene-49-controlled endonuclease. *Eur. J. Biochem.* **115**, 123–131.

Kemper, B., Garabett, M., and Courage, U. (1981). Studies on T4-head maturation. 2. Substrate specificity of gene-49-controlled endonuclease. *Eur. J. Biochem.* **115**, 133–141.

Kikuchi, Y., and Nash, H. A. (1978). The bacteriophage λ *int* gene product. A filter assay for genetic recombination, purification of Int, and specific binding to DNA. *J. biol. Chem.* **253**, 7147–7157.

Kikuchi, Y., and Nash, H. A. (1979). Nicking–closing activity associated with bacteriophage λ *int* gene product. *Proc. natn. Acad. Sci. U.S.A.* **76**, 3760–3764.

Kim, J.-S., Sharp, P. A., and Davidson, N. (1972). Electron microscope studies of heteroduplex DNA for a deletion mutant of bacteriophage φX174. *Proc. natn. Acad. Sci. U.S.A.* **69**, 1948–1952.

Kitani, Y., and Olive, L. S. (1967). Genetics of *Sordaria fimicola*. VI. Gene conversion at the *g* locus in mutant x wild-type crosses. *Genetics* **57**, 767–782.

Kitani, Y., and Olive, L. S. (1969). Genetics of *Sordaria fimicola*. VII. Gene conversion at the *g* locus in interallelic crosses. *Genetics* **62**, 23–66.

Kitani, Y., and Whitehouse, H. L. K. (1974). Aberrant ascus genotypes from crosses involving mutants at the *g* locus in *Sordaria fimicola*. *Genet. Res.* **24**, 229–250.

Kitani, Y., Olive, L. S., and El-Ani, A. S. (1961). Transreplication and crossing-over in *Sordaria fimicola*. *Science, N.Y.* **134**, 668–669.

Kitani, Y., Olive, L. S., and El-Ani, A. S. (1962). Genetics of *Sordaria fimicola*. V. Aberrant segregation at the *g* locus. *Am. J. Bot.* **49**, 697–706.

Kotewicz, M., Chung, S., Takeda, Y., and Echols, H. (1977). Characterization of the integration protein of bacteriophage λ as a site-specific DNA-binding protein. *Proc. natn. Acad. Sci. U.S.A.* **74**, 1511–1515.

Kozinski, A. W. (1961). Fragmentary transfer of P^{32}-labelled parental DNA to progeny phage. *Virology* **13**, 124–134.

Kozinski, A. W., and Felgenhauer, Z. Z. (1967). Molecular recombination in T4 bacteriophage deoxyribonucleic acid. II. Singe-strand breaks and exposure of uncomplemented areas as a prerequisite for recombination. *J. Virol.* **1**, 1193–1202.

Kozinski, A. W., and Kozinski, P. B. (1963). Fragmentary transfer of P^{32}-labelled DNA to progeny phage. II. The average size of the transferred parental fragment. Two-cycle transfer. Repair of the polynucleotide chain after fragmentation. *Virology* **20**, 213–229.

Kozinski, A. W., and Kozinski, P. B. (1964). Replicative fragmentation in T4 bacteriophage DNA. II. Biparental molecular recombination. *Proc. natn. Acad. Sci. U.S.A.* **52**, 211–218.

Kozinski, A. W., Kozinski, P. B., and Shannon, P. (1963). Replicative fragmentation in T4 phage: inhibition by chloramphenicol. *Proc. natn. Acad. Sci. U.S.A.* **50**, 746–753.

Kozinski, A. W., Kozinski, P. B., and James, R. (1967). Molecular recombination in T4 bacteriophage deoxyribonucleic acid. I. Tertiary structure of early replicative and recombining deoxyribonucleic acid. *J. Virol.* **1**, 758–770.

Krisch, H. M., Hamlett, N. V., and Berger, H. (1972). Polynucleotide ligase in bacteriophage T4D recombination. *Genetics* **72**, 187–203.

Kruszewska, A., and Gajewski, W. (1967). Recombination within the Y locus in *Ascobolus immersus*. *Genet. Res.* **9**, 159–177.

Kushner, S. R. (1974). Differential thermolability of exonuclease and endonuclease activities of the *recBC* nuclease isolated from thermosensitive *recB* and *recC* mutants. *J. Bact.* **120**, 1219–1222.

Kushner, S. R., Nagaishi, H., Templin, A., and Clark, A. J. (1971). Genetic recombination in *Escherichia coli*: the role of exonuclease I. *Proc. natn. Acad. Sci. U.S.A.* **68**, 824–827.

382

Kushner, S. R., Nagaishi, H., and Clark, A. J. (1972). Indirect suppression of *recB* and *recC* mutations by exonuclease I deficiency. *Proc. natn. Acad. Sci. U.S.A.* **69**, 1366–1370.

Kushner, S. R., Nagaishi, H., and Clark, A. J. (1974). Isolation of exonuclease VIII: the enzyme associated with the *sbcA* indirect suppressor. *Proc. natn. Acad. Sci. U.S.A.* **71**, 3593–3597.

Kutter, E. M., and Wiberg, J. S. (1968). Degradation of cytosine-containing bacterial and bacteriophage DNA after infection of *Escherichia coli* B with bacteriophage T4D wild type and with mutants defective in genes 46, 47 and 56. *J. molec. Biol.* **38**, 395–411.

Kvelland, I. (1969). The effect of homozygous deletions upon heterozygote formation in bacteriophage T4D. *Genet. Res.* **14**, 13–31.

Lacks, S. (1962). Molecular fate of DNA in genetic transformation of *Pneumococcus*. *J. molec. Biol.* **5**, 119–131.

Lacks, S. (1966). Integration efficiency and genetic recombination in pneumococcal transformation. *Genetics* **53**, 207–235.

Lacks, S. (1970). Mutants of *Diplococcus pneumoniae* that lack deoxyribonucleases and activities possibly pertinent to genetic transformation. *J. Bact.* **101**, 373–383.

Lacks, S., and Greenberg, B. (1976). Single-strand breakage on binding of DNA to cells in the genetic transformation of *Diplococcus pneumoniae*. *J. molec. Biol.* **101**, 255–275.

Lacks, S., and Hotchkiss, R. D. (1960*a*). A study of the genetic material determining an enzyme activity in *Pneumococcus*. *Biochim. biophys. Acta* **39**, 508–517.

Lacks, S., and Hotchkiss, R. D. (1960*b*). Formation of amylomaltase after genetic transformation of pnuemococcus. *Biochim. biophys. Acta* **45**, 155–163.

Lacks, S., Greenberg, B., and Neuberger, M. (1974). Role of deoxyribonuclease in the genetic transformation of *Diplococcus pneumoniae*. *Proc. natn. Acad. Sci. U.S.A.* **71**, 2305–2309.

Laipis, P. J., Olivera, B. M., and Ganesan, A. T. (1969). Enzymatic cleavage and repair of transforming DNA. *Proc. natn. Acad. Sci. U.S.A.* **62**, 289–296.

Lam, S. T., Stahl, M. M., McMilin, K. D., and Stahl, F. W. (1974). Rec-mediated recombinational hot spot activity in bacteriophage lambda. II. A mutation which causes hot spot activity. *Genetics* **77**, 425–433.

Lamb, B. C., and Helmi, S. (1978). A new type of genetic control of gene conversion, from *Ascobolus immersus*. *Genet. Res.* **32**, 67–78.

Landy, A., and Ross, W. (1977). Viral integration and excision: structure of the lambda *att* sites. *Science, N.Y.* **197**, 1147–1160.

Latt, S. A. (1973). Microfluorometric detection of deoxyribonucleic acid replication in human metaphase chromosomes. *Proc. natn. Acad. Sci. U.S.A.* **70**, 3395–3399.

Lawrence, C. W., Sherman, F., Jackson, M., and Gilmore, R. A. (1975). Mapping and gene conversion studies with the structural gene for iso-1-cytochrome *c* in yeast. *Genetics* **81**, 615–629.

Leblon, G. (1972*a*). Mechanism of gene conversion in *Ascobolus immersus*. I. Existence of a correlation between the origin of mutants induced by different mutagens and their conversion spectrum. *Molec. gen. Genet.* **115**, 36–48.

Leblon, G. (1972*b*). Mechanism of gene conversion in *Ascobolus immersus*. II. The relationships between the genetic alterations in b_1 or b_2 mutants and their conversion spectrum. *Molec. gen. Genet.* **116**, 322–335.

Leblon, G., and Rossignol, J.-L. (1973). Mechanism of gene conversion in *Ascobolus immersus*. III. The interaction of heteroalleles in the conversion process. *Molec. gen. Genet.* **122**, 165–182.

Lederberg, E. M. (1951). Lysogenicity in *E. coli* K-12. *Genetics* **36**, 560 (Abstr.).

Lederberg, E. M., and Lederberg, J. (1953). Genetic studies of lysogenicity in *Escherichia coli*. *Genetics* **38**, 51–64.

Lederberg, J., and Tatum, E. L. (1946). Gene recombination in *Escherichia coli*. *Nature, Lond.* **158**, 558.

Lederberg, J., Cavalli, L. L., and Lederberg, E. M. (1952). Sex compatibility in *Escherichia coli. Genetics* **37**, 720–730.

Lee, C. S., Davis, R. W., and Davidson, N. (1970). A physical study by electron microscopy of the terminally repetitious, circularly permuted DNA from the coliphage particles of *Escherichia coli* 15. *J. molec. Biol.* **48**, 1–22.

Lefevre, J. C., Claverys, J. P., and Sicard, A. M. (1979). Donor deoxyribonucleic acid length and marker effect in pneumococcal transformation. *J. Bact.* **138**, 80–86.

Lehman, I. R., and Nussbaum, A. L. (1964). The deoxyribonucleases of *Escherichia coli.* V. On the specificity of exonuclease I (phosphodiesterase). *J. biol. Chem.* **239**, 2628–2636.

Leupold, U. (1959). Studies on recombination in *Schizosaccharomyces pombe. Cold Spring Harb. Symp. quant. Biol.* **23**, 161–170.

Leupold, U. (1961). Intragene Rekombination und allele Komplementierung. *Arch. Julius Klaus-Stift. VererbForsch.* **36**, 89–117.

Levinthal, C. (1954). Recombination in phage T2: its relationship to heterozygosis and growth. *Genetics* **39**, 169–184.

Lieberman, R. P., and Oishi, M. (1973). Formation of the *recB–recC* DNase by *in vitro* complementation and evidence concerning its subunit nature. *Nature new Biol.* **243**, 75–77.

Lieberman, R. P., and Oishi, M. (1974). The *recBC* deoxyribonuclease of *Escherichia coli*: isolation and characterization of the subunit proteins and reconstitution of the enzyme. *Proc. natn. Acad. Sci. U.S.A.* **71**, 4816–4820.

Lin, P.-F., Bardwell, E., and Howard-Flanders, P. (1977). Initiation of genetic exchanges in λ phage–prophage crosses. *Proc. natn. Acad. Sci. U.S.A.* **74**, 291–295.

Lindegren, C. C. (1933). The genetics of *Neurospora* – III. Pure bred stocks and crossing-over in *N. crassa. Bull. Torrey bot. Club* **60**, 133–154.

Lindegren, C. C. (1953). Gene conversion in *Saccharomyces. J. Genet.* **51**, 625–637.

Lindegren, C. C., and Lindegren, G. (1942). Locally specific patterns of chromatid and chromosome interference in *Neurospora. Genetics* **27**, 1–24.

Lissouba, P. (1961). Mise en évidence d'une unité génétique polarisée et essai d'analyse d'un cas d'interférence négative. *Ann. Sci. nat. Bot. Biol vég.*, Sér. 12, **1**, 641–720.

Lissouba, P., and Rizet, G. (1960). Sur l'existence d'une génétique polarisée ne subissant que des échanges non réciproques. *C. r. hebd. Séanc. Acad. Sci., Paris* **250**, 3408–3410.

Lissouba, P., Mousseau, J., Rizet, G., and Rossignol, J.-L. (1962). Fine structure of genes in the ascomycete *Ascobolus immersus. Adv. Genet.* **11**, 343–380.

Little, J. W. (1967). An exonuclease induced by bacteriophage λ. II. Nature of the enzymatic reaction. *J. biol. Chem.* **242**, 679–686.

Little, J. W., Edmiston, S. H., Pacelli, L. Z., and Mount, D. W. (1980). Cleavage of the *Escherichia coli lexA* protein by the *recA* protease. *Proc. natn. Acad. Sci. U.S.A.* **77**, 3225–3229.

Lloyd, R. G. (1978). Hyper-recombination in *Escherichia coli* K-12 mutants constitutive for protein X synthesis. *J. Bact.* **134**, 929–935.

Lloyd, R. G., and Barbour, S. D. (1974). The genetic location of the *sbcA* gene of *Escherichia coli. Molec. gen. Genet.* **134**, 157–171.

Lloyd, R. G., and Low, B. (1976). Some genetic consequences of changes in the level of *recA* gene function in *Escherichia coli* K-12. *Genetics* **84**, 675–695.

Lloyd, R. G., Low, K. B., Godson, G. N., and Birge, E. A. (1974). Isolation and characterisation of a mutant of *Escherichia coli* K-12 with a temperature-sensitive RecA phenotype. *J. Bact.* **120**, 407–415.

Loeb, T. (1960). Isolation of a bacteriophage specific for the F+ and Hfr mating types of *Escherichia coli* K-12. *Science, N.Y.* **131**, 932–933.

Low, B. (1973). Restoration by the *rac* locus of recombinant forming ability in *recB⁻* and *recC⁻* merozygotes of *Escherichia coli* K-12. *Molec. gen. Genet.* **122**, 119–130.

384

Lucchesi, J. C., and Suzuki, D. T. (1968). The interchromosomal control of recombination. *A. Rev. Genet.* **2**, 53–86.

Luria, S. E. (1947). Reactivation of irradiated bacteriophage by transfer of self-reproducing units. *Proc. natn. Acad. Sci. U.S.A.* **33**, 253–264.

Luria, S. E., and Dulbecco, R. (1949). Genetic recombination leading to production of active bacteriophage from ultraviolet inactivated bacteriophage particles. *Genetics* **34**, 93–125.

Maccacaro, G. A., and Hayes, W. (1961). Pairing interaction as a basis for negative interference. *Genet. Res.* **2**, 406–413.

McCarron, M., Gelbart, W., and Chovnick, A. (1974). Intracistronic mapping of electrophoretic sites in *Drosophila melanogaster*: fidelity of information transfer by gene conversion. *Genetics* **76**, 289–299.

McClintock, B. (1965). The control of gene action in maize. *Brookhaven Symp. Biol.* **18**, 162–184.

McClintock, B. (1968). Genetic systems regulating gene expression during development. *Symp. Soc. devl. Biol.* **26**, 84–112.

MacDonald, M. V., and Whitehouse, H. L. K. (1979). A *buff* spore colour mutant in *Sordaria brevicollis* showing high-frequency conversion. I. Characteristics of the mutant. *Genet. Res.* **34**, 87–119.

McEntee, K. (1977). Protein X is the product of the *recA* gene of *Escherichia coli*. *Proc. natn. Acad. Sci. U.S.A.* **74**, 5275–5279.

McEntee, K., Hesse, J. E., and Epstein, W. (1976). Identification and radiochemical purification of the *recA* protein of *Escherichia coli* K-12. *Proc. natn. Acad. Sci. U.S.A.* **73**, 3979–3983.

McEntee, K., Weinstock, G. M., and Lehman, I. R. (1979). Initiation of general recombination catalyzed *in vitro* by the *recA* protein of *Escherichia coli*. *Proc. natn. Acad. Sci. U.S.A.* **76**, 2615–2619.

McEntee, K., Weinstock, G. M., and Lehman, I. R. (1980). *RecA* protein-catalyzed strand assimilation: stimulation by *Escherichia coli* single-stranded DNA-binding protein. *Proc. natn. Acad. Sci. U.S.A.* **77**, 857–861.

McGavin, S. (1971). Models of specifically paired like (homologous) nucleic acid structures. *J. molec. Biol.* **55**, 293–298.

MacHattie, L. A., Ritchie, D. A., Thomas, C. A., and Richardson, C. C. (1967). Terminal repetition in permuted T2 bacteriophage DNA molecules. *J. molec. Biol.* **23**, 355–363.

McKnight, G. L., Cardillo, T. S., and Sherman, F. (1981). An extensive deletion causing overproduction of yeast iso-2-cytochrome *c*. *Cell* **25**, 409–419.

McMilin, K. D. and Russo, V. E. A. (1972). Maturation and recombination of bacteriophage lambda DNA molecules in the absence of DNA duplication. *J. molec. Biol.* **68**, 49–55.

McMilin, K. D., Stahl, M. M., and Stahl, F. W. (1974). Rec-mediated recombinational hot spot activity in bacteriophage lambda. I. Hot spot activity associated with Spi deletions and *bio* substitutions. *Genetics* **77**, 409–423.

McPartland, A., Green, L., and Echols, H. (1980). Control of *recA* gene RNA in *E. coli*: regulatory and signal genes. *Cell* **20**, 731–737.

Malone, R. E., Chattoraj, D. K., Faulds, D. H., Stahl, M. M., and Stahl, F. W. (1978). Hotspots for generalized recombination in the *Escherichia coli* chromosome. *J. molec. Biol.* **121**, 473–491.

Marcou, D. (1969). Sur la nature des recombinaisons intracistroniques et sur leurs répercussions sur la ségrégation de marqueurs extérieurs chez le *Podospora anserina*. *C. r. hebd. Séanc. Acad. Sci., Paris, D*, **269**, 2362–2365.

Marcou, D., Masson, A., Simonet, J.-M., and Piquepaille, G. (1979). Evidence for non-random spatial distribution of meiotic exchanges in *Podospora anserina*: comparison between linkage groups 1 and 6. *Molec. gen. Genet.* **176**, 67–79.

Markham, P., and Whitehouse, H. L. K. (1982). A hypothesis for the initiation of genetic recombination in eukaryotes. *Nature, Lond.* **295**, 421–423.

Marvin, D. A., and Hoffmann-Berling, H. (1963). Physical and chemical properties of two new small bacteriophages. *Nature, Lond.* **197**, 517–518.

Mather, K. (1936). Competition between bivalents during chiasma formation. *Proc. R. Soc. B* **120** 208–227.

Maynard Smith, S., and Symonds, N. (1973). Involvement of bacteriophage T4 genes in radiation repair. *J. molec. Biol.* **74**, 33–44.

Meselson, M. (1964). On the mechanism of genetic recombination between DNA molecules. *J. molec. Biol.* **9**, 734–745.

Meselson, M. (1965). The duplication and recombination of genes. In *New Ideas in Biology*, Natural History Press, New York.

Meselson, M. (1967*a*). The molecular basis of genetic recombination. In *Heritage from Mendel* (R. A. Brink, ed.), pp. 81–104, University of Wisconsin Press, Madison, Wisc.

Meselson, M. (1967*b*). Reciprocal recombination in prophage lambda. *J. cell. comp. Physiol.* **70**, Suppl. 1, 113–118.

Meselson, M. (1972). Formation of hybrid DNA by rotary diffusion during genetic recombination. *J. molec. Biol.* **71**, 795–798.

Meselson, M. S., and Radding, C. M. (1975). A general model for genetic recombination. *Proc. natn. Acad. Sci. U.S.A.* **72**, 358–361.

Meselson, M., and Weigle, J. J. (1961). Chromosome breakage accompanying genetic recombination in bacteriophage. *Proc. natn. Acad. Sci. U.S.A.* **47**, 857–868.

Meselson, M., and Yuan, R. (1968). DNA restriction enzyme from *E. coli. Nature, Lond.* **217**, 1110–1114.

Meyer, R. R., Glassberg, J., and Kornberg, A. (1979). An *Escherichia coli* mutant defective in single-strand binding protein is defective in DNA replication. *Proc. natn. Acad. Sci. U.S.A.* **76**, 1702–1705.

Miao, R., and Guild, W. R. (1970). Competent *Diplococcus pneumoniae* accept both single- and double-stranded deoxyribonucleic acid. *J. Bact.* **101**, 361–364.

Michalke, W. (1967). Erhöhte Rekombinationshäufigkeit an den Enden des T1-Chromosoms. *Molec. gen. Genet.* **99**, 12–33.

Miller, H. I., Kikuchi, A., Nash, H. A., Weisberg, R. A., and Friedman, D. I. (1979). Site-specific recombination of bacteriophage λ: the role of host gene products. *Cold Spring Harb. Symp. quant. Biol.* **43**, 1121–1126.

Miller, L. K., Cooke, B. E., and Fried, M. (1976). Fate of mismatched base-pair regions in polyoma heteroduplex DNA during infection of mouse cells. *Proc. natn. Acad. Sci. U.S.A.* **73**, 3073–3077.

Miller, R. C., Kozinski, A. W., and Litwin, S. (1970). Molecular recombination in T4 bacteriophage deoxyribonucleic acid. III. Formation of long single strands during recombination. *J. Virol.* **5**, 368–380.

Mitchell, M. B. (1955*a*). Aberrant recombination of pyridoxine mutants of *Neurospora. Proc. natn. Acad. Sci. U.S.A.* **41**, 215–220.

Mitchell, M. B. (1955*b*). Further evidence of aberrant recombination in *Neurospora. Proc. natn. Acad. Sci. U.S.A.* **41**, 935–937.

Mitchell, M. B. (1956). A consideration of aberrant recombination in *Neurospora. C. r. trav. Lab. Carlsberg, Ser. Physiol.* **26**, 285–298.

Miura, A., and Tomizawa, J. (1968). Studies on radiation-sensitive mutants of *E. coli.* III. Participation of the Rec system in induction of mutation by ultraviolet irradiation. *Molec. gen. Genet.* **103**, 1–10.

Mizuuchi, K., and Nash, H. A. (1976). Restriction assay for integrative recombination of bacteriophage λ DNA *in vitro*: requirement for closed circular DNA substrate. *Proc. natn. Acad. Sci. U.S.A.* **73**, 3524–3528.

Mizuuchi, K., Gellert, M., and Nash, H. A. (1978). Involvement of supertwisted DNA in integrative recombination of bacteriophage lambda. *J. molec. Biol.* **121**, 375–392.

Moore, C. W., and Sherman, F. (1975). Role of DNA sequences in genetic recombination in the iso-1-cytochrome *c* gene of yeast. I. Discrepancies between physical distances and genetic distances determined by five mapping procedures. *Genetics* **79**, 397–418.

Moore, C. W., and Sherman, F. (1977). Role of DNA sequences in genetic recombination in the iso-1-cytochrome *c* gene of yeast. II. Comparison of mutants altered at the same and nearby base pairs. *Genetics* **85**, 1–22.

Moore, D. (1972). Genetic fine structure, site clustering and marker effect in the *ftr* cistron of *Coprinus*. *Genet. Res.* **19**, 281–303.

Moore, P. D. (1974). Recombination in higher organisms. *Comment. Pl. Sci.* **10**, 1–15, in *Curr. Adv. Pl. Sci.* **5**.

Moore, P. D., and Holliday, R. (1976). Evidence for the formation of hybrid DNA during mitotic recombination in Chinese hamster cells. *Cell* **8**, 573–579.

Morand, P., Blanco, M., and Devoret, R. (1977). Characterization of *lexB* mutations in *Escherichia coli* K-12. *J. Bact.* **131**, 572–582.

Morgan, T. H. (1911*a*). The application of the conception of pure lines to sex-limited inheritance and to sexual dimorphism. *Am. Nat.* **45**, 65–78.

Morgan, T. H. (1911*b*). An attempt to analyze the constitution of the chromosomes on the basis of sex-limited inheritance in *Drosophila*. *J. exp. Zool.* **11**, 365–414.

Morgan, T. H., and Cattell, E. (1912). Data for the study of sex-linked inheritance in *Drosophila*. *J. exp. Zool.* **13**, 79–101.

Morpurgo, G. (1963). Induction of mitotic crossing-over in *Aspergillus nidulans* by bifunctional alkylating agents. *Genetics* **48**, 1259–1263.

Morris, C. F., Sinha, N. K., and Alberts, B. M. (1975). Reconstruction of bacteriophage T4 DNA replication apparatus from purified components: rolling circle replication following *de novo* chain initiation on a single-stranded circular DNA template. *Proc. natn. Acad. Sci. U.S.A.* **72**, 4800–4804.

Morrison, D. A., and Guild, W. R. (1973). Structure of deoxyribonucleic acid on the cell surface during uptake by pneumococcus. *J. Bact.* **115**, 1055–1062.

Mosig, G., and Bock, S. (1976). Gene 32 protein of bacteriophage T4 moderates the activities of the T4 gene 46/47-controlled nuclease and of the *Escherichia coli recBC* nuclease *in vivo*. *J. Virol.* **17**, 756–761.

Mosig, G., and Breschkin, A. M. (1975). Genetic evidence for an additional function of phage T4 gene 32 protein: interaction with ligase. *Proc. natn. Acad. Sci. U.S.A.* **72**, 1226–1230.

Mosig, G., Ehring, R., Schliewen, W., and Bock, S. (1971). The patterns of recombination and segregation in terminal regions of T4 DNA molecules. *Molec. gen. Genet.* **113**, 51–91.

Mosig, G., Berquist, W., and Bock, S. (1977). Multiple interactions of a DNA-binding protein *in vivo*. III. Phage T4 gene-32 mutations differentially affect insertion-type recombination and membrane properties. *Genetics* **86**, 5–23.

Mount, D. W. (1971). Isolation and genetic analysis of a strain of *Escherichia coli* K-12 with an amber *recA* mutation. *J. Bact.* **107**, 388–389.

Mount, D. W. (1977). A mutant of *Escherichia coli* showing constitutive expression of the lysogenic induction and error-prone DNA repair pathways. *Proc. natn. Acad. Sci. U.S.A.* **74**, 300–304.

Mount, D. W., Low, K. B., and Edmiston, S. J. (1972). Dominant mutations (*lex*) in *Escherichia coli* K-12 which affect radiation sensitivity and frequency of ultraviolet light-induced mutations. *J. Bact.* **112**, 886–893.

Mousseau, J. (1966). Sur les variations de fréquence de conversion au niveau de divers sites d'un même locus. *C. r. hebd. Séanc. Acad. Sci., Paris, D,* **262**, 1254–1257.

Muller, H. J. (1916). The mechanism of crossing-over. *Am. Nat.* **50**, 193–221, 284–305, 350–366, 421–434.

Murray, N. E. (1960). Complementation and recombination between methionine-2 alleles in *Neurospora crassa*. *Heredity* **15**, 207–217.

Murray, N. E. (1963). Polarized recombination and fine structure within the *me*-2 gene of *Neurospora crassa*. *Genetics* **48**, 1163–1183.

Murray, N. E. (1968). Polarized intragenic recombination in chromosome rearrangements of *Neurospora*. *Genetics* **58**, 181–191.

Murray, N. E. (1969). Reversal of polarized recombination of alleles in *Neurospora* as a function of their position. *Genetics* **61**, 67–77.

Murray, N. E. (1970). Recombination events that span sites within neighbouring gene loci of *Neurospora*. *Genet. Res.* **15**, 109–121.

Murray, N. E., Batten, P. L., and Murray, K. (1973). Restriction of bacteriophage λ by *Escherichia coli* K. *J. molec. Biol.* **81**, 395–407.

Nash, H. A. (1975*a*). Integrative recombination in bacteriophage lambda: analysis of recombinant DNA. *J. molec. Biol.* **91**, 501–514.

Nash, H. A. (1975*b*). Integrative recombination of bacteriophage lambda DNA *in vitro*. *Proc. natn. Acad. Sci. U.S.A.* **72**, 1072–1076.

Nelson, O. E. (1962). The *waxy* locus in maize. I. Intralocus recombination frequency estimates by pollen and by conventional analyses. *Genetics* **47**, 737–742.

Nelson, O. E. (1968). The *waxy* locus in maize. II. The location of the controlling element alleles. *Genetics* **60**, 507–524.

Nelson, O. E. (1975). The *waxy* locus in maize. III. Effect of structural heterozygosity on intragenic recombination and flanking marker assortment. *Genetics* **79**, 31–44.

Nevers, P., and Saedler, H. (1977). Transposable genetic elements as agents of gene instability and chromosomal rearrangements. *Nature, Lond.* **268**, 109–115.

Nevers, P., and Spatz, H.-C. (1975). *Escherichia coli* mutants *uvrD* and *uvrE* deficient in gene conversion of λ-heteroduplexes. *Molec. gen. Genet.* **139**, 233–243.

Nomura, M., and Benzer, S. (1961). The nature of the 'deletion' mutants in the rII region of phage T4. *J. molec. Biol.* **3**, 684–692.

Nonn, E. M., and Bernstein, C. (1977). Multiplicity reactivation and repair of nitrous acid-induced lesions in bacteriophage T4. *J. molec. Biol.* **116**, 31–47.

Notani, N., and Goodgal, S. H. (1966). On the nature of recombinants formed during transformation in *Hemophilus influenzae*. *J. gen. Physiol.* **49**, no. 6, Suppl., 197–209.

Ohki, M., and Tomizawa, J. (1969). Asymmetric transfer of DNA strands in bacterial conjugation. *Cold Spring Harb. Symp. quant. Biol.* **33**, 651–658.

Olive, L. S. (1956). Genetics of *Sordaria fimicola*. I. Ascospore color mutants. *Am. J. Bot.* **43**, 97–107.

Olive, L. S. (1959). Aberrant tetrads in *Sordaria fimicola*. *Proc. natn. Acad. Sci. U.S.A.* **45**, 727–732.

Oliver, P. (1982). The effect of the *tif-1* mutation upon recombination in *Escherichia coli* K12. *J. gen. Microbiol.*, in press.

Olson, L. W., and Zimmermann, F. K. (1978*a*). Meiotic recombination and synaptonemal complexes in *Saccharomyces cerevisiae*. *Molec. gen. Genet.* **166**, 151–159.

Olson, L. W., and Zimmermann, F. K. (1978*b*). Mitotic recombination in the absence of synaptonemal complexes in *Saccharomyces cerevisiae*. *Molec. gen. Genet.* **166**, 161–165.

Oppenheim, A. B., and Riley, M. (1966). Molecular recombination following conjugation in *Escherichia coli*. *J. molec. Biol.* **20**, 331–357.

Oppenheim, A. B., and Riley, M. (1967). Covalent union of parental DNAs following conjugation in *Escherichia coli*. *J. molec. Biol.* **28**, 503–511.

Paquette, N. (1978). Detection of aberrant 4:4 asci in *Ascobolus immersus*. *Can. J. Genet. Cytol.* **20**, 9–17.

Paquette, N., and Rossignol, J.-L. (1978). Gene conversion spectrum of 15 mutants giving post-meiotic segregation in the *b2* locus of *Ascobolus immersus*. *Molec. gen. Genet.* **163**, 313–326.

Parag, Y., and Parag, G. (1975). Mutations affecting mitotic recombination frequency in haploids and diploids of the filamentous fungus *Aspergillus nidulans*. *Molec. gen. Genet.* **137**, 109–123.

Parker, J. S. (1975). Chromosome-specific control of chiasma formation. *Chromosoma* **49**, 391–406.

Parkinson, J. S., and Huskey, R. J. (1971). Deletion mutants of bacteriophage lambda. I. Isolation and initial characterization. *J. molec. Biol.* **56**, 369–384.

Paszewski, A. (1967). A study on simultaneous conversion in linked genes in *Ascobolus immersus*. *Genet. Res.* **10**, 121–126.

Peacock, W. J. (1970). Replication, recombination and chiasmata in *Goniaea australasiae* (Orthoptera: Acrididae). *Genetics* **65**, 593–617.

Pees, E. (1967). Genetic fine structure and polarized negative interference at the lys-51 (FL) locus of *Aspergillus nidulans*. *Genetica* **38**, 275–304.

Perkins, D. D. (1962). Crossing-over and interference in a multiply marked chromosome arm of *Neurospora*. *Genetics* **47**, 1253–1274.

Perry, P., and Wolff, S. (1974). New Giemsa method for the differential staining of sister chromatids. *Nature, Lond.* **253**, 156–158.

Pfeifer, D. (1961a). Genetic recombination in bacteriophage φX174. *Nature, Lond.* **189**, 422–423.

Pfeifer, D. (1961b). Genetische Untersuchungen am Bakteriophagen φX174. I. Aufbau eines selektiven Systems und nachweis genetischer Rekombination. *Z. VerebLehre* **92**, 317–329.

Piekarowicz, A., and Kunicki-Goldfinger, W. J. H. (1968). Mechanism of conjugation and recombination in bacteria. IV. Single-strandedness of donor DNA in mating bacteria. *Acta microbiol. pol.* **17**, 135–146.

Pittard, J., and Adelberg, E. A. (1964). Gene transfer by F' strains of *Escherichia coli* K-12. III. An analysis of the recombination events occurring in the F' male and in the zygotes. *Genetics* **49**, 995–1007.

Polani, P. E., Crolla, J. A., Seller, M. J., and Moir, F. (1979). Meiotic crossing-over exchange in the female mouse visualised by BUdR substitution. *Nature, Lond.* **278**, 348–349.

Pollock, T. J., and Abremski, K. (1979). DNA without supertwists can be an *in vitro* substrate for site-specific recombination of bacteriophage λ. *J. molec. Biol.* **131**, 651–654.

Pontecorvo, G., and Käfer, E. (1958). Genetic analysis based on mitotic recombination. *Adv. Genet.* **9**, 71–104.

Pontecorvo, G., and Roper, J. A. (1953). Diploids and mitotic recombination. *Adv. Genet.* **5**, 218–233.

Pontecorvo, G., Gloor, E. T., and Forbes, E. (1954). Analysis of mitotic recombination in *Aspergillus nidulans*. *J. Genet.* **52**, 226–237.

Postel, E. H., and Goodgal, S. H. (1966). Uptake of 'single-stranded' DNA in *Hemophilus influenzae* and its ability to transform. *J. molec. Biol.* **16**, 317–327.

Potter, H., and Dressler, D. (1978). *In vitro* system from *Escherichia coli* that catalyzes generalized genetic recombination. *Proc. natn. Acad. Sci. U.S.A.* **75**, 3698–3702.

Potter, H., and Dressler, D. (1979). Biochemical assay designed to detect formation of recombination intermediates *in vitro*. *Proc. natn. Acad. Sci. U.S.A.* **76**, 1084–1088.

Prashad, N., and Hosoda, J. (1972). Role of genes *46* and *47* in bacteriophage T4 reproduction. II. Formation of gaps on parental DNA of polynucleotide ligase defective mutants. *J. molec. Biol.* **70**, 617–635.

Pritchard, R. H. (1955). The linear arrangement of a series of alleles of *Aspergillus nidulans*. *Heredity* **9**, 343–371.

Pritchard, R. H. (1960). Localized negative interference and its bearing on models of gene recombination. *Genet. Res.* **1**, 1–24.

Pukkila, P. J. (1978). The recognition of mismatched base pairs in DNA by DNase I from *Ustilago maydis*. *Molec. gen. Genet.* **161**, 245–250.

Pukkila, P. J., Peterson, J. Herman, G., Modrich, P., and Meselson, M. (1982). Changes in the amount of DNA methylation influence mismatch correction. Manuscript.

Putrament, A. (1964). Mitotic recombination in the *paba1* cistron of *Aspergillus nidulans*. *Genet. Res.* **5**, 316–327.

Putrament, A., Rozbicka, T., and Wojciechowska, K. (1971). The highly polarized recombination pattern within the *methA* gene of *Aspergillus nidulans*. *Genet. Res.* **17**, 125–131.

Radding. C. M. (1970). The role of exonuclease and β protein of bacteriophage λ in genetic recombination. I. Effects of *red* mutants on protein structure. *J. molec. Biol.* **52**, 491–499.

Radding, C. M. (1973). Molecular mechanisms in genetic recombination. *A. Rev. Genet.* **7**, 87–111.

Radding, C. M., Szpirer, J., and Thomas, R. (1967). The structural gene for λ exonuclease. *Proc. natn. Acad. Sci. U.S.A.* **57**, 277–283.

Radman, M., Wagner, R. E., Glickman, B. W., and Meselson, M. (1980). DNA methylation, mismatch correction and genetic stability. In *Progress in Environmental Mutagenesis* (M. Alačevič, ed.), pp. 121–130, Elsevier/North-Holland Biomedical Press, Amsterdam.

Rasmussen, S. W., and Holm, P. B. (1980). Mechanics of meiosis. *Hereditas* **93**, 187–216.

Rees, H., and Naylor, B. (1960). Developmental variation in chromosome behaviour. *Heredity* **15**, 17–27.

Rizet, G., and Rossignol, J.-L. (1963). Sur la dissymétrie de certaines conversions et sur la dimension de l'erreur de copie chez l'*Ascobolus immersus*. *Revta Biol., Lisb.* **3**, 261–268.

Rizet, G., and Rossignol, J.-L. (1966). Sur la dimension probable des échanges réciproques au sein d'un locus complexe d'*Ascobolus immersus*. *C. r. hebd. Séanc. Acad. Sci., Paris, D*, **262**, 1250–1253.

Rizet, G., Lissouba, P., and Mousseau, J. (1960). Les mutations d'ascospores chez l'Ascomycète *Ascobolus immersus* et l'analyse de la structure fine des gènes. *Bull. Soc. fr. Physiol. vég.* **6**, 175–193.

Roberts, J. W., and Roberts, C. W. (1975). Proteolytic cleavage of bacteriophage lambda repressor in induction. *Proc. natn. Acad. Sci. U.S.A.* **72**, 147–151.

Roberts, J. W., Roberts, C. W., and Mount, D. W. (1977). Inactivation and proteolytic cleavage of phage λ repressor *in vitro* in an ATP-dependent reaction. *Proc. natn. Acad. Sci. U.S.A.* **74**, 2283–2287.

Roberts, J. W., Roberts, C. W., and Craig, N. L. (1978). *Escherichia coli recA* gene product inactivates phage λ repressor. *Proc. natn. Acad. Sci. U.S.A.* **75**, 4714–4718.

Rodarte-Ramón, U. S. (1972). Radiation-induced recombination in *Saccharomyces*: the genetic control of recombination in mitosis and meiosis. *Radiation Res.* **49**, 148–154.

Rodarte-Ramón, U. S., and Mortimer, R. K. (1972). Radiation-induced recombination in *Saccharomyces*: isolation and genetic study of recombination-deficient mutants. *Radiation Res.* **49**, 133–147.

Roger, M. (1972). Evidence for conversion of heteroduplex transforming DNAs to homoduplexes by recipient pneumococcal cells. *Proc. natn. Acad. Sci. U.S.A.* **69**, 466–470.

Roger, M. (1977). Mismatch excision and possible polarity effects result in preferred deoxyribonucleic acid strand of integration in pneumococcal transformation. *J. Bact.* **129**, 298–304.

Roman, H. (1957). Studies of gene mutation in *Saccharomyces*. *Cold Spring Harb. Symp. quant. Biol.* **21**, 175–185.

Roman, H. (1958). Sur les recombinaisons non réciproques chez *Saccharomyces cerevisiae* et sur les problèmes posés par ces phénomènes. *Ann. Génét.* **1**, 11–17.

Roman, H., and Jacob, F. (1957). Effet de la lumière ultraviolette sur la recombinaison génétique entre allèles chez la levure. *C. r. hebd. Séanc. Acad. Sci., Paris* **245**, 1032–1034.

Roman, H. L., and Jacob, F. (1959). A comparison of spontaneous and ultraviolet induced allelic recombination with reference to the recombination of outside markers. *Cold Spring Harb. Symp. quant. Biol.* **23**, 155–160.

Rommelaere, J., and Miller-Faurès, A. (1975). Detection by density equilibrium centrifugation of recombinant-like DNA molecules in somatic mammalian cells. *J. molec. Biol.* **98**, 195–218.

Rommelaere, J., Susskind, M., and Errera, M. (1973). Chromosome and chromatid exchanges in Chinese hamster cells. *Chromosoma* **41**, 243–257.

Roper, J. A., and Pritchard, R. H. (1955). Recovery of the complementary products of mitotic crossing-over. *Nature, Lond.* **175**, 639.

Ross, J. G., and Holm, G. (1960). Somatic segregation in tomato. *Heredity* **46**, 224–230.

Ross, P., and Howard-Flanders, P. (1977a). Initiation of *rec*A$^+$-dependent recombination in *Escherichia coli* (λ). I. Undamaged covalent circular lambda DNA molecules in *uvr*A$^+$ *rec*A$^+$ lysogenic host cells are cut following superinfection with psoralen-damaged lambda phages. *J. molec. Biol.* **117**, 137–158.

Ross, P., and Howard-Flanders, P. (1977b). Iniation of *rec*A$^+$-dependent recombination in *Escherichia coli* (λ). II. Specificity in the induction of recombination and strand cutting in undamaged covalent circular bacteriophage 186 and lambda DNA molecules in phage-infected cells. *J. molec. Biol.* **117**, 159–174.

Rossignol, J.-L. (1969). Existence of homogeneous categories of mutants exhibiting various conversion patterns in gene 75 of *Ascobolus immersus*. *Genetics* **63**, 795–805.

Rossignol, J.-L., and Haedens, V. (1978). The interaction during recombination between closely linked allelic frameshift mutant sites in *Ascobolus immersus*. I. A (or B) and C type mutant sites. *Heredity* **40**, 405–425.

Rossignol, J.-L., and Haedens, V. (1980). Relationship between asymmetrical and symmetrical hybrid DNA formation during meiotic recombination. *Curr. Genet.* **1**, 185–191.

Rossignol, J.-L., and Paquette, N. (1979). Disparity of gene conversion in frameshift mutants located in locus *b2* of *Ascobolus immersus*. *Proc. natn. Acad. Sci. U.S.A.* **76**, 2871–2875.

Rossignol, J.-L., Paquette, N., and Nicolas, A. (1979). Aberrant 4 :4 asci, disparity in the direction of conversion, and frequencies of conversion in *Ascobolus immersus*. *Cold Spring Harb. Symp. quant. Biol.* **43**, 1343–1352.

Roth, R., and Fogel, S. (1971). A system selective for yeast mutants deficient in meiotic recombination. *Molec. gen. Genet.* **112**, 295–305.

Rothman, J. L. (1965). Transduction studies on the relation between prophage and host chromosome. *J. molec. Biol.* **12**, 892–912.

Rupp, W. D., and Howard-Flanders, P. (1968). Discontinuities in the DNA synthesized in the excision-defective strain of *Escherichia coli* following ultraviolet irradiation. *J. molec. Biol.* **31**, 291–304.

Rupp, W. D., Wilde, C. E., Reno, D. L., and Howard-Flanders, P. (1971). Exchanges between DNA strands in ultraviolet-irradiated *Escherichia coli*. *J. molec. Biol.* **61**, 25–44.

Rush, M. G., and Warner, R. C. (1968). Multiple length rings of φX174 and S13 replicative forms. III. A possible intermediate in recombination. *J. biol. Chem.* **243**, 4821–4826.

Rush, M. G., Kleinschmidt, A. K., Hellmann, W., and Warner, R. C. (1967). Multiple-length rings in preparations of φX174 replicative form. *Proc. natn. Acad. Sci. U.S.A.* **58**, 1676–1683.

Russo, V. E. A. (1973). On the physical structure of λ recombinant DNA. *Molec. gen. Genet.* **122**, 353–366.

Rydberg, B. (1978). Bromouracil mutagenesis and mismatch repair in mutator strains of *Escherichia coli*. *Mutation Res.* **52**, 11–24.

Sadowski, P. D., and Hurwitz, J. (1969). Enzymatic breakage of deoxyribonucleic acid. I. Purification and properties of endonuclease II from T4 phage-infected *Escherichia coli*. *J. biol. Chem.* **244**, 6182–6191.

Sadowski, P. D., Warner, H. R., Hercules, K., Munro, J. L., Mendelsohn, S., and Wiberg, J. S. (1971). Mutants of bacteriophage T4 defective in the induction of T4 endonuclease II. *J. biol. Chem.* **246**, 3431–3433.

Saedler, H., and Starlinger, P. (1967). $O°$ mutations in the galactose operon in *E. coli*. I. Genetic characterization. *Molec. gen. Genet.* **100**, 178–189.

Sager, R. (1977). Genetic analysis of chloroplast DNA in *Chlamydomonas*. *Adv. Genet.* **19**, 287–340.

Sager, R., and Raminis, Z. (1963). The particulate nature of nonchromosomal genes in *Chlamydomonas*. *Proc. natn. Acad. Sci. U.S.A.* **50**, 260–268.

St Lawrence, P. (1965). The *q* locus of *Neurospora crassa*. *Proc. natn. Acad. Sci. U.S.A.* **42**, 189–194.

Sakaki, Y., Karu, A. E., Linn, S., and Echols, H. (1973). Purification and properties of the γ-protein specified by bacteriophage λ: an inhibitor of the host *recBC* recombination enzyme. *Proc. natn. Acad. Sci. U.S.A.* **70**, 2215–2219.

Salivar, W. O., Tzagoloff, H., and Pratt, D. (1964). Some physical-chemical and biological properties of the rod-shaped coliphage M13. *Virology* **24**, 359–371.

Sancar, A., Stachelek, C., Konigsberg, W., and Rupp, W. D. (1980). Sequences of the *recA* gene and protein. *Proc. natn. Acad. Sci. U.S.A.* **77**, 2611–2615.

Sandler, L. (1954). A genetic analysis of reversed acrocentric compound X chromosomes in *Drosophila melanogaster*. *Genetics* **39**, 923–942.

Sandler, L. (1957). The meiotic behavior of reversed compound ring X chromosomes in *Drosophila melanogaster*. *Genetics* **42**, 764–782.

Sandler, L., Lindsley, D. L., Nicoletti, B., and Trippa, G. (1968). Mutants affecting meiosis in natural populations of *Drosophila melanogaster*. *Genetics* **60**, 525–558.

Sang, H., and Whitehouse, H. L. K. (1979). Genetic recombination at the *buff* spore colour locus in *Sordaria brevicollis*. I. Analysis of flanking marker behaviour in crosses between *buff* mutants and wild type. *Molec. gen. Genet.* **174**, 327–334.

Sang, H., and Whitehouse, H. L. K. (1982). Genetic recombination at the *buff* spore colour locus in *Sordaria brevicollis*. II. Analysis of flanking marker behaviour in crosses between *buff* mutants. Manuscript.

Sarthy, P. V., and Meselson, M. (1976). Single burst study of *rec-* and *red-*mediated recombination in bacteriophage lambda. *Proc. natn. Acad. Sci. U.S.A.* **73**, 4613–4617.

Sato, K., and Sekiguchi, M. (1976). Studies on temperate-dependent ultraviolet light-sensitive mutants of bacteriophage T4: the structural gene for T4 endonuclease V. *J. molec. Biol.* **102**, 15–26.

Savage, E. A., and Hastings, P. J. (1981). Marker effects and the nature of the recombination event at the *his1* locus of *Saccharomyces cerevisiae*. *Curr. Genet.* **3**, 37–47.

Scaife, J., and Gross, J. D. (1963). The mechanism of chromosome mobilisation by an F-prime factor in *Escherichia coli* K-12. *Genet. Res.* **4**, 328–331.

Schultz, D. W., Swindle, J., and Smith, G. R. (1981). Clustering of mutations inactivating a Chi recombinational hotspot. *J. molec. Biol.* **146**, 275–286.

Schvartzman, J. B., Cortés, F., González-Fernández, A., Gutiérrez, C., and López-Sáez, J. F. (1979). On the nature of sister-chromatid exchanges in 5-bromodeoxyuridine-substituted chromosomes. *Genetics* **92**, 1251–1264.

Séchaud, J., Streisinger, G., Emrich, J., Newton, J., Lanford, H., Reinhold, H., and Stahl, M. M. (1965). Chromosome structure in phage T4, II. Terminal redundancy and heterozygosis. *Proc. natn. Acad. Sci. U.S.A.* **54**, 1333–1339.

Sekiguchi, M., Shimizu, K., Sato, K., Yasuda, S., and Ohshima, S. (1975). Enzymic mechanism of excision-repair in T4-infected cells. In *Molecular Mechanisms for the Repair of DNA* (P. C. Hanawalt and R. B. Setlow, eds), pp. 135–142, Plenum Press, New York.

Sertic, V., and Boulgakov, N. (1935a). Le groupement des bactériophages d'après leur type antigénique. *C. r. Soc. Biol.* **119**, 983–985.

Sertic, V., and Boulgakov, N. (1935b). Classification et identification des typhi-phages. *C. r. Soc. Biol.* **119**, 1270–1272.

Shah, D. B., and Berger, H. (1971). Replication of gene 46–47 amber mutants of bacteriophage T4D. *J. molec. Biol.* **57**, 17–34.

Shalitin, C., and Stahl, F. W. (1965). Additional evidence for two kinds of heterozygotes in phage T4. *Proc. natn. Acad. Sci. U.S.A.* **54**, 1340–1341.

Shanfield, B., and Käfer, E. (1969). UV-sensitive mutants increasing mitotic crossing-over in *Aspergillus nidulans. Mutation Res.* **7**, 485–487.

Shapiro, J. A. (1969). Mutations caused by the insertion of genetic material into the galactose operon of *Escherichia coli. J. molec. Biol.,* **40**, 93–105.

Shapiro, J. A. (1979). Molecular model for the transposition and replication of bacteriophage Mu and other transposable elements. *Proc. natn. Acad. Sci. U.S.A.* **76**, 1933–1937.

Sherman, F., and Roman, H. (1963). Evidence for two types of allelic recombination in yeast. *Genetics* **48**, 253–261.

Shibata, T., DasGupta, C., Cunningham, R. P., and Radding, C. M. (1979*a*). Purified *Escherichia coli recA* protein catalyzes homologous pairing of superhelical DNA and single-stranded fragments. *Proc. natn. Acad. Sci. U.S.A.* **76**, 1638–1642.

Shibata, T., Cunningham, R. P., DasGupta, C., and Radding, C. M. (1979*b*). Homologous pairing in genetic recombination: complexes of *recA* protein and DNA. *Proc. natn. Acad. Sci. U.S.A.* **76**, 5100–5104.

Shibata, T., DasGupta, C., Cunningham, R. P., and Radding, C. M. (1980). Homologous pairing in genetic recombination: formation of D loops by combined action of *recA* protein and a helix-destabilizing protein. *Proc. natn. Acad. Sci. U.S.A.* **77**, 2606–2610.

Shoemaker, N. B., and Guild, W. R. (1974). Destruction of low efficiency markers is a slow process occurring at the heteroduplex stage of transformation. *Molec. gen. Genet.* **128**, 283–290.

Shulman, M., and Gottesman, M. (1973). Attachment site mutants of bacteriophage lambda. *J. molec. Biol.* **81**, 461–482.

Sicard, A. M. (1964). A new synthetic medium for *Diplococcus pneumoniae*, and its use for the study of reciprocal transformations at the *amiA* locus. *Genetics* **50**, 31–44.

Sicard, A. M., and Ephrussi-Taylor, H. (1965). Genetic recombination in DNA-induced transformation of pneumococcus. II. Mapping the *amiA* region. *Genetics* **52**, 1207–1227.

Siddiqi, O. H. (1962). The fine genetic structure of the *pabal* region of *Aspergillus nidulans. Genet. Res.* **3**, 69–89.

Siddiqi, O. H. (1963). Incorporation of parental DNA into genetic recombinants of *E. coli. Proc. natn. Acad. Sci. U.S.A.* **49**, 589–592.

Siddiqi, O., and Fox, M. S. (1973). Integration of donor DNA in bacterial conjugation. *J. molec. Biol.* **77**, 101–123.

Sigal, N., and Alberts, B. (1972). Genetic recombination: the nature of a crossed strand-exchange between two homologous DNA molecules. *J. molec. Biol.* **71**, 789–793.

Sigal, N., Delius, H., Kornberg, T., Gefter, M. L., and Alberts, B. (1972). A DNA-unwinding protein isolated from *Escherichia coli*: its interaction with DNA and with DNA polymerases. *Proc. natn. Acad. Sci. U.S.A.* **69**, 3537–3541.

Signer, E. (1971). General recombination. In *The Bacteriophage Lambda* (E. D. Hershey, ed.). pp. 139–174, Cold Spring Harbor Laboratory, Cold Spring Harbor, N.Y.

Signer, E. R., and Weil, J. (1968). Recombination in bacteriophage λ. I. Mutants deficient in general recombination. *J. molec. Biol.* **34**, 261–271.

Signer, E., Echols, H., Weil, J., Radding, C., Schulman, M., Moore, L., and Manly, K. (1969). The general recombination system of bacteriophage λ. *Cold Spring Harb. Symp. quant. Biol.* **33**, 711–714.

Simchen, G., and Stamberg, J. (1969). Genetic control of recombination in *Schizophyllum commune*: specific and independent regulation of adjacent and non-adjacent chromosomal regions. *Heredity* **24**, 369–381.

Simonet, J.-M. (1973). Mutations affecting meiosis in *Podospora anserina*. II. Effect of *mei2* mutants on recombination. *Molec. gen. Genet.* **123**, 263–281.

Sinha, N. K., and Snustad, D. P. (1971). DNA synthesis in bacteriophage T4-infected *Escherichia coli*: evidence supporting a stoichiometric role for gene 32-product. *J. molec. Biol.* **62**, 267–271.

Sisco, K. L., and Smith, H. O. (1979). Sequence-specific DNA uptake in *Haemophilus* transformation. *Proc. natn. Acad. Sci. U.S.A.* **76**, 972–976.

Skalka, A. (1974). A replicator's view of recombination (and repair). In *Mechanisms in Recombination* (R. F. Grell, ed.), pp. 421–432, Plenum Press, New York and London.

Smith, B. R. (1966). Genetic controls of recombination. I. The recombination-2 gene of *Neurospora crassa. Heredity* **21**, 481–498.

Smith, G. R., Schultz, D. W., and Crasemann, J. M. (1980). Generalized recombination: nucleotide sequence homology between Chi recombinational hotspots. *Cell* **19**, 785–793.

Smith, P. D., Finnerty, V. G., and Chovnick, A. (1970). Gene conversion in *Drosophila*: non-reciprocal events at the maroon-like cistron. *Nature, Lond.* **228**, 442–444.

Smyth, D. R. (1971). Effect of *rec-3* on polarity of recombination in the *amination-1* locus of *Neurospora crassa. Aust. J. biol. Sci.* **24**, 97–106.

Smyth, D. R. (1973). A new map of the *amination-1* locus of *Neurospora crassa*, and the effect of the *recombination-3* gene. *Aust. J. biol. Sci.* **26**, 1355–1370.

Smyth, D. R., and Stern, H. (1973). Repeated DNA synthesized during pachytene in *Lilium henryi. Nature new Biol.* **245**, 94–96.

Snustad, D. P. (1968). Dominance interactions in *Escherichia coli* cells mixedly infected with bacteriophage T4D wild-type and *amber* mutants and their possible implications as to type of gene-product function: catalytic vs. stoichiometric. *Virology* **35**, 550–563.

Sobell, H. M. (1972). Molecular mechanism for genetic recombination. *Proc. natn. Acad. Sci. U.S.A.* **68**, 2483–2487.

Sobell, H. M. (1973). Symmetry in protein–nucleic acid interaction and its genetic implications. *Adv. Genet.* **17**, 411–490.

Sobell, H. M. (1974). Concerning the stereochemistry of strand equivalence in genetic recombination. In *Mechanisms in Recombination* (R. F. Grell, ed.), pp. 433–438, Plenum Press, New York and London.

Sobell, H. M. (1975). A mechanism to activate branch migration between homologous DNA molecules in genetic recombination. *Proc. natn. Acad. Sci. U.S.A.* **72**, 279–283.

Spatz, H. C., and Trautner, T. A. (1970). One way to do experiments on gene conversion? *Molec. gen. Genet.* **109**, 84–106.

Sprague, K. U., Faulds, D. H., and Smith, G. R. (1978). A single base-pair change creates a Chi recombinational hotspot in bacteriophage λ. *Proc. natn. Acad. Sci. U.S.A.* **75**, 6182–6186.

Stacey, K. A., and Lloyd, R. G. (1976). Isolation of Rec⁻ mutants from an F-prime merodiploid strain of *Escherichia coli* K-12. *Molec. gen. Genet.* **143**, 223–232.

Stadler, D. R. (1959). The relationship of gene conversion to crossing-over in *Neurospora. Proc. natn. Acad. Sci. U.S.A.* **45**, 1625–1629.

Stadler, D. R., and Towe, A. M. (1971). Evidence for meiotic recombination in *Ascobolus* involving only one member of a tetrad. *Genetics* **68**, 401–413.

Stadler, D. R., Towe, A. M., and Rossignol, J.-L. (1970). Intragenic recombination of ascospore color mutants in *Ascobolus* and its relationship to the segregation of outside markers. *Genetics* **66**, 429–447.

Stahl, F. W. (1979). *Genetic Recombination. Thinking about It in Phage and Fungi*, W. H. Freeman, San Francisco.

Stahl, F. W., and Stahl, M. M. (1971*b*). DNA synthesis associated with recombination. II. Recombination between repressed chromosomes. In *The Bacteriophage Lambda*

(A. D. Hershey, ed.), pp. 443–453, Cold Spring Harbor Laboratory, Cold Spring Harbor, N.Y.

Stahl, F. W., and Stahl, M. M. (1974). *Red*-mediated recombination in bacteriophage lambda. In *Mechanisms in Recombination* (R. F. Grell, ed.), pp. 407–419, Plenum Press, New York and London.

Stahl, F. W., and Stahl, M. M. (1975). Rec-mediated recombinational hot spot activity in bacteriophage λ. IV. Effect of heterology on Chi-stimulated crossing-over. *Molec. gen. Cenet.* **140**, 29–37.

Stahl, F. W., and Stahl, M. M. (1976). On recombination between close and distant markers in phage lambda. *Genetics* **82**, 577–593.

Stahl, F. W., and Stahl, M. M. (1977). Recombination pathway specificity of Chi. *Genetics* **86**, 715–725.

Stahl, F. W., McMilin, K., Stahl, M. M., Malone, R., Nozu, Y., and Russo, V. E. A. (1972*a*). A role for recombination in the production of 'free-loader' lambda bacteriophage particles. *J. molec. Biol.* **68**, 57–67.

Stahl, F. W., McMilin, K. D., Stahl, M. M., and Nozu, Y. (1972*b*). An enhancing role for DNA synthesis in formation of bacteriophage lambda recombinants. *Proc. natn. Acad. Sci. U.S.A.* **69**, 3598–3601.

Stahl, F. W., McMilin, K. D., Stahl, M. M., Crasemann, J. M., and Lam, S. (1974). The distribution of crossovers along unreplicated lambda bacteriophage chromosomes. *Genetics* **77**, 395–408.

Stahl, F. W., Crasemann, J. M., and Stahl, M. M. (1975). Rec-mediated recombinational hot spot activity in bacteriophage lambda. III. Chi mutations are site-mutations stimulating Rec-mediated recombination. *J. molec. Biol.* **94**, 203–212.

Stahl, F. W., Stahl, M. M., Malone, R. E., and Crasemann, J. M. (1980). Directionality and nonreciprocality of Chi-stimulated recombination in phage λ. *Genetics* **94**, 235–248.

Stahl, M. M., and Stahl, F. W. (1971*a*). DNA synthesis associated with recombination. I. Recombination in a DNA-negative host. In *The Bacteriophage Lambda* (A. D. Hershey, ed.), pp. 431–442. Cold Spring Harbor Laboratory, Cold Spring Harbor, N.Y.

Steinberg, C. M., and Edgar, R. S. (1961). On the absence of high negative interference in triparental crosses. *Virology* **15**, 511–513.

Stern, C. (1936). Somatic crossing-over and segregation in *Drosophila melanogaster*. *Genetics* **21**, 625–730.

Streisinger, G., and Bruce, V. (1960). Linkage of genetic markers in phages T2 and T4. *Genetics* **45**, 1289–1296.

Streisinger, G., and Franklin, N. C. (1957). Mutation and recombination at the host range genetic region of phage T2. *Cold Spring Harb. Symp. quant. Biol.* **21**, 103–111.

Streisinger, G., Edgar, R. S., and Denhardt, G. H. (1964). Chromosome structure in phage T4. I. Circularity of the linkage map. *Proc. natn. Acad. Sci. U.S.A.* **51**, 775–779.

Streisinger, G., Emrich, J., and Stahl, M. M. (1967). Chromosome structure in phage T4, III. Terminal redundancy and length determination. *Proc. natn. Acad. Sci. U.S.A.* **57**, 292–295.

Strickland, W. N. (1958*a*). Abnormal tetrads in *Aspergillus nidulans*. *Proc. R. Soc. B* **148**, 533–542.

Strickland, W. N. (1958*b*). An analysis of interference in *Aspergillus nidulans*. *Proc. R. Soc. B* **149**, 82–101.

Strickland, W. N. (1961). Tetrad analysis of short chromosome regions of *Neurospora crassa*. *Genetics* **46**, 1125–1141.

Sturtevant, A. H. (1913). The linear arrangement of six sex-linked factors in *Drosophila*, as shown by their mode of association. *J. exp. Zool.* **14**, 43–59.

Stuy, J. H. (1965). Fate of transforming DNA in the *Haemophilus influenzae* transformation system. *J. molec. Biol.* **13**, 554–570.

Symonds, N. (1953). A theorem on random-in-time mating. *Genetics* **38**, 32–33 (appendix to Visconti and Delbrück (1953)).

Symonds, N. (1962). The effect of pool size on recombination in phage. *Virology* **18**, 334–336.

Symonds, N. (1975). The repair of ultraviolet damage by phage T4: the role of the early phage genes. In *Molecular Mechanisms for the Repair of DNA* (P. C. Hanawalt and R. B. Setlow, eds), pp. 143–147, Plenum Press, New York.

Symonds, N., Van den Ende, P., Durston, A., and White, P. (1972). The structure of rII diploids of phage T4. *Molec. gen. Genet.* **116**, 223–238.

Symonds, N., Heindl, H., and White, P. (1973). Radiation sensitive mutants of phage T4. A comparative study. *Molec. gen. Genet.* **120**, 253–259.

Syvanen, M. (1974). *In vitro* genetic recombination of bacteriophage λ. *Proc. natn. Acad. Sci. U.S.A.* **71**, 2496–2499.

Szybalski, W. (1961). Molecular fate of transforming DNA. *J. Chim. phys. phys.-chim. Biol.* **58**, 1098–1099.

Takagi, J., Ando, T., and Ikeda, Y. (1968). Repair of single strand breaks in transforming DNA by polynucleotide ligase. *Biochem. biophys. Res. Commun.* **31**, 540–544.

Takano, T. (1966). Behaviour of some episomal elements in a recombination-deficient mutant of *Escherichia coli. Jap. J. Microbiol.* **10**, 201–210.

Taylor, A., and Smith, G. R. (1980). Unwinding and rewinding of DNA by the RecBC enzyme. *Cell* **22**, 447–457.

Taylor, J. H. (1965). Distribution of tritium-labeled DNA among chromosomes during meiosis. I. Spermatogenesis in the grasshopper. *J. Cell Biol.* **25** (2), 57–67.

Taylor, J. H., Woods, P. S., and Hughes, W. L. (1957). The organization and duplication of chromosomes as revealed by autoradiographic studies using tritium-labeled thymidine. *Proc. natn. Acad. Sci. U.S.A.* **43**, 122–128.

Tease, C. (1978). Cytological detection of crossing-over in BUdR substituted meiotic chromosomes using the fluorescent plus Giemsa technique. *Nature, Lond.* **272**, 823–824.

Templin, A., Kushner, S. R., and Clark, A. J. (1972). Genetic analysis of mutations indirectly suppressing *recB* and *recC* mutations. *Genetics* **72**, 205–215.

Tessman, E. S. (1965). Complementation groups in phage S13. *Virology* **25**, 303–321.

Tessman, E. S., and Shleser, R. (1963). Genetic recombination between phages S13 and φX174. *Virology* **19**, 239–240.

Tessman, E., and Tessman, I. (1959). Genetic recombination in phage S13. *Virology* **7**, 465–467.

Tessman, I. (1966). Genetic recombination of phage S13 in a recombination-deficient mutant of *Escherichia coli* K12. *Biochem. biophys. Res. Commun.* **22**, 169–174.

Tessman, I. (1968). Selective stimulation of one of the mechanisms for genetic recombination of bacteriophage S13. *Science, N.Y.* **16**, 481–482.

Theivendirarajah, K., and Whitehouse, H. L. K. (1982). The use of a spore colour mutant showing conversion with high frequency to test for symmetry in hybrid DNA formation. Manuscript.

Thomas, C. A., and MacHattie, L. A. (1964). Circular T2 DNA molecules. *Proc. natn. Acad. Sci. U.S.A.* **52**, 1297–1301.

Thomas, C. A., and Rubenstein, I. (1964). The arrangements of nucleotide sequences in T2 and T5 bacteriophage DNA molecules. *Biophys. J.* **4**, 93–106.

Thomas, D. Y., and Wilkie, D. (1968). Recombination of mitochondrial drug-resistance factors in *Saccharomyces cerevisiae. Biochem. biophys. Res. Commun.* **30**, 368–372.

Thompson, B. J., Escarmis, C., Parker, B., Slater, W. C., Doniger, J., Tessman, I., and Warner, R. C. (1975). Figure-8 configuration of dimers of S13 and φX174 replicative form DNA. *J. molec. Biol.* **91**, 409–419.

Thompson, B. J., Camien, M. N., and Warner, R. C. (1976). Kinetics of branch migration in double-stranded DNA. *Proc. natn. Acad. Sci. U.S.A.* **73**, 2299–2303.

Thuriaux, P., Minet, M., Munz, P., Ahmad, A., Zbaeren, D., and Leupold, U. (1980).

Gene conversion in nonsense suppressors of *Schizosaccharomyces pombe*. II. Specific marker effects. *Curr. Genet.* **1**, 89–95.

Tice, R., Windler, G., and Rary, J. M. (1978). Effect of cocultivation on sister chromatid exchange frequencies in Bloom's syndrome and normal fibroblast cells. *Nature, Lond.* **273**, 538–540.

Tiraby, J.-G., and Fox, M. S. (1973). Marker discrimination in transformation and mutation of pneumococcus. *Proc. natn. Acad. Sci, U.S.A.* **70**, 3541–3545.

Tiraby, J.-C., and Fox, M. S. (1974*a*). Marker discrimination and mutagen-induced alterations in pneumococcal transformation. *Genetics* **77**, 449–458.

Tiraby, J.-G., and Fox, M. S. (1974*b*). Marker effects in pneumococcal transformation. In *Mechanisms in Recombination* (R. F. Grell, ed.), pp. 225–236, Plenum Press, New York and London.

Tiraby, G., and Sicard, A. M. (1973*a*). Integration efficiency in DNA-induced transformation of pneumococcus. II. Genetic studies of a mutant integrating all the markers with a high efficiency. *Genetics* **75**, 35–48.

Tiraby, G., and Sicard, A. M. (1973*b*). Integration efficiencies of spontaneous mutant alleles of *amiA* locus in pneumococcal transformation. *J. Bact.* **116**, 1130–1135.

Tomizawa, J., and Anraku, N. (1964*a*). Molecular mechanisms of genetic recombination in bacteriophage. I. Effect of KCN on genetic recombination of phage T4. *J. molec. Biol.* **8**, 508–515.

Tomizawa, J., and Anraku, N. (1964*b*). Molecular mechanisms of genetic recombination in bacteriophage. II. Joining of parental DNA molecules of phage T4. *J. molec. Biol.* **8**, 516–540.

Tomizawa, J., and Ogawa, H. (1972). Structural genes of ATP-dependent deoxyribonuclease of *Escherichia coli*. *Nature new Biol.* **239**, 14–16.

Tomizawa, J., Anraku, N., and Iwama, Y. (1966). Molecular mechanisms of genetic recombination in bacteriophage. VI. A mutant defective in the joining of DNA molecules. *J. molec. Biol.* **21**, 247–253.

Unger, R. C., and Clark, A. J. (1972). Interaction of the recombination pathways of bacteriophage λ and its host *Escherichia coli* K12: effects on exonuclease V activity. *J. molec. Biol.* **70**, 539–548.

Unger, R. C., Echols, H., and Clark, A. J. (1972). Interaction of the recombination pathways of bacteriophage λ and host *Escherichia coli*: effects on λ recombination. *J. molec. Biol.* **70**, 531–537.

Van den Ende, P., and Symonds, N. (1972). The isolation and characterization of a T4 mutant partially defective in recombination. *Molec. gen. Genet.* **116**, 239–247.

Van de Putte, P., Zwenk, H., and Rörsch, A. (1966). Properties of four mutants of *Escherichia coli* defective in genetic recombination. *Mutation Res.* **3**, 381–392.

Verhoef, C., and de Haan, P. G. (1966). Genetic recombination in *Escherichia coli*. I. Relation between linkage of unselected markers and map distance. *Mutation Res.* **3**, 101–110.

Vielmetter, W., Bonhoeffer, F., and Schütte, A. (1968). Genetic evidence for transfer of a single DNA strand during bacterial conjugation. *J. molec. Biol.* **37**, 81–86.

Vigier, P. (1966). Rôle des hétérozygotes internes dans la formation de génômes double-recombinants chez le bactériophage T4. *C. r. hebd. Séanc. Acad. Sci. Paris, D*, **263**, 2010–2013.

Visconti, N., and Delbrück, M. (1953). The mechanism of genetic recombination in phage. *Genetics* **38**, 5–33.

Voll, M. J., and Goodgal, S. H. (1961). Recombination during transformation in *Haemophilus influenzae*. *Proc. natn. Acad. Sci. U.S.A.* **47**, 505–512.

Vovis, G. F., and Zinder, N. D. (1975). Methylation of f1 DNA by a restriction endonuclease from *Escherichia coli* B. *J. molec. Biol.* **95**, 557–568.

Vovis, G. F., Horiuchi, K., Hartman, N., and Zinder, N. D. (1973). Restriction endonuclease B and f1 heteroduplex DNA. *Nature new Biol.* **246**, 13–16.

Vovis, G. F., Horiuchi, K., and Zinder, N. D. (1974). Kinetics of methylation of DNA by a restriction endonuclease from *Escherichia coli* B. *Proc. natn. Acad. Sci. U.S.A.* **71**, 3810–3813.

Wackernagel, W., and Radding, C. M. (1974). Formation *in vitro* of infective joint molecules of λ DNA by T4 gene-32 protein. *Proc. natn. Acad. Sci. U.S.A.* **71**, 431–435.

Wagner, R., and Meselson, M. (1976). Repair tracts in mismatched DNA heteroduplexes. *Proc. natn. Acad. Sci. U.S.A.* **73**, 4135–4139.

Wagner, R. E., and Radman, M. (1975). A mechanism for initiation of genetic recombination. *Proc. natn. Acad. Sci. U.S.A.* **72**, 3619–3622.

Wake, C. T., and Wilson, J. H. (1979). Simian virus 40 recombinants are produced at high frequency during infection with genetically mixed oligomeric DNA. *Proc. natn. Acad. Sci. U.S.A.* **76**, 2876–2880.

Wang, J. C. (1971). Interaction between DNA and an *Escherichia coli* protein ω. *J. molec. Biol.* **55**, 523–533.

Watson, J. D. (1972). Origin of concatemeric T7 DNA. *Nature new Biol.* **239**, 197–201.

Watson, J. D., and Crick, F. H. C. (1953). A structure for deoxyribose nucleic acid. *Nature, Lond.* **171**, 737–738.

Weil, J. (1969). Reciprocal and non-reciprocal recombination in bacteriophage λ. *J. molec. Biol.* **43**, 351–355.

Weil, J., and Signer, E. R. (1968). Recombination in bacteriophage λ. II. Site-specific recombination promoted by the integration system. *J. molec. Biol.* **34**, 273–279.

Weil, J., and Terzaghi, B. (1970). The correlated occurrence of duplications and deletions in phage T4. *Virology* **42**, 234–237.

Weil, J., Terzaghi, B., and Crasemann, J. (1965). Partial diploidy in phage T4. *Genetics* **32**, 683–693.

Weinstock, G. M., McEntee, K., and Lehman, I. R. (1979). ATP-dependent renaturation of DNA catalyzed by the recA protein of *Escherichia coli*. *Proc. natn. Acad. Sci. U.S.A.* **76**, 126–130.

Weisbeek, P. J., and van de Pol, J. H. (1970). Biological activity of φX174 replicative form DNA fragments. *Biochim. biophys. Acta* **224**, 328–338.

Weisbeek, P. J., Borrias, W. E., Langeveld, S. A., Baas, P. D., and van Arkel, G. A. (1977). Bacteriophage φX174: gene *A* overlaps gene *B*. *Proc. natn. Acad. Sci. U.S.A.* **74**, 2504–2508.

Weiss, B., and Richardson, C. C. (1967). Enzymatic breakage and joining of deoxyribonucleic acid. I. Repair of single-strand breaks in DNA by an enzyme system from *Escherichia coli* infected with T4 bacteriophage. *Proc. natn. Acad. Sci. U.S.A.* **57**, 1021–1028.

West, S. C., Cassuto, E., Mursalim, J., and Howard-Flanders, P. (1980). Recognition of duplex DNA containing single-stranded regions by *recA* protein. *Proc. natn. Acad. Sci. U.S.A.* **77**, 2569–2573.

West, S. C., Cassuto, E., and Howard-Flanders, P. (1981*a*). Homologous pairing can occur before DNA strand separation in general genetic recombination. *Nature, Lond.* **290**, 29–33.

West, S. C., Cassuto, E., and Howard-Flanders, P. (1981*b*). recA protein promotes homologous-pairing and strand-exchange reactions between duplex DNA molecules. *Proc. natn. Acad. Sci. U.S.A.* **78**, 2100–2104.

West, S. C., Cassuto, E., and Howard-Flanders, P. (1981*c*). Heteroduplex formation by recA protein: polarity of strand exchanges. *Proc. natn. Acad. Sci. U.S.A.* **78**, 6149–6153.

White, R. L., and Fox, M. S. (1974). On the molecular basis of high negative interference. *Proc. natn. Acad. Sci. U.S.A.* **71**, 1544–1548.

White, R. L., and Fox, M. S. (1975*a*). Genetic heterozygosity in unreplicated bacteriophage λ recombinants. *Genetics* **81**, 33–50.

398

White, R. L., and Fox, M. S. (1975b). Genetic consequences of transfection with heteroduplex bacteriophage λ DNA. *Molec. gen. Genet.* **141**, 163–171.

Whitehouse, H. L. K. (1963). A theory of crossing-over by means of hybrid deoxyribonucleic acid. *Nature, Lond.* **199**, 1034–1040.

Whitehouse, H. L. K. (1965). A theory of crossing-over and gene conversion involving hybrid DNA. In *Genetics Today* (S. J. Geerts, ed.), **2**, 87–88, Pergamon, Oxford.

Whitehouse, H. L. K. (1966). An operator model of crossing-over. *Nature, Lond.* **211**, 708–713.

Whitehouse, H. L. K. (1967). Secondary crossing-over. *Nature, Lond.* **215**, 1352–1359.

Whitehouse, H. L. K. (1973a). *Towards an Understanding of the Mechanism of Heredity*, 3rd edn, Edward Arnold, London, and St Martin's Press, New York.

Whitehouse, H. L. K. (1973b). Hypothesis of post-recombination resynthesis of gene copies. *Nature, Lond.* **245**, 295–298.

Whitehouse, H. L. K. (1974). Genetic analysis of recombination at the *g* locus in *Sordaria fimicola. Genet. Res.* **24**, 251–279.

Whitehouse, H. L. K., and Hastings, P. J. (1965). The analysis of genetic recombination on the polaron hybrid DNA model. *Genet. Res.* **6**, 27–92.

Wiegand, R. C., Beattie, K. L., Holloman, W. K., and Radding, C. M. (1977). Uptake of homologous single-stranded fragments by superhelical DNA. III. The product and its enzymic conversion to a recombinant molecule. *J. molec. Biol.* **116**, 805–824.

Wildenberg, J. (1970). The relation of mitotic recombination to DNA replication in yeast pedigrees. *Genetics* **66**, 291–304.

Wildenberg, J., and Meselson, M. (1975). Mismatch repair in heteroduplex DNA. *Proc. natn. Acad. Sci. U.S.A.* **72**, 2202–2206.

Willetts, N. S., and Clark, A. J. (1969). Characteristics of some multiply recombination-deficient strains of *Escherichia coli. J. Bact.* **100**, 231–239.

Willetts, N. S., and Mount, D. W. (1969). Genetic analysis of recombination of *Escherichia coli* K-12 carrying *rec* mutations cotransducible with *thyA. J. Bact.* **100**, 923–934.

Wilson, J. H. (1979). Nick-free formation of reciprocal heteroduplexes: a simple solution to the topological problem. *Proc. natn. Acad. Sci. U.S.A.* **76**, 3641–3645.

Winkler, H. (1930). *Die Konversion der Gene*, Jena.

Witkin, E. M. (1969). The mutability toward ultraviolet light of recombination-deficient strains of *Escherichia coli. Mutation Res.* **8**, 9–14.

Wolff, S. (1977). Sister chromatid exchange. *A. Rev. Genet.* **11**, 183–201.

Wollman, E. L. (1953). Sur le determinisme génétique de la lysogénie. *Annls Inst. Pasteur, Paris* **84**, 281–293.

Wollman, E. L., and Jacob, F. (1954). Etude génétique d'un bactériophage tempéré d'*Escherichia coli*. II. Mécanisme de la recombinaison génétique. *Annls Inst. Pasteur, Paris* **87**, 674–690.

Wollman, E. L., Jacob, F., and Hayes, W. (1957). Conjugation and genetic recombination in *Escherichia coli* K-12. *Cold Spring Harb. Symp. quant. Biol.* **21**, 141–162.

Yajko, D. M., Valentine, M. C., and Weiss, B. (1974). Mutants of *Escherichia coli* with altered deoxyribonucleases. II. Isolation and characterization of mutants for exonuclease I. *J. molec. Biol.* **85**, 323–343.

Yasuda, S., and Sekiguchi, M. (1970). T4 endonuclease involved in repair of DNA. *Proc. natn. Acad. Sci. U.S.A.* **67**, 1839–1845.

Yu-Sun, C. C., Wickramaratne, M. R. T., and Whitehouse, H. L. K. (1977). Mutagen specificity in conversion pattern in *Sordaria brevicollis. Genet. Res.* **29**, 65–81.

Zahler, S. A. (1958). Some biological properties of bacteriophage S13 and φX-174. *J. Bact.* **75**, 310–315.

Zickler, H. (1934). Genetische Untersuchungen an einen heterothallischen Askomyzeten (*Bombardia lunata* nov. spec.). *Planta* **22**, 573–613.

Zimmermann, F. K. (1968). Enzyme studies on the products of mitotic gene conversion in *Saccharomyces cerevisiae. Molec. gen. Genet.* **101**, 171–184.

Zimmermann, F. K., and Schwaier, R. (1967). Induction of mitotic gene conversion with nitrous acid, 1-methyl-3-nitro-1-nitrosoguanidine and other alkylating agents in *Saccharomyces cerevisiae. Molec. gen. Genet.* **100**, 63–76.

Zinder, N. D. (1974). Recombination in bacteriophage f1. In *Mechanisms in Recombination* (R. F. Grell, ed.), pp. 19–28. Plenum Press, New York and London.

Zinder, N. D., Valentine, R. C., Roger, M., and Stoeckenius, W. (1963). f1, a rod-shaped male-specific bacteriophage that contains DNA. *Virology* **20**, 638–640.

Zissler, J. (1967). Integration-negative (*int*) mutants of phage λ. *Virology* **31**, 189.

Zissler, J., Signer, E., and Schaefer, F. (1971). The role of recombination in growth of bacteriophage lambda. I. The gamma gene. In *The Bacteriophage Lambda* (A. D. Hershey, ed.), pp. 455–468, Cold Spring Harbor Laboratory, Cold Spring Harbor, N.Y.

Index